Paul Wilmott on Quantitative Finance

Volume One

Paul Wilmott on Quantitative Finance

Volume One

www.wilmott.com

JOHN WILEY & SONS, LTD
Chichester • New York • Weinheim • Brisbane • Singapore • Toronto

Published by John Wiley & Sons Ltd., The Atrium, Southern Gate, Chichester,
West Sussex PO19 8SQ, England

Telephone (+44) 1243 779777

Email (for orders and customer service enquiries): cs-books@wiley.co.uk
Visit our Home Page on www.wileyeurope.com or www.wiley.com

Reprinted with corrections May and December 2003

Other Wiley Editorial Offices

John Wiley & Sons Inc., 111 River Street, Hoboken, NJ 07030, USA

Jossey-Bass, 989 Market Street, San Francisco, CA 94103-1741, USA

Wiley-VCH Verlag GmbH, Boschstr. 12, D-69469 Weinheim, Germany

John Wiley & Sons Australia Ltd, 33 Park Road, Milton, Queensland 4064, Australia

John Wiley & Sons (Asia) Pte Ltd, 2 Clementi Loop #02-01, Jin Xing Distripark, Singapore 129809

John Wiley & Sons Canada Ltd, 22 Worcester Road, Etobicoke, Ontario, Canada M9W 1L1

Library of Congress Cataloguing-in-Publication Data

Wilmott, Paul.
  Paul Wilmott on quantitative finance
    p. cm
  Rev. ed. of: Derivatives. 1998.
  Includes bibliographical references and index.
  ISBN 0-471-87438-8 (cloth: alk. Paper)
  1. Derivative securities — Mathematical models. 2. Options (Finance) — Mathematical
  models . 3. Options (Finance) — Prices — Mathematical models. I. Wilmott, Paul.
  Derivatives. II. Title.

HG6024.A3 W555 2000
332.64'5—dc21

                                        99-089739

British Library Cataloguing in Publication Data

A catalogue record for this book is available from the British Library

ISBN 0-471-87438-8

Typeset in 10/12pt Times by Laser Words, Madras India.
Printed and bound in Great Britain by Antony Rowe Ltd, Chippenham, Wiltshire.
This book is printed on acid-free paper responsibly manufactured from sustainable forestry
in which at least two trees are planted for each one used for paper production.

To my young alien friend, welcome home!

"I really must say that you are an ignorant person, friend Greybeard, if you know nothing of this enigmatic business which is at once the fairest and most deceitful in Europe, the noblest and the most infamous in the world, the finest and most vulgar on earth. It is a quintessence of academic learning and a paragon of fraudulence; it is a touchstone for the intelligent and a tombstone for the audacious, a treasury of usefulness and a source of disaster, and finally a counterpart of Sisyphus who never rests as also of Ixion who is chained to a wheel that turns perpetually."

Joseph de la Vega in *Confusión de Confusiones*, 1688.

contents of this volume

contents of volume two

prolog

Despite having written a book on quantitative finance with more pages than *Seven Pillars of Wisdom*, my view on math modeling in finance is very cynical. But normally, I'm the least cynical person you could wish to meet. This is how I think of math finance: it's like a safety net across Niagara Falls. Without the safety net only a fool would try to walk across the falls on a tightrope, but with a safety net, we'll all have a go. The catch is that the safety net isn't really there, it's a hologram. What does this mean? Math modeling gives us the confidence to trade and invent new products. The modeling is probably very inaccurate, and in times of crisis is completely useless. But without the models there'd be much less business being done.

One of the problems of quantitative finance is the stability of parameters. The mass of an electron is 9.11×10^{-28} g. The ratio of the circumference of a circle to its diameter, denoted by π, is 3.1415926.... The speed of light in vacuum is 3×10^8 m/s. However, I can't think of a single parameter in finance that is meaningful and stable. Hell, there are numerous financial conventions for the number of days in January.

Another problem is that quantities we'd like to measure don't even exist. Here's an example. I was speaking to an ex-academic recently about the subject of correlation between financial quantities. I said I didn't believe in the existence of correlation in finance. The pompous old fool laughed to himself (you know the type[1]) and said that, of course, it exists. Being uncharacteristically polite, I didn't say what was on my mind, that you can *measure* the correlation between being born in the year of the rat and becoming a lawyer, but that may not *mean* much. Correlation doesn't exist; deal with it. To some extent, that's what this book is about, dealing with the realities of the financial markets.

As the Marquis de Sade said in *Justine*, 'Such are the thoughts which caused me to take up my pen; and it is in consideration of such motives that I beg the indulgence of my readers for the untrue philosophies placed in the mouths of several of my characters, and for the sometimes rather painful situations which, for truth's sake, I am obliged to bring before his eyes.' I describe most of the quant finance models in this book, even those I don't believe in. There are icons of me scattered throughout the book, telling which bits I believe in and which bits I don't. Broadly speaking, I like those models for which the assumption of no arbitrage and one price is dropped. A model that gives a range for prices is OK with me.

The last time I wrote a book it got some really bad reviews. The word 'shame' stands out from one, another said I had 'unsavory personality traits'. True, but how did *he* know? I don't remember sleeping with him. Occasionally, people said nice things. I was compared to Aristotle, Carl Sagan, Richard Feynman and God. Some of the best comments came via email. One Dutch girl emailed 'Love you for ever, baby. xxx M' and that was after just reading my

---

[1] Email me if you want to know who it was.

last prolog. (I communicated with this girl for a while before meeting her... and when I did... Phwoar!) Hey, the *Financial Times* even called me a 'cult derivatives lecturer', but that was probably just a spelling mistake.

This book is essentially a second edition of *Derivatives*, but given a catchier title because of the broader subject matter. It's also a more arrogant title, which suits me just fine.[2] It has been expanded enormously. There are many new chapters, more details, more spreadsheets, more termsheets, more of everything. Still, many chapters remain untouched, since I did such a good job first time around ;-)

I have two lists of people to thank, one specifically for people who helped with the book and one for those special people in my life. Here's the first list. A big thank you to Hyungsok Ahn, Richard Arkell, David Bakstein, Bafkam Bim, Mauricio Bouabci, Daniel Bruno, Eric Cartman, Keesup Choe, Elsa Cortina, Grace Davies, Sam Ehrlichman, David Epstein, Simon Gould, Rich Haber, Einar Holstad, Arefin Huq, Stephen Jefferies, Varqa Khadem, Eli Lilly, Jenny Matthews, Chris McCoy, Kent Osband, Henrik Rassmussen, Howard Roark, Aram Schlosberg, Philipp Schönbucher, Richard Sherry, Phoebus Theologites, Dave Thomson and David Wilson.

The following is a list of people that I'd like to thank for adding immeasurably to my life; some helped with this book, but most were just 'there'. Asli Oztukel, Barbara Passy, Buddy Holly, Colin Atkinson, David Herring, Edna Hepburn-Ruston, Elisabeth Keck, Fouad Khennach, Glen Matlock, Henriette Prast, Ingrid Blauer, Jean Laidlaw, Jeff Dewynne, John Lydon, John Ockendon, Karen Mason, Malcolm McLaren, Marieke Bos, Maureen Doherty, Nabeela Khan, Oscar Wilmott, Patricia Sadro, Paul Cook, Peter Jäckel, Philip Hua, Quentin Crisp, Rowan Douglas, Sandra Maler, Sara Statman, Simon Ritchie, Steve Jones, Terri Colpi, Truman Capote, Veronika Guggenbichler, Zachary Wilmott. Unusually for one of my lists of acknowledgements, these are all real people.

I simply love my public so, please, please email me at paul@wilmott.com. I am especially keen to hear from good looking, skinny brunettes in their 20s. Keep an eye on www.wilmott.com for news, research papers, software and updates on errors in this book, maybe even some gossip.

ABOUT THE AUTHOR

Paul Wilmott is still 'immensely talented', and the author of even more books and research articles on finance than the last time he wrote an 'About the Author'. Nor has his star sign changed... Sunday November 8th 1959, 12:40 a.m., St Catherine's Hospital, Tranmere, Merseyside.[3]

He has two children, Oscar and Zachary. He is married to Andrea. They live in London where they can usually be found in the fashionable restaurants, nightclubs and bars, hanging out with the other beautiful people.

Having written a best-selling finance textbook, his only remaining ambition is to meet and swap parenting tips with his hero, Homer Simpson.

---

[2] *Paul Wilmott on 20th Century Gay and Lesbian Fiction* will be in the shops soon.

[3] Conceived 'the day the music died'. (At least that's what I estimate, I haven't dared ask.)

Specific info about the particular meaning of an icon is contained in its 'speech box'.

Quiz: in the background of many of the icons you will find a depiction of a scene from a movie or a movie poster. Email paul@wilmott.com the name of each movie and the page number. Correct entries will be put into a draw for an as yet undecided prize.

PART ONE
basic theory of derivatives

The first part of the book contains the fundamentals of derivatives theory and practice. It only deals with the equity, currency and commodity worlds since these are simpler than the fixed-income world, for technical reasons. I introduce the important concepts of hedging and no arbitrage, on which most sophisticated finance theory is based.

The assumptions, key concepts and results in Part One make up what is loosely known as the 'Black–Scholes world', named for Fischer Black and Myron Scholes who, together with Robert Merton, first conceived them. Their original work was published in 1973, after some resistance (the famous equation was first written down in 1969). In October 1997 Myron Scholes and Robert Merton were awarded the Nobel Prize for Economics for their work, Fischer Black having died in August 1995. The *New York Times* of Wednesday, 15th October 1997 wrote: 'Two North American scholars won the Nobel Memorial Prize in Economic Science yesterday for work that enables investors to price accurately their bets on the future, a breakthrough that has helped power the explosive growth in financial markets since the 1970's and plays a profound role in the economics of everyday life'.[1]

Part One is self contained, requiring little knowledge of finance or any more than elementary calculus.

Chapter 1: Products and Markets An overview of the workings of the financial markets and their products. A chapter such as this is obligatory. However, my readers will fall into one of two groups. Either they will know everything in this chapter and much, much more besides. Or they will know little, in which case what I write will not be enough.

Chapter 2: Derivatives An introduction to options, options markets, market conventions. Definitions of the common terms, simple no arbitrage, put-call parity and elementary trading strategies.

Chapter 3: The Random Behavior of Assets An examination of data for various financial quantities, leading to a model for the random behavior of prices. Almost all of sophisticated finance theory assumes that prices are random, the question is how to model that randomness.

Chapter 4: Elementary Stochastic Calculus We'll need a little bit of theory for manipulating our random variables. I keep the requirements down to the bare minimum. The key concept is Itô's lemma which I will try to introduce in as accessible a manner as possible.

---

[1] We'll be hearing more about these two in Chapter 59 on 'Derivatives ****Ups'.

Chapter 5: The Black–Scholes Model   I present the classical model for the fair value of options on stocks, currencies and commodities. This is the chapter in which I describe delta hedging and no arbitrage and show how they lead to a unique price for an option. This is the foundation for most quantitative finance theory and I will be building on this foundation for much, but by no means all, of the book.

Chapter 6: Partial Differential Equations   Partial differential equations play an important role in most physical applied mathematics. They also play a role in finance. Most of my readers trained in the physical sciences, engineering and applied mathematics will be comfortable with the idea that a partial differential equation is almost the same as 'the answer', the two being separated by at most some computer code. If you are not sure of this connection I hope that you will persevere with the book. This requires some faith on your part, you may have to read the book through twice: I have necessarily had to relegate the numerics, the real 'answer', to the last few chapters.

Chapter 7: The Black–Scholes Formulae and the 'Greeks'   From the Black–Scholes partial differential equation we can find formulae for the prices of some options. Derivatives of option prices with respect to variables or parameters are important for hedging. I will explain some of the most important such derivatives and how they are used.

Chapter 8: Simple Generalizations of the Black–Scholes World   Some of the assumptions of the Black–Scholes world can be dropped or stretched with ease. I will describe several of these. Later chapters are devoted to more extensive generalizations.

Chapter 9: Early Exercise and American Options   Early exercise is of particular importance financially. It is also of great mathematical interest. I will explain both of these aspects.

Chapter 10: Probability Density Functions and First Exit Times   The random nature of financial quantities means that we cannot say with certainty what the future holds in store. For that reason we need to be able to describe that future in a probabilistic sense.

Chapter 11: Multi-asset Options   Another conceptually simple generalization of the basic Black–Scholes world is to options on more than one underlying asset. Theoretically simple, this extension has its own particular problems in practice.

Chapter 12: The Binomial Model   One of the reasons that option theory has been so successful is that the ideas can be explained and implemented very easily with no complicated mathematics. This chapter is a slight digression, but it's hard not to include it.

Chapter 13: Predicting the Markets?   Although almost all sophisticated finance theory assumes the random movement of assets, many traders rely on technical indicators to predict the future direction of assets. These indicators may be simple geometrical constructs of the asset price path or quite complex algorithms. The hypothesis is that information about short-term future asset price movements are contained within the past history of prices. All traders use technical indicators at some time. In this chapter I describe some of the more common techniques.

Chapter 14: A Trading Game   Many readers of this book will never have traded anything more sophisticated than baseball cards. To get them into the swing of the subject from a practical point of view I include some suggestions on how to organize your own trading game based on the buying and selling of derivatives. I had a lot of help with this chapter from David Epstein who has been running such games for several years.

CHAPTER I
products and markets

In this Chapter...

- the time value of money
- an introduction to equities, commodities, currencies and indices
- fixed and floating interest rates
- futures and forwards
- no arbitrage, one of the main building blocks of finance theory

1.1 INTRODUCTION

This first chapter is a very gentle introduction to the subject of finance, and is mainly just a collection of definitions and specifications concerning the financial markets in general. There is little technical material here, and the one technical issue, the 'time value of money', is extremely simple. I will give the first example of 'no arbitrage'. This is important, being one part of the foundation of derivatives theory. Whether you read this chapter thoroughly or just skim it will depend on your background; mathematicians new to finance may want to spend more time on it than practitioners, say.

1.2 THE TIME VALUE OF MONEY

The simplest concept in finance is that of the **time value of money**; $1 today is worth more than $1 in a year's time. This is because of all the things we can do with $1 over the next year. At the very least, we can put it under the mattress and take it out in one year. But instead of putting it under the mattress we could invest it in a gold mine, or a new company. If those are too risky, then lend the money to someone who is willing to take the risks and will give you back the dollar with a little bit extra, the **interest**. That is what banks do, they borrow your money and invest it in various risky ways, but by spreading their risk over many investments they reduce their overall risk. And by borrowing money from many people they can invest in ways that the average individual cannot. The banks compete for your money by offering high interest rates. Free markets and the ability to quickly and cheaply change banks ensure that interest rates are fairly consistent from one bank to another.

I am going to denote interest rates by r. Although rates vary with time I am going to assume for the moment that they are constant. We can talk about several types of interest. First of all there is **simple** and **compound interest**. Simple interest is when the interest you receive is based only on the amount you initially invest, whereas compound interest is when you also get interest on your interest. Compound interest is the only case of relevance. And compound interest comes in two forms, **discretely compounded** and **continuously compounded**. Let me illustrate how they each work.

Suppose I invest \$1 in a bank at a discrete interest rate of r paid once *per annum*. At the end of one year my bank account will contain

$$1 \times (1 + r).$$

If the interest rate is 10% I will have one dollar and ten cents. After two years I will have

$$1 \times (1 + r) \times (1 + r) = (1 + r)^2$$

or one dollar and twenty-one cents. After n years I will have $(1 + r)^n$. That is an example of discrete compounding.

Now suppose I receive m interest payments at a rate of r/m *per annum*. After one year I will have

$$\left(1 + \frac{r}{m}\right)^m. \tag{1.1}$$

Now I am going to imagine that these interest payments come at increasingly frequent intervals, but at an increasingly smaller interest rate: I am going to take the limit $m \to \infty$. This will lead to a rate of interest that is paid continuously. Expression (1.1) becomes

$$\left(1 + \frac{r}{m}\right)^m = e^{m \log(1 + r/m)} \sim e^r.$$

That is how much money I will have in the bank after one year if the interest is continuously compounded. And similarly, after a time t I will have an amount

$$e^{rt} \tag{1.2}$$

in the bank. Almost everything in this book assumes that interest is compounded continuously.

Another way of deriving the result (1.2) is via a differential equation. Suppose I have an amount $M(t)$ in the bank at time t, how much does this increase in value from one day to the next? If I look at my bank account at time t and then again a short while later, time $t + dt$, the amount will have increased by

$$M(t + dt) - M(t) \approx \frac{dM}{dt} dt + \cdots,$$

where the right-hand side comes from a Taylor series expansion. But I also know that the interest I receive must be proportional to the amount I have, M, the interest rate, r, and the timestep, dt. Thus

$$\frac{dM}{dt} dt = rM(t) dt.$$

Dividing by dt gives the ordinary differential equation

$$\frac{dM}{dt} = rM(t)$$

OUR FIRST (AND SIMPLEST) DIFFERENTIAL EQUATION

the solution of which is

$$M(t) = M(0)\,e^{rt}.$$

If the initial amount at $t = 0$ was \$1 then I get (1.2) again.

This equation relates the value of the money I have now to the value in the future. Conversely, if I know I will get one dollar at time T in the future, its value at an earlier time t is simply

$$e^{-r(T-t)}.$$

I can relate cashflows in the future to their **present value** by multiplying by this factor. As an example, suppose that r is 5% i.e. $r = 0.05$, then the present value of \$1,000,000 to be received in two years is

$$\$1,000,000 \times e^{-0.05 \times 2} = \$904,837.$$

The present value is clearly less than the future value.

Interest rates are a very important factor determining the present value of future cashflows. For the moment I will only talk about one interest rate, and that will be constant. In later chapters I will generalize.

Important Aside

What mathematics have we seen so far? To get to (1.2) all we needed to know about are the two functions e (or exp) and log, and Taylor series. Believe it or not, you can appreciate almost all finance theory by knowing these three things together with 'expectations'. I'm going to build up to the basic Black–Scholes and derivatives theory assuming that you know all four of these. Don't worry if you don't know about these things yet, take a

I BELIEVE THAT 99% OF FINANCE REQUIRES ONLY BASIC MATH

look at Appendix A where I review these requisites and show how to interpret finance theory and practice in terms of the most elementary mathematics.

Just because you *can* understand derivatives theory in terms of basic math doesn't mean that you *should*. I hope that there's enough in the book to please the PhDs[1] as well.

1.3 **EQUITIES**

The most basic of financial instruments is the **equity, stock** or **share**. This is the ownership of a small piece of a company. If you have a bright idea for a new product or service then you

[1] And Nobel laureates.

could raise capital to realize this idea by selling off future profits in the form of a stake in your new company. The investors may be friends, your Aunt Joan, a bank, or a venture capitalist. The investor in the company gives you some cash, and in return you give him a contract stating how much of the company he owns. The **shareholders** who own the company between them then have some say in the running of the business, and technically the directors of the company are meant to act in the best interests of the shareholders. Once your business is up and running, you could raise further capital for expansion by issuing new shares.

This is how small businesses begin. Once the small business has become a large business, your Aunt Joan may not have enough money hidden under the mattress to invest in the next expansion. At this point shares in the company may be sold to a wider audience or even the general public. The investors in the business may have no link with the founders. The final point in the growth of the company is with the quotation of shares on a regulated stock exchange so that shares can be bought and sold freely, and capital can be raised efficiently and at the lowest cost.

Figures 1.1 and 1.2 show screens from Bloomberg giving details of Microsoft stock, including price, high and low, names of key personnel, weighting in various indices (see below) etc. There is much, much more info available on Bloomberg for this and all other stocks. We'll be seeing many Bloomberg screens throughout this book.

In Figure 1.3 I show an excerpt from the *Wall Street Journal Europe* of 5th January 2000. This shows a small selection of the many stocks traded on the New York Stock Exchange. The listed information includes highs and lows for the day as well as the change since the previous day's close.

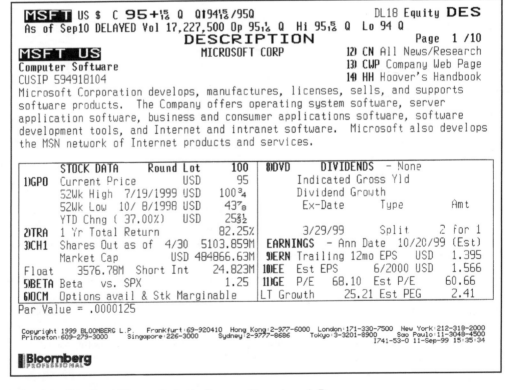

Figure 1.1 Details of Microsoft stock. Source: Bloomberg L.P.

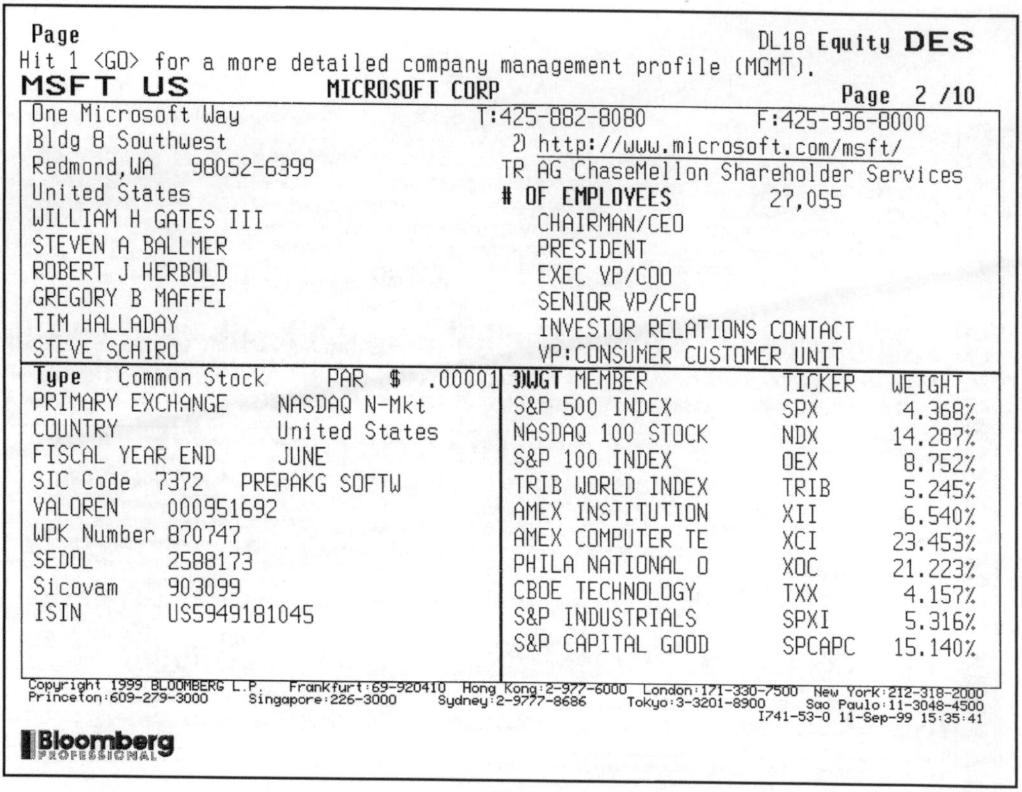

```
 Page                                             DL18 Equity DES
 Hit 1 <GO> for a more detailed company management profile (MGMT).
 MSFT  US            MICROSOFT CORP                    Page  2 /10
 ┌──────────────────────────────────────────────────────────────────┐
 │ One Microsoft Way         T:425-882-8080      F:425-936-8000       │
 │ Bldg 8 Southwest          2) http://www.microsoft.com/msft/        │
 │ Redmond,WA   98052-6399   TR AG ChaseMellon Shareholder Services   │
 │ United States             # OF EMPLOYEES         27,055            │
 │ WILLIAM H GATES III        CHAIRMAN/CEO                            │
 │ STEVEN A BALLMER           PRESIDENT                               │
 │ ROBERT J HERBOLD           EXEC VP/COO                             │
 │ GREGORY B MAFFEI           SENIOR VP/CFO                           │
 │ TIM HALLADAY               INVESTOR RELATIONS CONTACT             │
 │ STEVE SCHIRO               VP:CONSUMER CUSTOMER UNIT              │
 ├──────────────────────────────────────────────────────────────────┤
 │ Type   Common Stock    PAR $  .00001 3)WGT MEMBER    TICKER WEIGHT │
 │ PRIMARY EXCHANGE    NASDAQ N-Mkt    S&P 500 INDEX    SPX    4.368% │
 │ COUNTRY            United States    NASDAQ 100 STOCK NDX   14.287% │
 │ FISCAL YEAR END    JUNE             S&P 100 INDEX    OEX    8.752% │
 │ SIC Code  7372   PREPAKG SOFTW      TRIB WORLD INDEX TRIB   5.245% │
 │ VALOREN   000951692                 AMEX INSTITUTION  XII    6.540% │
 │ WPK Number 870747                   AMEX COMPUTER TE  XCI   23.453% │
 │ SEDOL     2588173                   PHILA NATIONAL O  XOC   21.223% │
 │ Sicovam   903099                    CBOE TECHNOLOGY   TXX    4.157% │
 │ ISIN      US5949181045              S&P INDUSTRIALS   SPXI   5.316% │
 │                                     S&P CAPITAL GOOD  SPCAPC 15.140%│
 └──────────────────────────────────────────────────────────────────┘
 Copyright 1999 BLOOMBERG L.P.  Frankfurt:69-920410  Hong Kong:2-977-6000  London:171-330-7500  New York:212-318-2000
 Princeton:609-279-3000   Singapore:226-3000   Sydney:2-9777-8686   Tokyo:3-3201-8900   Sao Paulo:11-3048-4500
                                                                I741-53-0 11-Sep-99 15:35:41
 ║Bloomberg
 ║PROFESSIONAL
```

Figure 1.2 Details of Microsoft stock continued. Source: Bloomberg L.P.

The behavior of the quoted prices of stocks is far from being predictable. In Figure 1.4 I show the Dow Jones Industrial Average over the period from August 1964 to February 1999. In Figure 1.5 is a time series of the Glaxo–Wellcome share price, as produced by Bloomberg.

If we could predict the behavior of stock prices in the future then we could become very rich. Although many people have claimed to be able to predict prices with varying degrees of accuracy, no one has yet made a completely convincing case. In this book I am going to take the point of view that prices have a large element of randomness. This does *not* mean that we cannot model stock prices, but it does mean that the modeling must be done in a probabilistic sense. No doubt the reality of the situation lies somewhere between complete predictability and perfect randomness, not least because there have been many cases of market manipulation where large trades have moved stock prices in a direction that was favorable to the person doing the moving.

To whet your appetite for the mathematical modeling later, I want to show you a simple way to simulate a random walk that looks something like a stock price. One of the simplest random processes is the tossing of a coin. I am going to use ideas related to coin tossing as a model for the behavior of a stock price. As a simple experiment start with the number 100 which you should think of as the price of your stock, and toss a coin. If you throw a head multiply the number by 1.01, if you throw a tail multiply by 0.99. After one toss your number will be either 99 or 101. Toss again. If you get a head multiply your *new* number by 1.01 or by 0.99 if you

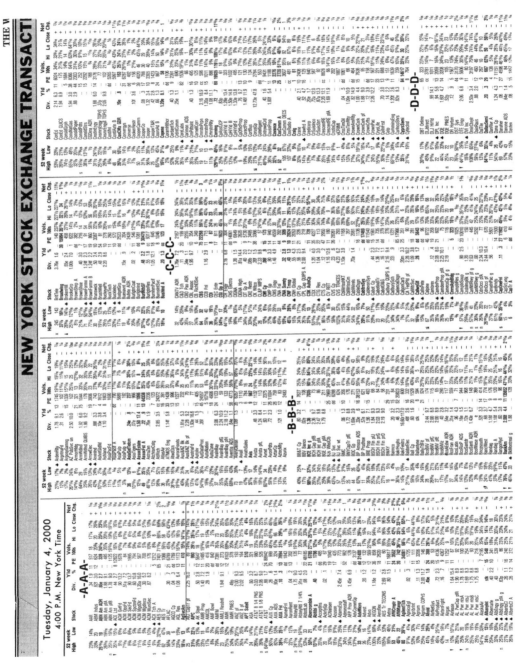

Figure 1.3 The *Wall Street Journal Europe* of 5th January 2000. Reproduced by permission of Dow Jones & Company, Inc.

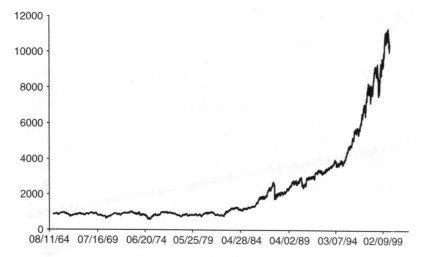

Figure 1.4 A time series of the Dow Jones Industrial Average from August 1964 to February 1999.

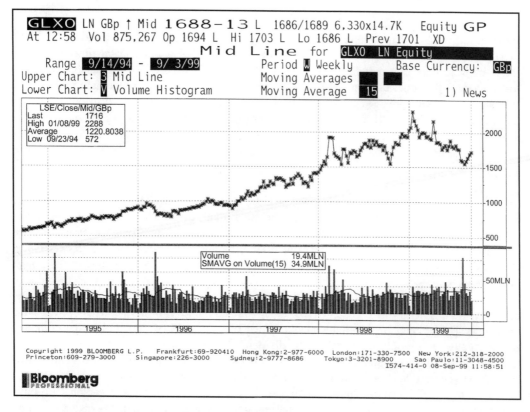

Figure 1.5 Glaxo–Wellcome share price (volume below). Source: Bloomberg L.P.

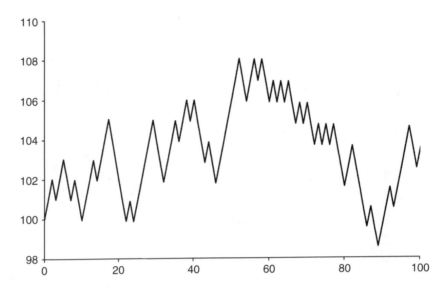

Figure 1.6 A simulation of an asset price path?

throw a tail. You will now have either $1.01^2 \times 100$, $1.01 \times 0.99 \times 100 = 0.99 \times 1.01 \times 100$ or $0.99^2 \times 100$. Continue this process and plot your value on a graph each time you throw the coin. Results of one particular experiment are shown in Figure 1.6. Instead of physically tossing a coin, the series used in this plot was generated on a spreadsheet like that in Figure 1.7. This uses the Excel spreadsheet function RAND() to generate a uniformly distributed random number between 0 and 1. If this number is greater than one half it counts as a 'head' otherwise a 'tail'.

1.3.1 Dividends

The owner of the stock theoretically owns a piece of the company. This ownership can only be turned into cash if he owns so many of the stock that he can take over the company and keep all the profits for himself. This is unrealistic for most of us. To the average investor the value in holding the stock comes from **dividends** and any growth in the stock's value. Dividends are lump sum payments, paid out every quarter or every six months, to the holder of the stock.

The amount of the dividend varies from year to year depending on the profitability of the company. As a general rule companies like to try to keep the level of dividends about the same each time. The amount of the dividend is decided by the board of directors of the company and is usually set a month or so before the dividend is actually paid.

When the stock is bought it either comes with its entitlement to the next dividend (**cum**) or not (**ex**). There is a date at around the time of the dividend payment when the stock goes from cum to ex. The original holder of the stock gets the dividend but the person who buys it obviously does not. All things being equal a stock that is cum dividend is better than one that is ex dividend. Thus at the time that the dividend is paid and the stock goes ex dividend there will be a drop in the value of the stock. The size of this drop in stock value offsets the disadvantage of not getting the dividend.

	A	B	C	D	E
1	Initial stock price	100		**Stock**	
2	Up move	1.01		100	
3	Down move	0.99		99	
4	Probability of up	0.5		98.01	
5				97.0299	
6		=B1		96.0596	
7				97.0202	
8				97.9904	
9		=D6*IF(RAND()>1-$B$4,$B$2,$B$3)		9	
10				97.9806	
11				98.96041	
12				99.95001	
13				98.95051	
14				97.961	
15				98.94061	
16				99.93002	
17				100.9293	
18				99.92003	
19				100.9192	
20				101.9284	
21				100.9091	
22				99.90004	
23				98.90104	
24				99.89005	
25				100.889	
26				99.88007	
27				98.88127	
28				97.89245	
29				96.91353	
30				95.94439	
31				96.90384	

Figure 1.7 Simple spreadsheet to simulate the coin-tossing experiment.

This jump in stock price is in practice more complex than I have just made out. Often capital gains due to the rise in a stock price are taxed differently from a dividend, which is often treated as income. Some people can make a lot of risk-free money by exploiting tax 'inconsistencies'.

I discuss dividends in depth in Chapter 8 and again in Chapter 37.

1.3.2 Stock Splits

Stock prices in the US are usually of the order of magnitude of $100. In the UK they are typically around £1. There is no real reason for the popularity of the number of digits, after all, if I buy a stock I want to know what percentage growth I will get, the absolute level of the stock is irrelevant to me, it just determines whether I have to buy tens or thousands of the stock to invest a given amount. Nevertheless there is some psychological element to the stock size. Every now and then a company will announce a **stock split**. For example, the company with a stock price of $900 announces a three-for-one stock split. This simply means that instead of holding one stock valued at $900, I hold three valued at $300 each.[2]

---

[2] In the UK this would be called a two-for-one split.

```
 <HELP> for explanation, <MENU> for similar functions.     DL18 Equity DVD
 Hit # <GO> to view details.
           DIVIDEND/SPLIT SUMMARY                      Page  1/ 1
 MSFT US      MICROSOFT CORP                  Currency █
 ┌──────────────────────────────┐
 │ 12 Month Yield    n.a.        │
 │ Indicated Yield   n.a.        │
 └──────────────────────────────┘

   Graph Selections                 GRAPH NOT AVAILABLE
   B-Both
   G-Gross Yield
   Y-Adjust for Splits

   Range  1990  to  1999  Type  1-All           Frequency  Irregular
   ┌───────────────────────────────────────────────────────────────────┐
   │ Declared Ex-Date  Record   Payable      Amount    Type             │
   │ 1)  1/25/99 3/29/99  3/12/99 3/26/99     2 for 1   Stock Split      │
   │ 2)  1/26/98 2/23/98  2/ 6/98 2/20/98     2 for 1   Stock Split      │
   │ 3) 11/12/96 12/ 9/96 11/22/96 12/ 6/96   2 for 1   Stock Split      │
   │ 4)  4/25/94 5/23/94  5/ 6/94 5/20/94     2 for 1   Stock Split      │
   │ 5)  6/ 3/92 6/15/92  6/ 3/92 6/12/92     3 for 2   Stock Split      │
   │ 6)  5/ 8/91 6/27/91  6/18/91 6/26/91     3 for 2   Stock Split      │
   │ 7)  3/13/90 4/16/90  3/26/90 4/13/90     2 for 1   Stock Split      │
   └───────────────────────────────────────────────────────────────────┘

 Copyright 1999 BLOOMBERG L.P.  Frankfurt:69-920410  Hong Kong:2-977-6000  London:171-330-7500  New York:212-318-2000
 Princeton:609-279-3000  Singapore:226-3000  Sydney:2-9777-8686  Tokyo:3-3201-8900  Sao Paulo:11-3048-4500
                                                           I741-53-0 11-Sep-99 15:36:49
 ▌Bloomberg
 ▌PROFESSIONAL
```

Figure 1.8 Stock split info for Microsoft. Source: Bloomberg L.P.

1.4 COMMODITIES

Commodities are usually raw products such as precious metals, oil, food products etc. The prices of these products are unpredictable but often show seasonal effects. Scarcity of the product results in higher prices. Commodities are usually traded by people who have no need of the raw material. For example they may just be speculating on the direction of gold without wanting to stockpile it or make jewellery. Most trading is done on the futures market, making deals to buy or sell the commodity at some time in the future. The deal is then closed out before the commodity is due to be delivered. Futures contracts are discussed below.

Figure 1.9 shows a time series of the price of pulp, used in paper manufacture.

1.5 CURRENCIES

Another financial quantity we shall discuss is the **exchange rate**, the rate at which one currency can be exchanged for another. This is the world of **foreign exchange**, or **Forex** or **FX** for short. Some currencies are pegged to one another, and others are allowed to float freely. Whatever the exchange rates from one currency to another, there must be consistency throughout. If it is possible to exchange dollars for pounds and then the pounds for yen, this implies a relationship between the dollar/pound, pound/yen and dollar/yen exchange rates. If this relationship moves out of line it is possible to make **arbitrage profits** by exploiting the mispricing.

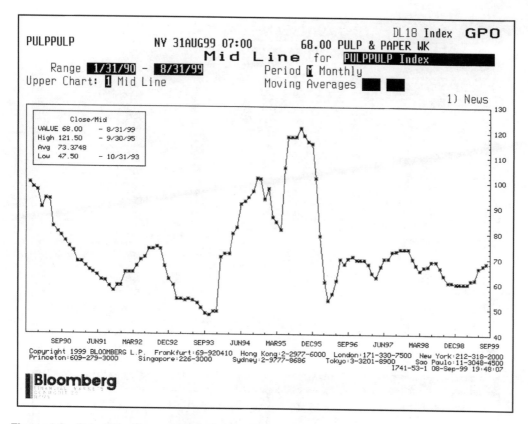

Figure 1.9 Pulp price. Source: Bloomberg L.P.

Figure 1.10 is an excerpt from the *Wall Street Journal Europe* of 5th January 2000. At the bottom of this excerpt is a matrix of exchange rates. A similar matrix is shown in Figure 1.11 from Bloomberg.

Although the fluctuation in exchange rates is unpredictable, there is a link between exchange rates and the interest rates in the two countries. If the interest rate on dollars is raised while the interest rate on pounds sterling stays fixed we would expect to see sterling depreciating against the dollar for a while. Central banks can use interest rates as a tool for manipulating exchange rates, but only to a degree.

At the start of 1999 Euroland currencies were fixed at the rates shown in Figure 1.12.

1.6 INDICES

For measuring how the stock market/economy is doing as a whole, there have been developed the stock market **indices**. A typical index is made up from the weighted sum of a selection or **basket** of representative stocks. The selection may be designed to represent the whole market, such as the Standard & Poor's 500 (S&P500) in the US or the Financial Times Stock Exchange index (FTSE100) in the UK, or a very special part of a market. In Figure 1.4 we saw the DJIA, representing major US stocks. In Figure 1.13 is shown JP Morgan's Emerging Market Bond Index. The EMBI+ is an index of emerging market debt instruments, including external-currency-denominated Brady bonds, Eurobonds and US dollar local markets instruments. The

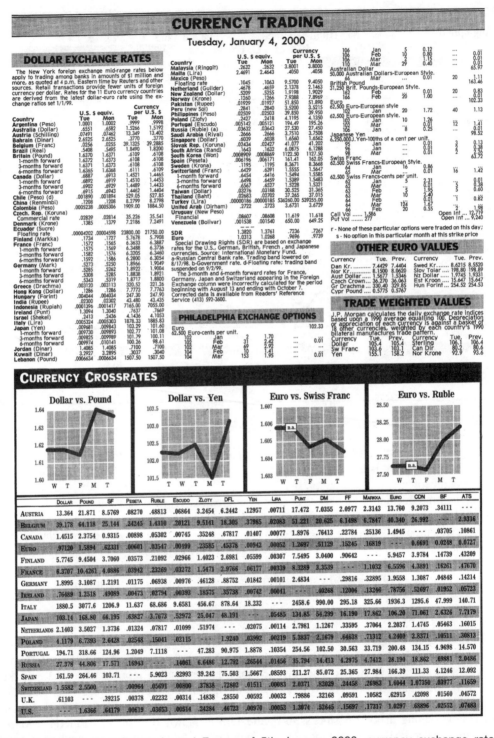

Figure 1.10 The *Wall Street Journal Europe* of 5th January 2000, currency exchange rates. Reproduced by permission of Dow Jones & Company, Inc.

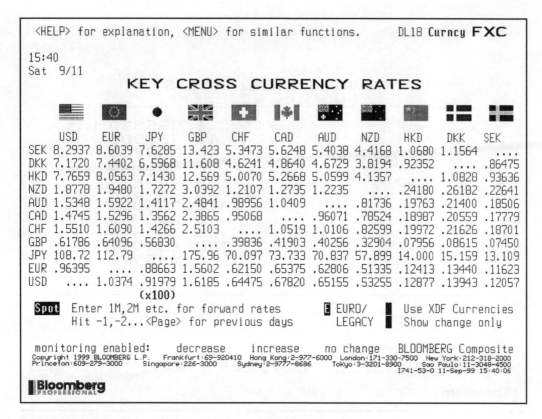

Figure 1.11   Key cross currency rates. Source: Bloomberg L.P.

main components of the index are the three major Latin American countries, Argentina, Brazil and Mexico. Bulgaria, Morocco, Nigeria, the Philippines, Poland, Russia and South Africa are also represented.

Figure 1.14 shows a time series of the MAE All Bond Index which includes Peso and US dollar denominated bonds sold by the Argentine Government.

1.7 **FIXED-INCOME SECURITIES**

In lending money to a bank you may get to choose for how long you tie your money up and what kind of interest rate you receive. If you decide on a fixed-term deposit the bank will offer to lock in a fixed rate of interest for the period of the deposit, a month, six months, a year, say. The rate of interest will not necessarily be the same for each period, and generally the longer the time that the money is tied up the higher the rate of interest, although this is not always the case. Often, if you want to have immediate access to your money then you will be exposed to interest rates that will change from time to time, as interest rates are not constant.

These two types of interest payments, **fixed** and **floating**, are seen in many financial instruments. **Coupon-bearing bonds** pay out a known amount every six months or year etc. This is the **coupon** and would often be a fixed rate of interest. At the end of your fixed term you get a final coupon and the return of the **principal**, the amount on which the interest was calculated.

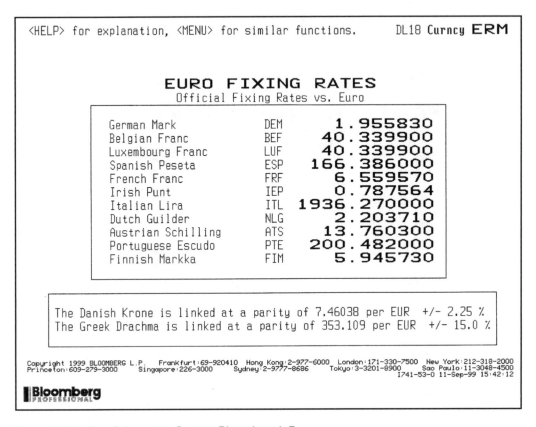

```
<HELP> for explanation, <MENU> for similar functions.     DL18 Curncy ERM


                    EURO  FIXING  RATES
                   Official Fixing Rates vs. Euro

        German Mark          DEM        1.955830
        Belgian Franc        BEF       40.339900
        Luxembourg Franc     LUF       40.339900
        Spanish Peseta       ESP      166.386000
        French Franc         FRF        6.559570
        Irish Punt           IEP        0.787564
        Italian Lira         ITL     1936.270000
        Dutch Guilder        NLG        2.203710
        Austrian Schilling   ATS       13.760300
        Portuguese Escudo    PTE      200.482000
        Finnish Markka       FIM        5.945730


     The Danish Krone is linked at a parity of 7.46038 per EUR  +/- 2.25 %
     The Greek Drachma is linked at a parity of 353.109 per EUR  +/- 15.0 %


Copyright 1999 BLOOMBERG L.P.   Frankfurt:69-920410  Hong Kong:2-977-6000  London:171-330-7500  New York:212-318-2000
Princeton:609-279-3000      Singapore:226-3000     Sydney:2-9777-8686     Tokyo:3-3201-8900     Sao Paulo:11-3048-4500
                                                                          I741-53-0 11-Sep-99 15:42:12
```

Bloomberg PROFESSIONAL

Figure 1.12 Euro fixing rates. Source: Bloomberg L.P.

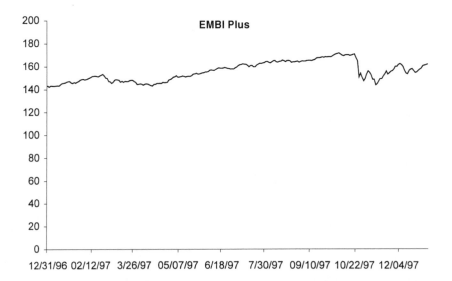

Figure 1.13 JP Morgan's EMBI Plus.

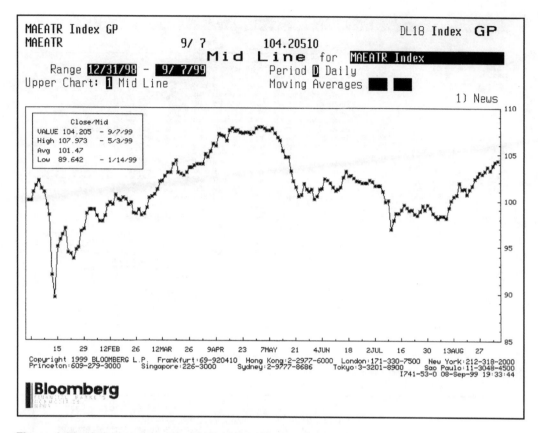

Figure 1.14 A time series of the MAE All Bond Index. Source: Bloomberg L.P.

Interest rate swaps are an exchange of a fixed rate of interest for a floating rate of interest. Governments and companies issue bonds as a form of borrowing. The less creditworthy the issuer, the higher the interest that they will have to pay out. Bonds are actively traded, with prices that continually fluctuate.

Fixed-income modeling and products are the subject of the whole of Part Four.

1.8 INFLATION-PROOF BONDS

A very recent addition to the list of bonds issued by the US government is the **index-linked bond**. These have been around in the UK since 1981, and have provided a very successful way of ensuring that income is not eroded by inflation.

In the UK inflation is measured by the **Retail Price Index** or **RPI**. This index is a measure of year-on-year inflation, using a 'basket' of goods and services including mortgage interest payments. The index is published monthly. The coupons and principal of the index-linked bonds are related to the level of the RPI. Roughly speaking, the amounts of the coupon and principal are scaled with the increase in the RPI over the period from the issue of the bond to the time of the payment. There is one slight complication in that the actual RPI level used in these calculations is set back *eight months*. Thus the base measurement is eight months before issue and the scaling of any coupon is with respect to the increase in the RPI from this base

UK GILTS PRICES

Shorts" (Lives up to Five Years)	Notes	Yield Int	Yield Red	Price £	+ or −	52 week High	52 week Low
Treas $8\frac{1}{2}$pc 2000	✠	8.49	5.54	100.13	......	103.51	100.13
Conv 9pc 2000		8.96	5.56	100.47	−.01	104.36	100.40
Treas 13pc 2000		12.57	6.03	103.44xd	−.01	111.61	103.44
Treas 8pc 2000		7.89	6.35	101.42	+.01	105.82	101.41
Treas Fltg Rate 2001		−	0.24	100.16	......	100.81	100.11
Treas 10pc 2001		9.63	6.43	103.80	+.02	110.36	103.77
Conv $9\frac{1}{2}$pc 2001	✠	9.11	6.43	104.33xd	+.03	110.88	104.28
Conv $9\frac{3}{4}$pc 2001	✠	9.29	6.43	104.91	+.03	111.79	104.85
Treas 7pc 2001		6.94	6.44	100.93	+.08	106.51	100.75
Conv 10pc 2002	✠	9.34	6.57	107.04	+.11	116.41	106.82
Treas 7pc 2002		6.92	6.45	101.19	+.11	108.13	100.95
Conv $9\frac{1}{2}$pc 2002	✠	8.92	6.57	106.45	+.12	115.64	106.21
Treas $9\frac{3}{4}$pc 2002		9.07	6.57	107.54	+.13	117.29	107.28
Exch 9pc 2002	✠	8.47	6.57	106.21	+.14	115.62	105.92
Conv $9\frac{3}{4}$pc 2003	✠	8.92	6.57	109.33	+.17	120.19	108.99
Treas 8pc 2003		7.64	6.42	104.78	+.17	114.74	104.45
Treas 10pc 2003		8.98	6.46	111.34	+.19	123.52	110.99
Treas $13\frac{3}{4}$pc 2000–3	✠	13.22	6.03	104.02xd	−.01	112.93	104.02
Treas $6\frac{1}{2}$pc 2003		6.47	6.34	100.53	+.19	110.21	100.15
Treas $11\frac{1}{2}$pc 2001–4		10.89	6.49	105.62	+.04	113.56	105.58
Treas 10pc 2004	✠	8.83	6.46	113.22	+.23	126.55	112.80
Five to Fifteen Years							
Treas 5pc 2004		5.25	6.24	95.28	+.23	98.80	94.53
Funding $3\frac{1}{2}$pc 1999–4		3.89	6.06	90.03xd	+.19	98.95	89.66
Conv $9\frac{1}{2}$pc 2004	✠	8.40	6.28	113.12	+.26	126.43	112.66
Treas $6\frac{3}{4}$pc 2004		6.61	6.22	102.19	+.25	113.28	101.39
Conv $9\frac{1}{2}$pc 2005		8.30	6.23	114.46	+.27	128.16	113.68
Exch $10\frac{1}{2}$pc 2005	✠	8.74	6.23	120.16	+.30	135.46	119.38
Treas $12\frac{1}{2}$pc 2003–5	✠	10.39	6.46	120.32	+.21	135.55	119.95
Treas $8\frac{1}{2}$pc 2005		7.63	6.16	111.41	+.32	125.31	110.35
Conv $9\frac{3}{4}$pc 2006	✠	8.10	6.06	120.41	+.43	136.21	119.10
Treas $7\frac{3}{4}$pc 2006		7.13	6.14	108.69	+.38	122.40	107.42
Treas 8pc 2002–6		7.75	6.66	103.29	+.12	111.73	103.05
Treas $7\frac{1}{2}$pc 2006		6.95	6.08	107.89	+.40	121.62	106.44
Treas $11\frac{3}{4}$pc 2003–7	✠	10.31	6.57	114.02xd	+.16	126.29	113.72
Treas $8\frac{1}{2}$pc 2007		7.42	6.06	114.56xd	+.45	129.86	112.98
Treas $7\frac{1}{4}$pc 2007		6.72	5.98	107.93	+.46	122.55	106.17
Treas $13\frac{1}{2}$pc 2004–8	✠	10.67	6.22	126.56	+.24	142.61	126.15
Treas 9pc 2008		7.46	5.94	120.68	+.55	138.52	118.77
Treas 8pc 2009	✠	6.86	5.73	116.70	+.55	132.91	114.73
Treas $5\frac{3}{4}$pc 2009		5.70	5.63	100.93	+.58	114.67	99.10
Treas $6\frac{1}{4}$pc 2010		5.92	5.57	105.51	+.59	118.76	103.35
Conv 9pc Ln 2011		6.99	5.58	128.75	+.69	145.31	126.38
Treas 9pc 2012		6.85	5.50	131.48	+.75	147.74	128.45
Treas $5\frac{1}{2}$pc 2008–12		5.63	5.76	97.70	+.39	112.26	95.68
Treas 8pc 2013		6.34	5.29	126.13	+.60	139.64	122.43
Treas $7\frac{3}{4}$pc 2012–15		6.54	5.61	118.59	+.64	133.56	116.18
Over Fifteen Years							
Treas 8pc 2015		6.05	5.04	132.13	+.81	144.12	127.70
Treas $8\frac{3}{4}$pc 2017		6.03	4.90	145.04	+1.00	156.08	139.82
Exch 12pc 2013–17	✠	7.44	5.60	161.31	+.96	183.80	158.27
Treas 8pc 2021		5.57	4.73	143.69	+1.45	153.21	137.64
Treas 6pc 2028		4.84	4.50	124.06	+1.55	132.62	116.63
Undated							
Consols 4pc	✠	5.03	−	79.50	+1.30	87.19	73.95
War Loan $3\frac{1}{2}$pc		4.82	−	72.66	+1.17	79.83	67.14
Conv $3\frac{1}{2}$pc '61 Aft.	✠	4.13	−	84.66	+1.17	95.68	79.14
Treas 3pc '66 Aft.	✠	5.38	−	55.72	+.80	61.93	51.78
Consols $2\frac{1}{2}$pc	✠	4.93	−	50.67	+.79	56.93	46.78
Treas. $2\frac{1}{2}$pc		4.96	−	50.41	+.51	55.17	46.70

Index–Linked	Notes (b)	Yield (1)	Yield (2)	Price £	+ or −	52 week High	52 week Low
$2\frac{1}{2}$pc '01	(78.3)	3.72	4.35	203.28	+.09	206.02	201.90
$2\frac{1}{2}$pc '03	(78.8)	3.17	3.48	201.98	+.18	207.83	201.24
$4\frac{3}{8}$pc '04	(135.6)	2.70	2.93	128.72	+.13	134.77	127.26
2pc '06	(69.5)	1.85	2.02	235.96xd	+.29	239.80	230.16
$2\frac{1}{2}$pc '09	(78.8)	1.85	1.97	217.78	+.39	221.45	208.87
$2\frac{1}{2}$pc '11	(74.6)	1.96	2.06	229.97	+.62	235.62	219.11
$2\frac{1}{2}$pc '13	(89.2)	1.94	2.03	194.34	+.66	199.07	183.57
$2\frac{1}{2}$pc '16	(81.6)	1.91	1.99	215.37	+.67	221.10	203.03
$2\frac{1}{2}$pc '20	(83.0)	1.83	1.89	218.00	+.88	221.68	203.17
$2\frac{1}{2}$pc '24	(97.7)	1.72	1.78	192.00xd	+.88	195.05	176.32
$4\frac{1}{8}$pc '30	(135.1)	1.65	1.70	191.40xd	+.78	194.33	174.35

Prospective real redemption rate on projected inflation of (1) 5% and (2) 3%. (b) Figures in parentheses show RPI base for indexing (ie 8 months prior to issue) and have been adjusted to reflect rebasing of RPI to 100 in February 1987. Conversion factor 3.945. RPI for April 1999: 165.2 and for November1999: 166.7.

Other Fixed Interest

	Notes	Yield Int	Yield Red	Price £	+ or −	52 week High	52 week Low
Asian Dev $10\frac{1}{4}$pc 2009		8.23	6.60	$124\frac{1}{2}$	......	$140\frac{23}{32}$	$124\frac{1}{2}$
B'ham $11\frac{1}{2}$pc 2012		8.04	6.40	143	......	$158\frac{3}{4}$	142
Leeds $13\frac{1}{2}$pc 2006		9.93	6.80	136	......	152	136
Liverpool $3\frac{1}{2}$pc Irred.		5.38	5.40	65	......	73	55
LCC 3pc '20 Aft.		5.33	5.30	$56\frac{1}{4}$	......	64	50
Manchester $11\frac{1}{2}$pc 2007		8.97	6.70	$128\frac{1}{4}$	......	$151\frac{3}{4}$	$128\frac{1}{4}$
Met. Wtr. 3pc 'B'		3.33	6.40	90	......	$93\frac{1}{8}$	85
N'wide Anglia $3\frac{7}{8}$pc IL 2021		−	2.90	$196\frac{3}{4}$	......	$200\frac{1}{2}$	$174\frac{31}{32}$
$4\frac{1}{4}$pc IL 2024		−	2.90	$192\frac{1}{2}$	......	$196\frac{3}{4}$	169

● 'Tap' stock. All UK Gilts are tax-free to non-residents on application. E Auction basis. xd Ex dividend. Closing mid-prices are shown in pounds per £100 nominal of stock. Prospective real Index-Linked redemption yields are calculated by HSBC Bank plc from Gemma closing prices. ✠ Indicative price.

Figure 1.15 UK gilts prices from the *Financial Times* of 11th January 2000. Reproduced by permission of the *Financial Times*.

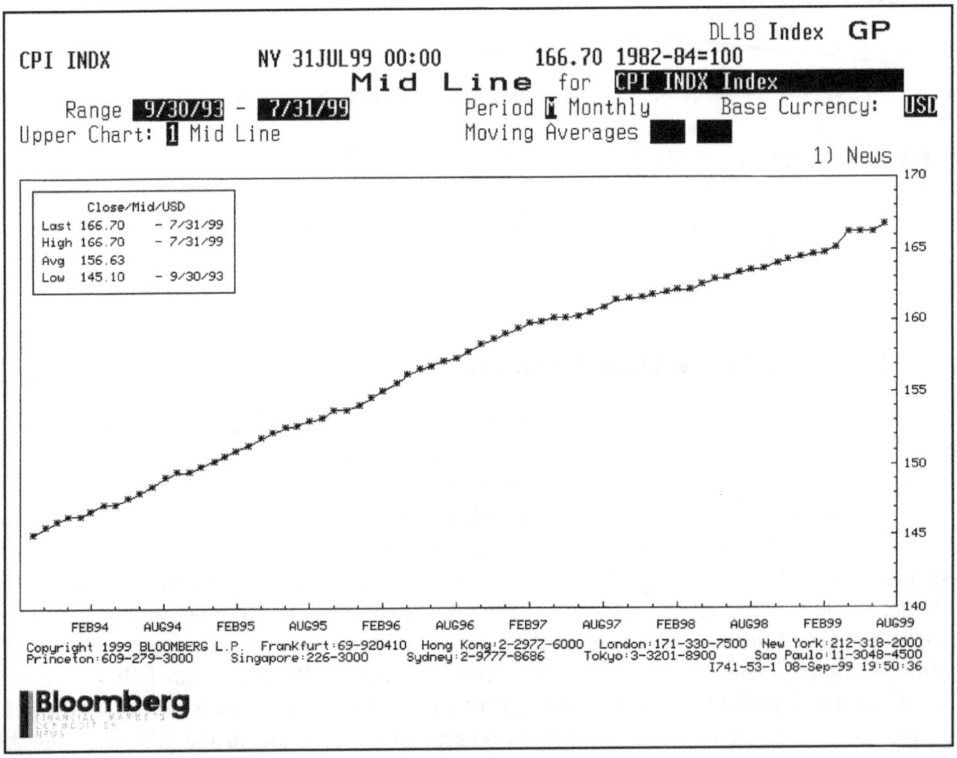

Figure 1.16 The CPI index. Source: Bloomberg L.P.

measurement to the level of the RPI eight months before the coupon is paid. One of the reasons for this complexity is that the initial estimate of the RPI is usually corrected at a later date.

Figure 1.15 shows the UK gilts prices published in the *Financial Times* of 11th January 2000. The index-linked bonds are on the right. The figures in parentheses give the base for the index, the RPI eight months prior to the issue of the gilt.

In the US the inflation index is the **Consumer Price Index (CPI)**. A time series of this index is shown in Figure 1.16.

I will not pursue the modeling of inflation or index-linked bonds in this book. I would just like to say that the dynamics of the relationship between inflation and short-term interest rates is particularly interesting. Clearly the level of interest rates will affect the rate of inflation directly through mortgage repayments, but also interest rates are often used by central banks as a tool for keeping inflation down.

1.9 **FORWARDS AND FUTURES**

A **forward contract** is an agreement where one party promises to buy an asset from another party at some specified time in the future and at some specified price. No money changes hands until the **delivery date** or **maturity** of the contract. The terms of the contract make it an obligation to buy the asset at the delivery date, there is no choice in the matter. The asset could be a stock, a commodity or a currency.

A **futures contract** is very similar to a forward contract. Futures contracts are usually traded through an exchange, which standardizes the terms of the contracts. The profit or loss from the futures position is calculated every day and the change in this value is paid from one party to the other. Thus with futures contracts there is a gradual payment of funds from initiation until maturity.

Forwards and futures have two main uses, in speculation and in hedging. If you believe that the market will rise you can benefit from this by entering into a forward or futures contract. If your market view is right then a lot of money will change hands (at maturity or every day) in your favor. That is speculation and is very risky. Hedging is the opposite, it is avoidance of risk. For example, if you are expecting to get paid in yen in six months' time, but you live in America and your expenses are all in dollars, then you could enter into a futures contract to lock in a guaranteed exchange rate for the amount of your yen income. Once this exchange rate is locked in you are no longer exposed to fluctuations in the dollar/yen exchange rate. But then you won't benefit if the yen appreciates.

1.9.1 A First Example of No Arbitrage

Although I won't be discussing futures and forwards very much they do provide us with our first example of the **no-arbitrage** principle. I am going to introduce mathematical notation now for the first time in the book; it will be fairly consistent throughout. Consider a forward contract that obliges us to hand over an amount $\$F$ at time T to receive the underlying asset. Today's date is t and the price of the asset is currently $\$S(t)$, this is the **spot price**, the amount for which we could get immediate delivery of the asset. When we get to maturity we will hand over the amount $\$F$ and receive the asset, then worth $\$S(T)$. How much profit we make cannot be known until we know the value $S(T)$, and we can't know this until time T. From now on I am going to drop the '$' sign from in front of monetary amounts.

We know all of F, $S(t)$, t and T, is there any relationship between them? You might think not, since the forward contract entitles us to receive an amount $S(T) - F$ at expiry and this is unknown. However, by entering into a special portfolio of trades *now* we can eliminate all randomness in the future. This is done as follows.

Enter into the forward contract. This costs us nothing up front but exposes us to the uncertainty in the value of the asset at maturity. Simultaneously sell the asset. It is called **going short** when you sell something you don't own. This is possible in many markets, but with some timing restrictions. We now have an amount $S(t)$ in cash due to the sale of the asset, a forward contract, and a short asset position. But our net position is zero. Put the cash in the bank, to receive interest.

When we get to maturity we hand over the amount F and receive the asset, this cancels our short asset position regardless of the value of $S(T)$. At maturity we are left with a guaranteed $-F$ in cash as well as the bank account. The word 'guaranteed' is important because it emphasizes that it is independent of the value of the asset. The bank account contains the initial investment of an amount $S(t)$ with added interest, which has a value at maturity of

$$S(t)e^{r(T-t)}.$$

Our net position at maturity is therefore

$$S(t)e^{r(T-t)} - F.$$

Since we began with a portfolio worth zero and we end up with a predictable amount, that predictable amount should also be zero. We can conclude that

$$F = S(t)e^{r(T-t)}. \tag{1.3}$$

OUR FIRST EXAMPLE OF NO ARBITRAGE

This is the relationship between the spot price and the forward price. It is a linear relationship, the forward price is proportional to the spot price.

The cashflows in this special hedged portfolio are shown in Table 1.1.

In Figure 1.17 is a path taken by the spot asset price and its forward price. As long as interest rates are constant, these two are related by (1.3).

TODAY'S CASHFLOW IS ZERO, THE FUTURE CASHFLOW IS KNOWN, SO.....

Table 1.1 Cashflows in a hedged portfolio of asset and forward.

Holding	Worth today (t)	Worth at maturity (T)
Forward	0	$S(T) - F$
− Stock	$-S(t)$	$-S(T)$
Cash	$S(t)$	$S(t)e^{r(T-t)}$
Total	0	$S(t)e^{r(T-t)} - F$

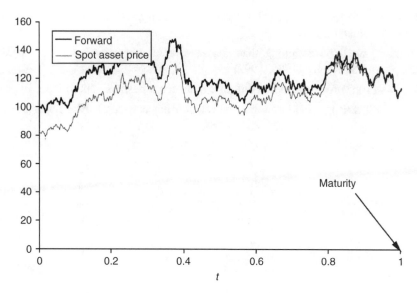

Figure 1.17 A time series of a spot asset price and its forward price.

If this relationship is violated then there will be an arbitrage opportunity. To see what is meant by this, imagine that F is less than $S(t)e^{r(T-t)}$. To exploit this and make a riskless arbitrage profit, enter into the deals as explained above. At maturity you will have $S(t)e^{r(T-t)}$ in the bank, a short asset and a long forward. The asset position cancels when you hand over the amount F, leaving you with a profit of $S(t)e^{r(T-t)} - F$. If F is greater than that given by (1.3) then you enter into the opposite positions, going short the forward. Again you make a riskless profit. The standard economic argument then says that investors will act quickly to exploit the opportunity, and in the process prices will adjust to eliminate it.

1.10 SUMMARY

The above descriptions of financial markets are enough for this introductory chapter. Perhaps the most important point to take away with you is the idea of no arbitrage. In the example here, relating spot prices to futures prices, we saw how we could set up a very simple portfolio which completely eliminated any dependence on the future value of the stock. When we come to value derivatives, in the way we just valued a future, we will see that the same principle can be applied albeit in a far more sophisticated way.

FURTHER READING

- For general financial news visit www.bloomberg.com and www.reuters.com. CNN has online financial news at www.cnnfn.com. There are also online editions of the *Wall Street Journal Europe*, www.wsj.com, the *Financial Times*, www.ft.com and *Futures and Options World*, www.fow.com.
- For more information about futures see the Chicago Board of Trade website www.cbot.com.
- Many, many financial links can be found at Wahoo!, www.io.com/~gibbonsb/wahoo.html.

- See Bloch (1995) for an empirical analysis of inflation data and a theoretical discussion of pricing index-linked bonds.
- In the main, we'll be assuming that markets are random. For insight about alternative hypotheses see Schwager (1990, 1992).
- See Brooks (1967) for how the raising of capital for a business might work in practice.
- Cox, Ingersoll & Ross (1981) discuss the relationship between forward and future prices.

CHAPTER 2
derivatives

In this Chapter...

- the definitions of basic derivative instruments
- option jargon
- no arbitrage and put-call parity
- how to draw payoff diagrams
- simple option strategies

2.1 INTRODUCTION

The previous chapter dealt with some of the basics of financial markets. I didn't go into any detail, just giving the barest outline and setting the scene for this chapter. Here I introduce the theme that is central to the book, the subject of options, a.k.a. derivatives or contingent claims. This chapter is non-technical, being a description of some of the most common option contracts, and explaining the market-standard jargon. It is in later chapters that I start to get technical.

Options have been around for many years, but it was only on 26th April 1973 that they were first traded on an exchange. It was then that The Chicago Board Options Exchange (CBOE) first created standardized, listed options. Initially there were just calls on 16 stocks. Puts weren't even introduced until 1977. In the US options are traded on CBOE, the American Stock Exchange, the Pacific Stock Exchange and the Philadelphia Stock Exchange. Worldwide, there are over 50 exchanges on which options are traded.

2.2 OPTIONS

If you are reading the book in a linear fashion, from start to finish, then the last topics you read about will have been futures and forwards. The holder of future or forward contracts is *obliged* to trade at the maturity of the contract. Unless the position is closed before maturity the holder must take possession of the commodity, currency or whatever is the subject of the contract, regardless of whether the asset has risen or fallen. Wouldn't it be nice if we only had to take possession of the asset if it had risen?

The simplest **option** gives the holder the *right* to trade in the future at a previously agreed price but takes away the obligation. So if the stock falls, we don't have to buy it after all.

> A **call option** is the right to
> buy a particular asset for an
> agreed amount at a specified
> time in the future

As an example, consider the following call option on Amazon stock. It gives the holder the right to buy one of Amazon stock for an amount $95 in six months' time. Today's stock price is $86. The amount '95' which we can pay for the stock is called the **exercise price** or **strike price**. The date on which we must **exercise** our option, if we decide to, is called the **expiry** or **expiration date**. The stock on which the option is based is known as the **underlying asset**.

Let's consider what may happen over the next six months, up until expiry. Suppose that nothing happens, that the stock price remains at $86. What do we do at expiry? We could exercise the option, handing over $95 to receive the stock. Would that be sensible? No, because the stock is only worth $86, either we wouldn't exercise the option or if we really wanted the stock we would buy it in the stock market for the $86. But what if the stock price rises to $100? Then we'd be laughing, we would exercise the option, paying $95 for a stock that's worth $100, a profit of $5.

We would exercise the option at expiry if the stock is above the strike and not if it is below. If we use S to mean the stock price and E the strike then at expiry the option is worth

$$\max(S - E, 0).$$

This function of the underlying asset is called the **payoff function**. The 'max' function represents the optionality.

Why would we buy such an option? Clearly, if you own a call option you want the stock to rise as much as possible. The higher the stock price the greater will be your profit. I will discuss this below, but your decision whether to buy it will depend on how much it costs; the option is valuable, there is no downside to it unlike a future. In our example the option was valued at $20.5. Where did this number come from? The valuation of options is the subject of this book, and I'll be showing you how to find this value later on.

What if you believe that the stock is going to fall, is there a contract that you can buy to benefit from the fall in a stock price?

> A **put option** is the right to
> *sell* a particular asset for an
> agreed amount at a specified
> time in the future

The holder of a put option wants the stock price to fall so that he can sell the asset for more than it is worth. The payoff function for a put option is

$$\max(E - S, 0).$$

Now the option is only exercised if the stock falls below the strike price.

Figure 2.1 is an excerpt from the *Wall Street Journal Europe* of 5th January 2000 showing options on various stocks. The table lists closing prices of the underlying stocks and the last

U.S. LISTED OPTIONS QUOTATIONS

Tuesday, January 4, 2000

Volume and close for actively traded equity options with results for corresponding put or call contract as of 3 p.m. Volume figures are unofficial. Open interest is total outstanding for all exchanges and reflects previous trading day. Close when possible is shown for the underlying stock on primary market. **CB**-Chicago Board Options Exchange. **AM**-American Stock Exchange. **PB**-Philadelphia Stock Exchange. **PC**-Pacific Stock Exchange. **NY**-New York Stock Exchange. **XC**-Composite. **c**-Call. **p**-Put.

MOST ACTIVE CONTRACTS

Option	Strike	Vol.	Exch.	Last	Net Chg	3 pm Close	Open Int.
Micsft	Jan 100	p 13,675	XC	9/16	+ 1/16	115 3/8	109,572
Disney	Jan 27 1/2	13,388	XC	3 7/8	+ 1 5/16	31 1/16	95,848
AmOnline	Jan 80	11,888	XC	5 1/8	− 2 1/4	78 1/2	211,536
Micsft	Jan 90	p 10,448	XC	3/16	...	115 3/8	140,712
Intel	Jan 70	p 9,805	XC	7/16	+ 1/16	85 3/8	158,564
DellCptr	Jan 45	p 8,784	XC	1 1/8	+ 1/2	48	81,720
Disney	Feb 30	6,982	XC	2 9/16	+ 13/16	31 1/16	4,152
Intel	Jan 90	6,457	XC	2 3/16	− 7/16	85 3/8	148,840
Cisco	Jan 90	p 6,344	XC	3/4	+ 1/4	104 3/16	53,396
Bk of Am	Jan 47 1/2	6,196	XC	1 3/8	− 1 5/16	45 5/16	16,968
Compaq	Jan 30	6,161	XC	1 3/8	− 11/16	28 13/16	256,144
Intel	Jan 80	6,053	XC	7 1/2	− 1 1/8	85 3/8	169,968
Qualcom	Jan 77 1/2	p 6,049	XC	1/16	...	162 3/8	31,856
AmOnline	Jan 90	6,039	XC	2 1/8	− 1 1/4	78 1/2	211,516
Yahoo	Jan 02 135	p 6,002	XC	6 1/8	− 1	481	160
Yahoo	Jan 450	5,857	XC	66	+ 2 1/8	481	27,424
Disney	Jan 30	p 5,619	XC	9/16	− 9/16	31 1/16	32,984
AmOnline	Jan 100	5,383	XC	1 1/16	− 7/16	78 1/2	261,988
Micsft	Jan 125	5,052	XC	1 1/16	− 3/16	115 3/8	65,676
CBS Cp	Feb 60	5,010	XC	2 5/16	− 1 13/16	57	2,211
DellCptr	Feb 45	p 4,772	XC	2 1/2	+ 3/4	48	46,540
Intel	Jan 85	4,571	XC	4 1/2	− 5/8	85 3/8	148,724
CMGI Inc	Jan 320	4,504	XC	28	− 11	307	16,305
Atl R	Feb 85	4,500	XC	4 3/4	− 2 1/4	80 5/8	740
DellCptr	Jan 50	4,456	XC	1 1/2	− 1 3/8	48	187,664
Cmpuwr	Jan 30	4,385	XC	6 7/8	− 1/8	36 9/16	15,858
MCI Wrld	Jan 46 5/8	p 4,332	XC	1	+ 1/4	80	49,432
Compaq	Apr 20	4,227	XC	10	− 3/4	28 13/16	52,660
Caterp	Aug 50	4,154	XC	6	+ 1	48 7/8	240
Disney	Jan 30	4,004	XC	1 13/16	+ 13/16	31 1/16	100,672
Citigrp	Jan 55	3,907	XC	1/2	− 9/16	50 1/16	87,340
LoralSp	Feb 22 1/2	3,811	XC	2 3/8	+ 3/16	22	777
Cendant	Feb 25	3,797	XC	1 7/16	− 7/16	23 3/4	64,845
Intel	Jan 95	3,706	XC	1	− 3/16	85 3/8	70,272
GMagic	Feb 5	3,687	XC	1 5/8	+ 1	5	156,147
Compaq	Feb 30	3,647	XC	2 3/8	− 5/8	28 13/16	34,532
DellCptr	Jan 55	3,610	XC	7/16	− 5/8	48	133,160
SunMicro	Apr 45	p 3,572	XC	1	+ 1/16	73	92,732
ETradeGr	Jan 30	3,564	XC	2	− 1/16	28 5/16	78,180
MerrLyn	Jan 80	3,534	XC	2 5/8	− 1 3/8	77 1/8	48,848

Option	Strike	Exp.	-Call- Vol.	-Call- 3 pm	-Put- Vol.	-Put- 3pm
ACTV	35	Jan	186	5 3/8	2503	2 3/16
38 5/8	45	Jan	92	1 1/4	2380	8 1/2
AT&T	45	Jan	1144	7	95	1/4
51 13/16	50	Jan	147	2 7/8	596	1 1/16
51 13/16	55	Jan	665	13/16	185	4 1/8
Abbt L	35	May	1725	3	3	3 1/4
A M D	15	Jan	14	14 3/4	500	1/16
29 5/8	25	Jan	1868	5 1/2	143	15/16
29 5/8	25	Feb	1720	6 3/4	10	1 5/8
29 5/8	30	Jan	1557	2 11/16	356	2 13/16
AdvRdio	22 1/2	Feb	485	1 11/16	...	...
Alcatl	35	Jan	...	...	600	3/8
AlterraHl	5	Feb	...	...	500	5/8
6	7 1/2	Feb	350	5/16	500	2 1/4
Amazon	65	Jan	66	20 3/4	785	1 3/4
86	80	Jan	385	10 1/2	626	6 3/8
86	85	Jan	1203	8 1/8	295	8 1/4
86	90	Jan	1331	6	316	12
86	95	Jan	685	4 1/2	10	12
86	95	Jul	21	20 1/2	500	26 1/8
86	100	Jan	1043	3	131	18 5/8
AmOnline	57 1/2	Jan	5	21 5/8	580	1/2
78 1/2	65	Jan	118	14 3/4	702	1 1/4
78 1/2	70	Jan	847	11	1810	2 5/16
78 1/2	75	Jan	2564	7 1/2	1113	4
78 1/2	75	Feb	808	11 1/8	208	7
108 1/16	100	Jan	1023	10 3/8	415	2 1/2
108 1/16	105	Jan	520	6 7/8	224	4 1/4
Enron	40	Feb	1006	3 1/2	...	...
41 7/16	45	Apr	533	2 3/4	...	...
Equant	115	Jan	715	5 1/4	...	...
EricTel	60	Jan	573	6 1/4	1315	1 1/2
65 1/8	65	Feb	517	5 1/2	605	5 5/8
65 1/8	65	Apr	525	8 3/4	527	7 1/2
eToys	25	Jan	200	3 7/8	578	2 11/16
26 3/8	35	Jan	883	15/16	20	9 5/8
26 3/8	40	Jan	528	1/2	2	13 1/2
ExodsCm	90	Jan	504	8 3/8	64	9 3/4
87 5/16	95	Feb	558	10 5/8	...	...
Exxon	70	Apr	500	9 3/8	10	1 1/2
FEMSA	40	Jul	8	8 7/8	1000	4 5/8
F N M	50	Jan	837	7 3/8	225	3/8
FUnion	30	Feb	1819	2 1/2	193	1 1/2
30 11/16	35	Feb	1749	3/4	22	4 5/8
Firstar	20	Jan	2460	15/16	20	1 1/16
Gateway	50	Mar	...	...	520	2 1/2
65 1/2	60	Jan	46	6 7/8	573	2
65 1/2	60	Mar	25	10 1/2	533	6
65 1/2	65	Jan	2354	3 5/8	516	4 1/4
Gen El	135	Jan	62	11 7/8	1137	1 1/8
145 3/4	140	Jan	503	8 1/8	873	2 1/4
145 3/4	145	Jan	884	[illegible]	479	4
92 1/2	95	Jan	...	...	700	5
OceanEgy	7 1/2	Feb	1010	3/4	...	...
Oracle o	30	Mar	1510	87	...	...
Oracle	70	Jan	58	39 1/2	565	1/8
108	75	Jan	53	33 1/2	1254	5/16
108	115	Mar	563	12 7/8	127	17 7/8
108	120	Jan	2574	4	81	14 1/2
108	120	Feb	504	8	7	18 1/8
PRI Auto	65	Feb	500	7 1/8	...	...
ParmTc	20	Jan	1011	1 7/8	701	1 3/4
19 3/8	22 1/2	Jan	2147	13/16	29	3 7/8
19 3/8	25	Feb	689	1 1/4	206	5 1/8
19 3/8	35	Feb	2140	5/16	...	...
PepsiCo	32 1/2	Apr	...	...	507	7/8
36 5/16	37 1/2	Jan	697	1/2	85	7/16
PetrlGeo	20	Feb	...	...	550	3 5/8
Pfizer	30	Jan	438	1 3/4	546	3/4
31	30	Feb	161	2 1/2	818	1 3/8
31	30	Jun	63	4	1238	2 9/16
31	35	Jan	733	1/4	92	4 1/4
31	35	Feb	498	5/8	6	4 3/8
31	35	Mar	497	1 1/16	1037	4 7/8
Ph Mor	25	Jan	717	9/16	1307	1 3/4
23 7/8	25	Feb	1051	1 7/16	110	2 9/16
23 7/8	25	Mar	275	2 1/16	1130	3 1/4

Figure 2.1 The *Wall Street Journal Europe* of 5th January 2000, Stock Options. Reproduced by permission of Dow Jones & Company, Inc.

INDEX OPTIONS TRADING

Tuesday, January 4, 2000

Volume, close, net change and open interest for all contracts. Volume figures are unofficial. Open interest reflects previous trading day. p-Put. c-Call. The totals for call and put volume and open interest are midday figures.

CHICAGO

CB MEXICO INDEX(MEX)

Strike			Vol.	3 pm Close	Net. Chg.	Open Int.
Jun	90	c	5	19 3/8	− 1 3/8	13
Mar	110	c	10	4 1/4	− 1 1/8	46
Call Vol.			**15**	**Open Int.**		**116**
Put Vol.			**0**	**Open Int.**		**338**

CB TECHNOLOGY(TXX)

Strike			Vol.	3 pm Close	Net. Chg.	Open Int.
Jan	650	p	10	1/8	− 6 1/8	26
Feb	820	p	10	8 1/8	+ 1 3/8	10
Feb	900	p	60	17 1/4	+ 3 1/2	30
Call Vol.			**0**	**Open Int.**		**319**
Put Vol.			**80**	**Open Int.**		**381**

DJ INDUS AVG(DJX)

Strike			Vol.	3 pm Close	Net. Chg.	Open Int.
Jan	90	p	540	1/16	...	9,881
Jan	92	p	180	1/16	...	1,546
Feb	92	p	150	3/16	...	150
Mar	92	c	1	21 1/2	− 1	1
Jun	92	p	5	1 1/2	+ 5/16	9,117
Jan	96	p	13	1/8	+ 1/16	23,136
Feb	96	p	23	9/16	+ 1/4	1,340
Mar	96	p	50	1	+ 3/16	10,585
Jun	96	p	795	1 7/8	+ 1/4	6,974
Jan	100	p	35	5/16	+ 3/16	9,841
Feb	100	p	112	1 1/16	+ 1/4	2,983
Mar	100	c	82	13 1/4	− 3 3/4	258
Mar	100	p	15	1 3/8	+ 7/16	9,170
Jun	100	p	5	2 3/4	+ 3/4	5,023
Jan	102	p	1,088	5/16	+ 1/8	4,358
Feb	102	p	365	1 1/16	+ 1/4	1,750
Mar	102	p	30	1 13/16	+ 3/8	3,147
Jan	104	c	111	9 1/8	− 7/8	2,654
Jan	104	p	449	1/2	+ 1/4	3,671
Feb	104	p	4	1 3/16	...	227
Mar	104	c	2	10 1/4	− 1 5/8	1,051
Mar	104	p	8	1 5/8	+ 3/8	2,367
Jun	104	p	5	3 1/2	+ 3/4	1,632
Jan	105	p	1,281	3/4	+ 3/8	959
Jan	106	c	10	7 1/8	− 1	377
Jan	106	p	28	3/4	+ 3/8	1,999
Mar	106	c	8	9 3/8	− 2 1/4	783
Mar	106	p	1	2 5/16	+ 5/16	757
Jun	106	c	2	12 1/2	+ 1/8	3
Jan	108	c	23	4 1/2	− 3 5/8	618
Jan	108	p	159	1 3/8	+ 9/16	2,157
Feb	108	p	25	2 1/2	+ 5/8	272
Mar	108	c	8	7 3/8	− 3	3,594
Mar	108	p	8	3 1/8	+ 9/16	4,769
Jun	108	c	12	10 1/4	− 2 3/4	311
Jan	109	c	6	3 7/8	− 3 3/8	30
Jan	109	p	43	1 1/2	+ 11/16	689
Jan	110	c	143	2 3/4	− 3 1/8	1,519
Jan	110	p	213	1 13/16	+ 13/16	5,035
Feb	110	c	13	4 5/8	− 2	63
Feb	110	p	58	3	+ 15/16	514
Mar	110	c	34	5 7/8	− 2 5/8	7,029

Strike			Vol.	3 pm Close	Net. Chg.	Open Int.
Feb	470	c	2	27 1/8	− 4	8
Feb	470	p	1	8	+ 3/8	615
Jan	480	c	15	11	− 14 3/4	19
Mar	480	p	1	15 1/2	− 1/8	2,364
Jan	485	c	2	10	− 6 7/8	9
Jan	490	c	4	9 3/8	− 1 3/4	75
Feb	490	c	86	12 3/8	− 6 3/8	219
Mar	490	c	3	23	− 2	100
Mar	490	p	3	18 1/2	+ 1/2	100
Feb	500	c	400	8 1/4	− 5	186
Feb	500	p	100	20 3/4	+ 1 1/4	195
Jan	510	c	5	2	− 2 1/4	5
Feb	510	c	10	5 3/4	− 3 1/4	21
Mar	520	c	3	6 1/2	− 3 3/4	5
Call Vol.			**533**	**Open Int.**		**8,978**
Put Vol.			**266**	**Open Int.**		**14,765**

S & P 100 INDEX(OEX)

Strike			Vol.	3 pm Close	Net. Chg.	Open Int.
Mar	540	p	33	1 1/8	+ 1/4	486
Jan	550	p	796	1/16	− 1/16	6,551
Jan	560	p	60	3/16	+ 1/16	1,498
Feb	560	p	10	5/8	...	272
Jan	580	p	226	1/4	+ 1/8	1,195
Feb	580	p	32	3/4	+ 1/8	231
Jan	600	p	400	3/16	...	3,531
Feb	600	c	1	187	− 17	1
Feb	600	p	14	1	+ 1/8	1,088
Jan	610	c	1	172	− 18	25
Jan	610	p	251	5/16	+ 1/16	1,200
Jan	620	p	95	3/8	+ 1/8	2,490
Feb	620	c	1	165	− 19	10
Feb	620	p	43	2	...	496
Mar	620	p	81	4 3/8	+ 1 1/8	712
Jan	630	p	29	3/8	+ 1/16	1,623
Jan	640	c	3	144	− 12 1/2	199
Jan	640	p	50	7/16	− 1/16	2,022
Feb	640	c	1	146	− 19	31
Feb	640	p	60	3 1/8	+ 3/4	838
Mar	640	p	41	5 1/8	+ 1 3/4	315
Apr	640	p	6	6 1/4	+ 1/8	32
Jan	650	p	36	9/16	− 3/16	4,315
Jan	660	p	90	1	+ 1/2	3,318
Feb	660	c	1	125	− 7 1/4	5
Feb	660	p	42	4 1/8	+ 1	1,083
Mar	660	c	4	132	− 18	59
Mar	660	p	90	7	+ 1	789
Apr	660	p	2	8 7/8	+ 2 1/8	4
Jan	670	c	18	99 1/2	− 31 1/2	137
Jan	670	p	413	1 1/4	+ 1/4	2,349
Feb	670	p	44	4 7/8	+ 7/8	125
Jan	680	c	706	102	− 7 3/8	1,309
Jan	680	p	244	1 9/16	+ 7/16	3,464
Feb	680	p	92	6	+ 2 1/8	223
Mar	680	p	76	10 1/4	+ 3 1/2	1,657
Jan	690	p	291	1 3/4	+ 3/8	4,583
Feb	690	c	4	91	− 14 1/4	11
Feb	690	p	171	6 1/2	+ 2 1/4	1,977
Jan	695	p	283	2 3/8	+ 7/8	1,415
Jan	700	c	124	73	− 18 7/8	6,023
Jan	700	p	968	2 1/4	+ 5/8	12,174
Feb	700	p	278	7 1/2	+ 2	1,305
Mar	700	p	7	10	+ 1/2	2,731
Apr	700	p	3	18 1/2	+ 5 1/2	98
Jan	705	p	245	2 5/8	+ 5/8	1,569
Jan	710	c	74	61 1/2	− 19 1/2	2,386
Jan	710	p	726	3 1/4	+ 1 3/8	7,712
Feb	710	p	6	9 3/8	+ 2 3/4	580
Jan	715	p	237	3 1/4	+ 1 1/4	3,720

Strike			Vol.	3 pm Close	Net. Chg.	Open Int.
Feb	1500	c	669	12	− 12 1/4	5,484
Feb	1500	p	55	78 1/2	+ 9	116
Mar	1500	c	1,001	32 1/2	− 4 1/2	13,524
Mar	1500	p	17	83	+ 8	740
Jan	1525	c	541	3/4	− 3 1/4	5,628
Jan	1525	p	203	97	+ 16	278
Feb	1525	c	181	7 3/4	− 8 1/4	1,861
Mar	1525	c	1,606	15	− 25 1/2	7,549
Mar	1525	p	500	114	+ 28	218
Jan	1550	c	789	3/8	− 5/8	13,957
Feb	1550	c	221	3 5/8	− 4 7/8	3,059
Feb	1550	p	2	111	+ 25	77
Mar	1550	c	1,397	11	− 5 1/2	9,956
Jan	1575	c	10	3/8	+ 1/8	2,102
Feb	1575	c	6	2	− 3 5/8	1,779
Mar	1575	c	209	6 3/4	− 6	619
Mar	1575	p	10	152 1/2	+ 38 1/2	11
Feb	1600	c	155	1	− 1 7/16	1,925
Mar	1600	c	502	3 3/4	− 3 1/4	9,199
Mar	1650	c	55	1 1/4	− 1 3/4	2,052
Jan	1700	p	1	267	+ 15	373
Mar	1700	c	100	1	− 1/8	1,416
Mar	1700	p	3	264	+ 22	32
Call Vol.			**36,803**	**Open Int.**		**820,987**
Put Vol.			**45,134**	**Open Int.**		**975,013**

AMERICAN

COMP TECH(XCI)

Strike			Vol.	3 pm Close	Net. Chg.	Open Int.
Jan	860	p	3	16 1/4	...	...
Jan	1120	c	22	259 3/4	+ 160 1/8	22
Jan	1170	p	3	7	+ 2 7/8	3
Call Vol.			**22**	**Open Int.**		**71**
Put Vol.			**6**	**Open Int.**		**60**

JAPAN INDEX(JPN)

Strike			Vol.	3 pm Close	Net. Chg.	Open Int.
Mar	170	c	2	25 1/4	+ 3 3/4	651
Mar	175	c	9	21	− 1	39
Jan	180	p	30	11/16	− 1/8	125
Mar	180	c	8	17 1/2	+ 3 1/2	3,390
Jan	185	c	10	9 5/8	+ 1/8	10
Jan	185	p	10	11/16	+ 1/4	91
Feb	185	p	10	3	− 3/4	10
Mar	185	c	1	14 3/8	− 1 1/4	99
Jan	190	p	35	2 1/8	+ 1/16	145
Mar	190	c	156	10 3/4	− 1 1/2	252
Jan	195	c	1	3 5/8	− 7/8	42
Feb	195	p	35	5 3/4	− 4 1/8	39
Mar	195	c	4	8 1/4	− 1/8	101
Call Vol.			**556**	**Open Int.**		**23,806**
Put Vol.			**120**	**Open Int.**		**7,561**

MS CYCLICAL(CYC)

Strike			Vol.	3 pm Close	Net. Chg.	Open Int.
Jan	520	p	125	3 1/8	...	125
Feb	520	p	2,000	7 1/2	− 1/2	5,000
Feb	560	c	2,000	22	− 4	5,000
Jan	590	c	20	4 1/2	− 7 1/2	400
Call Vol.			**2,020**	**Open Int.**		**12,676**
Put Vol.			**2,125**	**Open Int.**		**11,224**

MS HITECH 35(MSH)

Strike			Vol.	3 pm Close	Net. Chg.	Open Int.
Jan	1560	c	50	310 1/4	+ 10 1/4	39
Jan	1570	c	850	303 3/4	− 36 1/4	865
Jan	1580	c	50	291 1/4	+ 23 1/2	14
Feb	1600	p	275	45	+ 10	260
Jan	1610	p	60	14 3/4	− 8 1/8	21
Jan	1630	c	4	218 3/8	− 11 7/8	3

Figure 2.2 The *Wall Street Journal Europe* of 5th January 2000, Index Options. Reproduced by permission of Dow Jones & Company, Inc.

traded prices of the options on the stocks. To understand how to read this let us examine the prices of options on Gateway. Go to 'Gateway' in the list. The closing price on 4th January 2000 was $65\frac{1}{2}$, and is written beneath 'Gateway' several times. Calls and puts are quoted here with strikes of $60 and $65, others may exist but are not mentioned in the newspaper for want of space. The available expiries are January and March. Part of the information included here is the volume of the transactions in each series, we won't worry about that but some people use option volume as a trading indicator. From the data, we can see that the January calls with a strike of $60 were worth $6\frac{7}{8}$. The puts with same strike and expiry were worth $2. The March calls with a strike of $60 were worth $10\frac{1}{2}$ and the puts with same strike and expiry were worth $6. Note that the higher the strike, the lower the value of the calls but the higher the value of the puts. This makes sense when you remember that the call allows you to buy the underlying for the strike, so that the lower the strike price the more this right is worth to you. The opposite is true for a put since it allows you to sell the underlying for the strike price.

There are more strikes and expiries available for options on indices, so let's now look at the Index Options section of the *Wall Street Journal Europe* of 5th January 2000, shown in Figure 2.2.

In Figure 2.3 are the quoted prices of the March and June DJIA calls against the strike price. Also plotted is the payoff function *if the underlying were to finish at its current value at expiry*, the current closing price of the DJIA was 10997.93.

This plot reinforces the fact that the higher the strike the lower the value of a call option. It also appears that the longer time to maturity the higher the value of the call. Is it obvious that this should be so? As the time to expiry decreases what would we see happen? As there is less and less time for the underlying to move, so the option value must converge to the payoff function.

One of the most interesting feature of calls and puts is that they have a nonlinear dependence on the underlying asset. This contrasts with futures which have a linear dependence on the underlying. This nonlinearity is very important in the pricing of options, the randomness in the underlying asset and the curvature of the option value with respect to the asset are intimately related.

Calls and puts are the two simplest forms of option. For this reason they are often referred to as **vanilla** because of the ubiquity of that flavor. There are many, many more kinds of options,

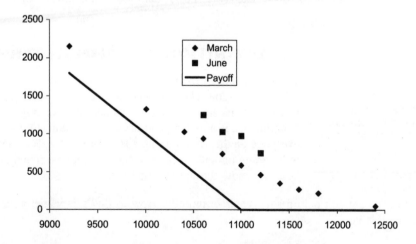

Figure 2.3   Option prices versus strike, March and June series of DJIA.

		BID m	ASK m	LAST m	1CHG m	IVBD	IVAS	BEST	DEBS	GABS	VEBS	THEO	7DEC
GLXO LN CALLS		Bid Price	Ask Price	Last Trade	1 Day Net Change	Imp Volat Bid	Imp Volat Ask	Best Price	Delta Best Price	Gamma Best Price	Vega Best Price	Theo. Value	7 Day Decay
GLXOCT99		1686.0	1689.0	1688.0	-13.0			1687					
1)	1200	489.50	504.50	509.50	unch	N.A.	69.97	504.50	.942	.0003	.674	494.09	4.6870
2)	1250	440.00	455.00	460.00	unch	N.A.	63.58	455.00	.936	.0003	.689	444.92	4.6853
3)	1300	390.50	405.50	410.50	unch	N.A.	57.36	405.50	.928	.0004	.837	396.33	4.6828
4)	1350	342.00	357.00	362.00	unch	N.A.	52.29	357.00	.915	.0005	.853	348.72	4.8888
5)	1400	294.50	309.50	314.50	unch	N.A.	48.07	309.50	.895	.0007	1.018	302.62	5.2385
6)	1450	249.00	264.00	268.50	unch	29.45	45.11	264.00	.864	.0008	1.194	258.66	5.8316
7)	1500	203.00	218.00	224.00	unch	30.67	42.27	220.00	.823	.0011	1.538	217.53	6.3538
8)	1600	125.00	137.50	136.00	-6.00	29.86	37.59	136.00	.706	.0017	2.013	146.02	7.0423
9)	1700	69.00	76.00	80.00	unch	30.95	34.02	76.00	.516	.0020	2.280	90.95	7.4785
10)	1800	32.00	38.00	40.00	unch	30.62	33.12	37.00	.319	.0019	2.005	52.71	36.2390
11)	1900	16.00	20.00	21.50	unch	32.84	35.47	20.00	.190	.0013	1.552	28.38	64.8611
12)	2000	6.00	9.00	9.00	unch	32.53	35.83	9.00	.099	.0008	1.041	14.27	32.9660
13)	2100	2.00	4.00	3.50	unch	32.32	36.52	3.50	.044	.0005	.581	6.728	1.4568
14)	2200		2.00	1.00	unch	N.A.	38.08	1.00	.015	.0002	.232	2.968	.5272
15)	2300		1.50	.50	unch	N.A.	41.58	.50	.008	.0001	.132	1.262	.2929
16)	2400		1.00	.50	unch	N.A.	43.98	.50	.007	.0001	.126	.502	.2977
17)	2500		1.00	.50	unch	N.A.	48.40	.50	.007	.0001	.101	.195	.3010

GLXO LN GBp ↑   1688 −13   L 5s L 1686/1689 L Trd Equity **OCM**
At 12:50 Vol 854,194 Op 1694 L Hi 1703 L Lo 1686 L Prev 1701
OPTION MONITOR 3   COMP   Center: 1687   1 〈GO〉 to Edit Spreadsheet

Figure 2.4   Prices for Glaxo–Wellcome calls expiring in October. Source: Bloomberg L.P.

some of which will be described and examined later on. Other terms used to describe contracts with some dependence on a more fundamental asset are **derivatives** or **contingent claims**.

Figure 2.4 shows the prices of call options on Glaxo–Wellcome for a variety of strikes. All these options are expiring in October. The table shows many other quantities that we will be seeing later on.

WE'LL BE USING THESE TERMS ALL THE TIME, THEY ARE STANDARD THROUGHOUT THE INDUSTRY

2.3 **DEFINITION OF COMMON TERMS**

The subjects of mathematical finance and derivatives theory are filled with jargon. The jargon comes from both the mathematical world and the financial world. Generally speaking the jargon from finance is aimed at simplifying communication, and to put everyone on the same footing.[1] Here are a few loose definitions to be going on with, some you have already seen and there will be many more throughout the book.

- **Premium**: The amount paid for the contract initially. How to find this value is the subject of much of this book.

---

[1] I have serious doubts about the purpose of most of the math jargon.

- **Underlying (asset)**: The financial instrument on which the option value depends. Stocks, commodities, currencies and indices are going to be denoted by S. The option payoff is defined as some function of the underlying asset at expiry.

- **Strike (price)** or **exercise price**: The amount for which the underlying can be bought (call) or sold (put). This will be denoted by E. This definition only really applies to the simple calls and puts. We will see more complicated contracts in later chapters and the definition of strike or exercise price will be extended.

- **Expiration (date)** or **expiry (date)**: Date on which the option can be exercised or date on which the option ceases to exist or give the holder any rights. This will be denoted by T.

- **Intrinsic value**: The payoff that would be received if the underlying is at its current level when the option expires.

- **Time value**: Any value that the option has above its intrinsic value. The uncertainty surrounding the future value of the underlying asset means that the option value is generally different from the intrinsic value.

- **In the money**: An option with positive intrinsic value. A call option when the asset price is above the strike, a put option when the asset price is below the strike.

- **Out of the money**: An option with no intrinsic value, only time value. A call option when the asset price is below the strike, a put option when the asset price is above the strike.

- **At the money**: A call or put with a strike that is close to the current asset level.

- **Long position**: A positive amount of a quantity, or a positive exposure to a quantity.

- **Short position**: A negative amount of a quantity, or a negative exposure to a quantity. Many assets can be sold short, with some constraints on the length of time before they must be bought back.

2.4 **PAYOFF DIAGRAMS**

The understanding of options is helped by the visual interpretation of an option's value at expiry. We can plot the value of an option at expiry as a function of the underlying in what is known as a **payoff diagram**. At expiry the option is worth a known amount. In the case of a call option the contract is worth $\max(S - E, 0)$. This function is the bold line in Figure 2.5.

Figure 2.6 shows Bloomberg's standard option valuation screen and Figure 2.7 shows the value against the underlying and the payoff.

The payoff for a put option is $\max(E - S, 0)$, this is the bold line plotted in Figure 2.8.

Figure 2.6 shows Bloomberg's option valuation screen and Figure 2.7 shows the value against the underlying and the payoff.

These payoff diagrams are useful since they simplify the analysis of complex strategies involving more than one option.

Make a note of the thin lines in all of these figures. The meaning of these will be explained very shortly.

2.4.1 Other Representations of Value

The payoff diagrams shown above only tell you about what happens at expiry, how much money your option contract is worth at that time. They make no allowance for how much premium you had to pay for the option. To adjust for the original cost of the option, sometimes

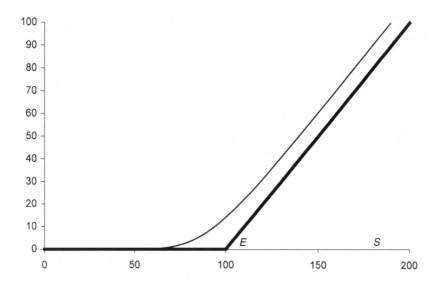

Figure 2.5   Payoff diagram for a call option.

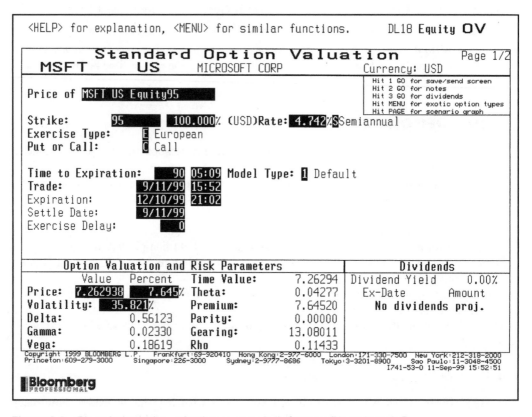

Figure 2.6   Bloomberg option valuation screen, call. Source: Bloomberg L.P.

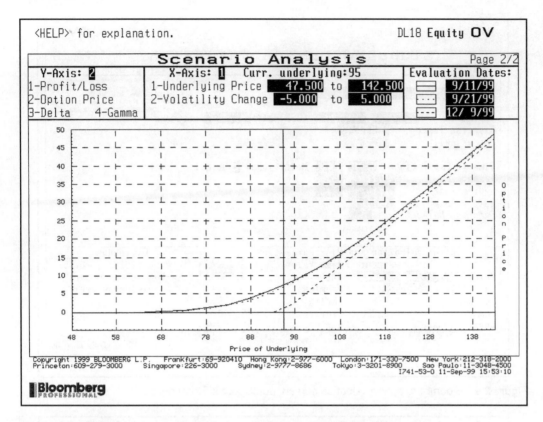

Figure 2.7 Bloomberg scenario analysis, call. Source: Bloomberg L.P.

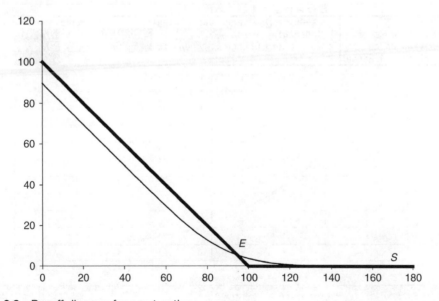

Figure 2.8 Payoff diagram for a put option.

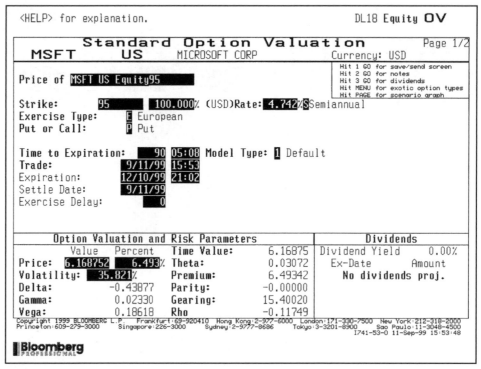

Figure 2.9 Bloomberg option valuation screen, put. Source: Bloomberg L.P.

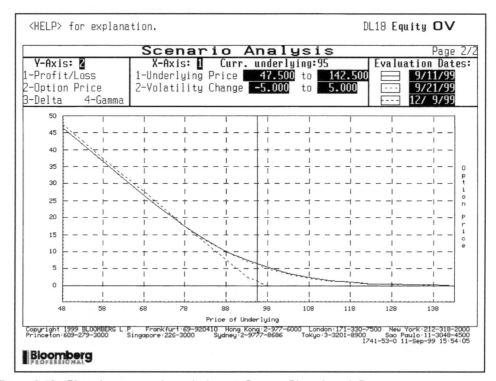

Figure 2.10 Bloomberg scenario analysis, put. Source: Bloomberg L.P.

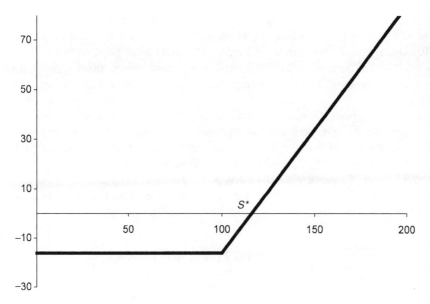

Figure 2.11 Profit diagram for a call option.

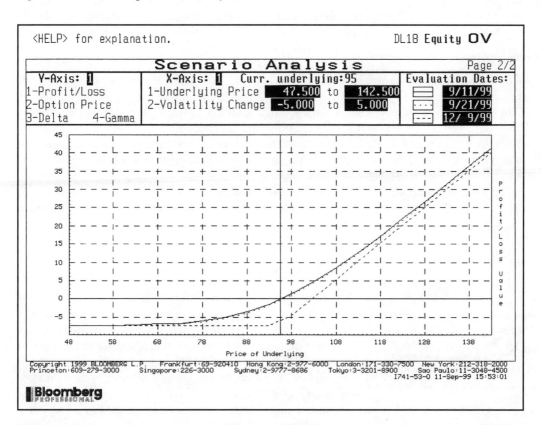

Figure 2.12 Profit diagram for a call. Source: Bloomberg L.P.

one plots a diagram such as that shown in Figure 2.11. In this **profit diagram** for a call option I have subtracted from the payoff the premium originally paid for the call option. This figure is helpful because it shows how far into the money the asset must be at expiry before the option becomes profitable. The asset value marked S^* is the point which divides profit from loss; if the asset at expiry is above this value then the contract has made a profit, if below the contract has made a loss.

As it stands, this profit diagram takes no account of the time value of money. The premium is paid up front but the payoff, if any, is only received at expiry. To be consistent one should either discount the payoff by multiplying by $e^{-r(T-t)}$ to value everything at the present, or multiply the premium by $e^{r(T-t)}$ to value all cashflows at expiry.

Figure 2.12 shows Bloomberg's call option profit diagram. Note that the profit today is zero; if we buy the option and immediately sell it we make neither a profit nor a loss (this is subject to issues of transaction costs).

2.5 **WRITING OPTIONS**

WRITING AN OPTION IS NOT THE SAME AS SELLING IT

I have talked above about the rights of the purchaser of the option. But for every option that is sold, someone somewhere must be liable if the option is exercised. If I hold a call option entitling me to buy a stock some time in the future, who do I buy this stock from? Ultimately, the stock must be delivered by the person who **wrote** the option. The **writer** of an option is the person who promises to deliver the underlying asset, if the option is a call, or buy it, if the option is a put. The writer is the person who receives the premium.

In practice, most simple option contracts are handled through an exchange so that the purchaser of an option does not know who the writer is. The holder of the option can even sell the option on to someone else via the exchange to close his position. However, regardless of who holds the option, or who has handled it, the writer is the person who has the obligation to deliver or buy the underlying.

The asymmetry between owning and writing options is now clear. The purchaser of the option hands over a premium in return for special rights, and an uncertain outcome. The writer receives a guaranteed payment up front, but then has obligations in the future.

2.6 **MARGIN**

SOME CONTRACTS HAVE MARGIN REQUIREMENTS OTHERS DO NOT

Writing options is very risky. The downside of buying an option is just the initial premium, the upside may be unlimited. The upside of writing an option is limited, but the downside could be huge. For this reason, to cover the risk of default in the event of an unfavorable outcome, the **clearing houses** that register and settle options insist on the deposit of a **margin** by the writers of options. Clearing houses act as counterparty to each transaction.

Margin comes in two forms, the **initial margin** and the **maintenance margin**. The initial margin is the amount deposited at the initiation of the contract. The total amount held as margin must stay above a prescribed maintenance margin. If it ever falls below this level then more money (or equivalent in bonds, stocks etc.) must be deposited. The levels of these margins vary from market to market.

Margin has been much neglected in the academic literature. But a poor understanding of the subject has led to a number of famous financial disasters, most notably Metallgesellschaft and Long Term Capital Management. We'll discuss the details of these cases in Chapter 59, and we'll also be seeing how to model margin and how to margin hedge.

2.7 **MARKET CONVENTIONS**

Most of the simpler options contracts are bought and sold through exchanges. These exchanges make it simpler and more efficient to match buyers with sellers. Part of this simplification involves the conventions about such features of the contracts as the available strikes and expiries. For example, simple calls and puts come in **series**. This refers to the strike and expiry dates. Typically a stock has three choices of expiries trading at any time. Having standardized contracts traded through an exchange promotes liquidity of the instruments.

Some options are an agreement between two parties, often brought together by an intermediary. These agreements can be very flexible and the contract details do not need to satisfy any conventions. Such contracts are known as **over the counter** or **OTC** contracts. I give an example at the end of this chapter.

2.8 **THE VALUE OF THE OPTION BEFORE EXPIRY**

We have seen how much calls and puts are worth at expiry, and drawn these values in payoff diagrams. The question that we can ask, and the question that is central to this book, is 'How much is the contract worth *now*, before expiry?' How much would you pay for a contract, a piece of paper, giving you rights in the future? You may have no idea what the stock price will do between now and expiry in six months, say, but clearly the contract has value. At the very least you know that there is no downside to owning the option, the contract gives you specific rights but no *obligations*. Two things are clear about the contract value before expiry: the value will depend on how high the asset price is today and how long there is before expiry.

The higher the underlying asset today, the higher we might expect the asset to be at expiry of the option and therefore the more valuable we might expect a call option to be. On the other hand a put option might be cheaper by the same reasoning.

The dependence on time to expiry is more subtle. The longer the time to expiry, the more time there is for the asset to rise or fall. Is that good or bad if we own a call option? Furthermore, the longer we have to wait until we get any payoff, the less valuable will that payoff be simply because of the time value of money.

I will ask you to suspend disbelief for the moment (it won't be the last time in the book) and trust me that we will be finding a 'fair value' for these options contracts. The aspect of finding the 'fair value' that I want to focus on now is the dependence on the asset price and time. I am going to use V to mean the value of the option, and it will be a function of the value of the underlying asset S at time t. Thus we can write $V(S, t)$ for the value of the contract.

We know the value of the contract *at expiry*. If I use T to denote the expiry date then at $t = T$ the function V is known, it is just the payoff function. For example if we have a call option then

$$V(S, T) = \max(S - E, 0).$$

This is the function of S that I plotted in the earlier payoff diagrams. Now I can tell you what the fine lines are in Figures 2.5 and 2.8, they are the values of the contracts $V(S, t)$ *at some time before expiry*, plotted against S. I have not specified how long before expiry, since the plot is for explanatory purposes only. I will spend a lot of time showing you how to find these values for a wide variety of contracts.

2.9 FACTORS AFFECTING DERIVATIVE PRICES

The two most important factors affecting the prices of options are the value of the underlying asset S and the time to expiry t. These quantities are **variables** meaning that they inevitably change during the life of the contract; if the underlying did not change then the pricing would be trivial. This contrasts with the **parameters** that affect the price of options.

Examples of parameters are the interest rate and strike price. The interest rate will have an effect on the option value via the time value of money since the payoff is received in the future. The interest rate also plays another role which we will see later. Clearly the strike price is important; the higher the strike in a call, the lower the value of the call.

If we have an equity option then its value will depend on any dividends that are paid on the asset during the option's life. If we have an FX option then its value will depend on the interest rate received by the foreign currency.

There is one important parameter that I have not mentioned, and which has a major impact

on the option value. That parameter is the **volatility**. Volatility is a measure of the amount of fluctuation in the asset price, a measure of the randomness. Figure 2.13 shows two asset price paths, the more jagged of the two has the higher volatility. The technical definition of volatility is the 'annualized standard deviation of the asset returns'. I will show how to measure this parameter in Chapter 3.

Volatility is a particularly interesting parameter because it is so hard to estimate. And having estimated it, one finds that it

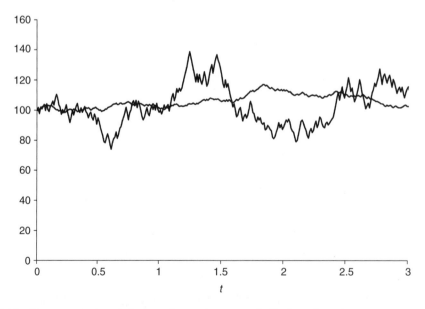

Figure 2.13   Two asset price paths, one is much more volatile than the other.

never stays constant and is unpredictable. Once you start to think of the volatility as varying in a random fashion then it becomes natural to treat it as a variable also. We will see examples of this later in the book.

The distinction between parameters and variables is very important. I shall be deriving equations for the value of options, partial differential equations. These equations will involve differentiation with respect to the variables, but the parameters, as their name suggests, remain as parameters in the equations.

2.10 **SPECULATION AND GEARING**

If you buy a far out-of-the-money option it may not cost very much, especially if there is not very long until expiry. If the option expires worthless, then you also haven't lost very much. However, if there is a dramatic move in the underlying, so that the option expires in the money, you may make a large profit relative to the amount of the investment. Let me give an example.

Example

Today's date is 14th April and the price of Wilmott Inc. stock is $666. The cost of a 680 call option with expiry 22nd August is $39. I expect the stock to rise significantly between now and August, how can I profit if I am right?

Buy the Stock

Suppose I buy the stock for $666. And suppose that by the middle of August the stock has risen to $730. I will have made a profit of $64 per stock. More importantly my investment will have risen by

$$\frac{730 - 666}{666} \times 100 = 9.6\%.$$

Buy the Call

If I buy the call option for $39, then at expiry I can exercise the call, paying $680 to receive something worth $730. I have paid $39 and I get back $50. This is a profit of $11 per option, but in percentage terms I have made

$$\frac{\text{value of asset at expiry} - \text{strike} - \text{cost of call}}{\text{cost of call}} \times 100 = \frac{730 - 680 - 39}{39} \times 100 = 28\%.$$

This is an example of **gearing** or **leverage**. The out-of-the-money option has a high gearing, a possible high payoff for a small investment. The downside of this leverage is that the call option is more likely than not to expire completely worthless and you will lose all of your investment. If Wilmott Inc. remains at $666 then the stock investment has the same value but the call option experiences a 100% loss.

Highly-leveraged contracts are very risky for the writer of the option. The buyer is only risking a small amount; although he is very likely to lose, his downside is limited to his initial premium. But the writer is risking a large loss in order to make a probable small profit. The writer is likely to think twice about such a deal unless he can offset his risk by buying other contracts. This offsetting of risk by buying other related contracts is called **hedging**.

Gearing explains one of the reasons for buying options. If you have a strong view about the direction of the market then you can exploit derivatives to make a better return, if you are right, than buying or selling the underlying.

2.11 EARLY EXERCISE

The simple options described above are examples of **European options** because exercise is only permitted *at expiry*. Some contracts allow the holder to exercise *at any time* before expiry, and these are called **American options**. American options give the holder more rights than their European equivalent and can therefore be more valuable, and they can never be less valuable. The main point of interest with American-style contracts is deciding *when* to exercise. In Chapter 9 I will discuss American options in depth, and show how to determine when it is *optimal* to exercise, so as to give the contract the highest value.

Note that the terms 'European' and 'American' do not in any way refer to the continents on which the contracts are traded.

Finally, there are **Bermudan options**. These allow exercise on specified dates, or in specified periods. In a sense they are half-way between European and American since exercise is allowed on some days and not on others.

2.12 PUT-CALL PARITY

Imagine that you buy one European call option with a strike of E and an expiry of T and that you write a European put option with the same strike and expiry. Today's date is t. The payoff you receive at T for the call will look like the line in the first plot of Figure 2.14. The payoff for the put is the line in the second plot in the figure. Note that the sign of the payoff is negative, you *wrote* the option and are liable for the payoff. The payoff for the portfolio of the two options is the sum of the individual payoffs, shown in the third plot. The payoff for this portfolio of options is

$$\max(S(T) - E, 0) - \max(E - S(T), 0) = S(T) - E,$$

where $S(T)$ is the value of the underlying asset at time T.

The right-hand side of this expression consists of two parts, the asset and a fixed sum E. Is there another way to get exactly this payoff? If I buy the asset today it will cost me $S(t)$ and be worth $S(T)$ at expiry. I don't know what the value $S(T)$ will be but I do know how to guarantee to get that amount, and that is to buy the asset. What about the E term? To lock in a payment of E at time T involves a cashflow of $Ee^{-r(T-t)}$ at time t. The conclusion is that the portfolio of a long call and a short put gives me exactly the same payoff as a long asset, short cash position. The equality of these cashflows is independent of the future behavior of the stock and is model independent:

$$C - P = S - Ee^{-r(T-t)},$$

ANOTHER EXAMPLE OF NO ARBITRAGE. FROM NOW ON THEY WON'T BE SO EASY

where C and P are today's values of the call and the put respectively. This relationship holds at any time up to expiry and is known as **put-call parity**. If this relationship did not hold then there would be riskless arbitrage opportunities.

In Table 2.1 I show the cashflows in the perfectly hedged portfolio. In this table I have set up the cashflows to have a guaranteed value of zero at expiry.

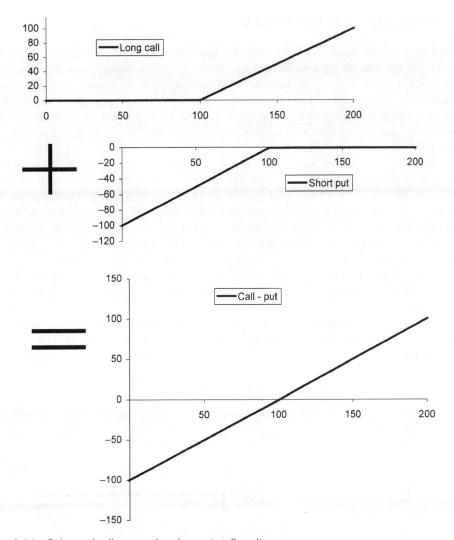

Figure 2.14 Schematic diagram showing put-call parity.

Table 2.1 Cashflows in a hedged portfolio of options and asset.

Holding	Worth today (t)	Worth at expiry (T)
Call	C	$\max(S(T) - E, 0)$
−Put	$-P$	$-\max(E - S(T), 0)$
−Stock	$-S(t)$	$-S(T)$
Cash	$Ee^{-r(T-t)}$	E
Total	$C - P - S(t) + Ee^{-r(T-t)}$	0

FUTURE CASHFLOW IS ZERO, SO TODAY'S MUST BE AS WELL

2.13 BINARIES OR DIGITALS

The original and still most common contracts are the vanilla calls and puts. Increasingly important are the **binary** or **digital options**. These contracts have a payoff at expiry that is discontinuous in the underlying asset price. An example of the payoff diagram for one of these options, a **binary call**, is shown in Figure 2.15. This contract pays $1 at expiry, time T, if the asset price is then greater than the exercise price E. Again, and as with the rest of the figures in this chapter, the bold line is the payoff and the fine line is the contract value some time before expiry.

Why would you invest in a binary call? If you think that the asset price will rise by expiry, to finish above the strike price then you might choose to buy either a vanilla call or a binary call. The vanilla call has the best upside potential, growing linearly with S beyond the strike. The binary call, however, can never pay off more than the $1. If you expect the underlying to rise dramatically then it may be best to buy the vanilla call. If you believe that the asset rise will be less dramatic then buy the binary call. The gearing of the vanilla call is greater than that for a binary call if the move in the underlying is large.

Figure 2.16 shows the payoff diagram for a **binary put**, the holder of which receives $1 if the asset is *below* E at expiry. The binary put would be bought by someone expecting a modest fall in the asset price.

There is a particularly simple binary put-call parity relationship. What do you get at expiry if you hold both a binary call and a binary put with the same strikes and expiries? The answer is that you will always get $1 regardless of the level of the underlying at expiry. Thus

$$\text{Binary call} + \text{Binary put} = e^{-r(T-t)}.$$

What would the table of cashflows look like for the perfectly hedged digital portfolio?

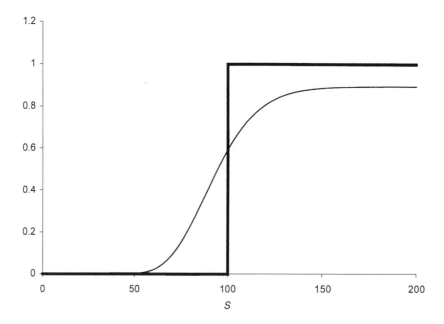

Figure 2.15 Payoff diagram for a binary call option.

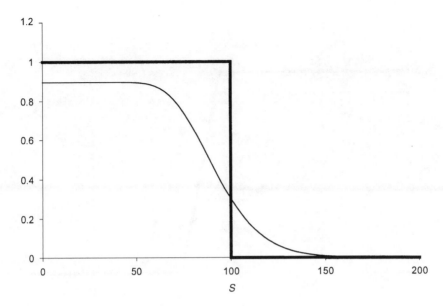

Figure 2.16 Payoff diagram for a binary put option.

2.14 **BULL AND BEAR SPREADS**

A payoff that is similar to a binary option can be made up with vanilla calls. This is our first example of a **portfolio of options** or an **option strategy**.

Suppose I buy one call option with a strike of 100 and write another with a strike of 120 and with the same expiration as the first then my resulting portfolio has a payoff that is shown in Figure 2.17. This payoff is zero below 100, 20 above 120 and linear in between. The payoff

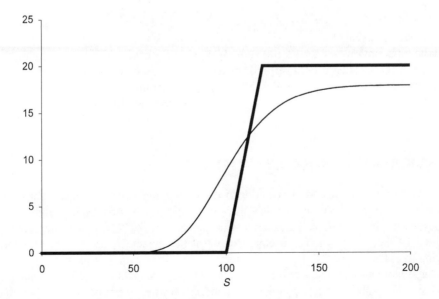

Figure 2.17 Payoff diagram for a bull spread.

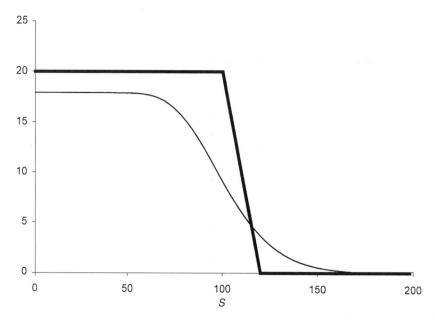

Figure 2.18 Payoff diagram for a bear spread.

is continuous, unlike the binary call, but has a payoff that is superficially similar. This strategy is called a **bull spread** because it benefits from a bull, i.e. rising, market.

The payoff for a general bull spread, made up of calls with strikes E_1 and E_2, is given by

$$\frac{1}{E_2 - E_1}(\max(S - E_1, 0) - \max(S - E_2, 0)),$$

where $E_2 > E_1$. Here I have bought/sold $(E_2 - E_1)^{-1}$ of each of the options so that the maximum payoff is scaled to 1.

If I write a put option with strike 100 and buy a put with strike 120 I get the payoff shown in Figure 2.18. This is called a **bear spread**, benefitting from a bear, i.e. falling, market. Again, it is very similar to a binary put except that the payoff is continuous.

Because of put-call parity it is possible to build up these payoffs using other contracts.

A strategy involving options of the same type (i.e. calls or puts) is called a **spread**.

2.15 **STRADDLES AND STRANGLES**

If you have a precise view on the behavior of the underlying asset you may want to be more precise in your choice of option; simple calls, puts, and binaries may be too crude.

The **straddle** consists of a call and a put with the same strike. The payoff diagram is shown in Figure 2.19. Such a position is usually bought at the money by someone who expects the underlying to either rise or fall, but not to remain at the same level. For example, just before an anticipated major news item stocks often show a 'calm before the storm'. On the announcement the stock suddenly moves either up or down depending on whether or not the news was favorable to the company.

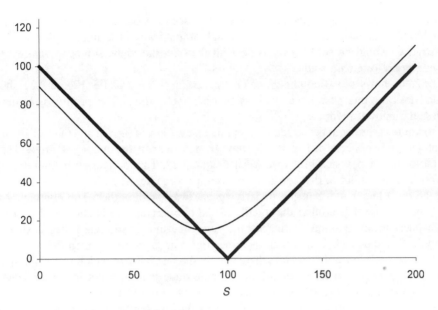

Figure 2.19 Payoff diagram for a straddle.

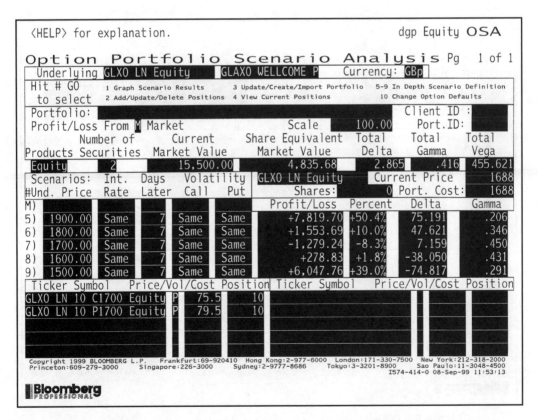

Figure 2.20 A portfolio of two options making up a straddle. Source: Bloomberg L.P.

They may also be bought by technical traders who see the stock at a key support or resistance level and expect the stock to either break through dramatically or bounce back.

The straddle would be sold by someone with the opposite view, someone who expects the underlying price to remain stable.

Figure 2.20 shows the Bloomberg screen for setting up a straddle. Figure 2.21 shows the profit and loss for this position at various times before expiry. The profit/loss is the option value less the upfront premium.

The **strangle** is similar to the straddle except that the strikes of the put and the call are different. The contract can be either an **out-of-the-money strangle** or an **in-the-money strangle**. The payoff for an out-of-the money strangle is shown in Figure 2.22. The motivation behind the purchase of this position is similar to that for the purchase of a straddle. The difference is that the buyer expects an even larger move in the underlying one way or the other. The contract is usually bought when the asset is around the middle of the two strikes and is cheaper than a straddle. This cheapness means that the gearing for the out-of-the-money strangle is higher than that for the straddle. The downside is that there is a much greater range over which the strangle has no payoff at expiry, while for the straddle there is only the one point at which there is no payoff.

There is another reason for a straddle or strangle trade that does not involve a view on the direction of the underlying. These contracts are bought or sold by those with a view on the direction of volatility, they are one of the simplest **volatility trades**. Because of the relationship between the price of an option and the volatility of the asset one can speculate on the direction

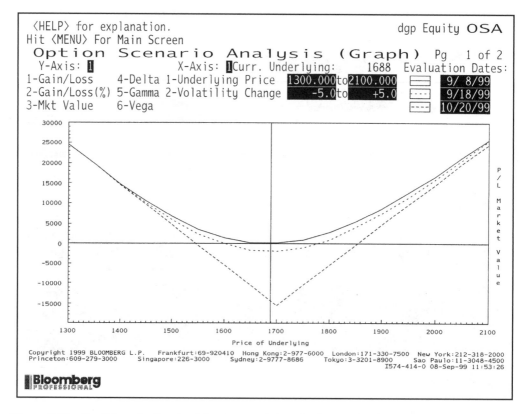

Figure 2.21 Profit/loss for the straddle at several times before expiry. Source: Bloomberg L.P.

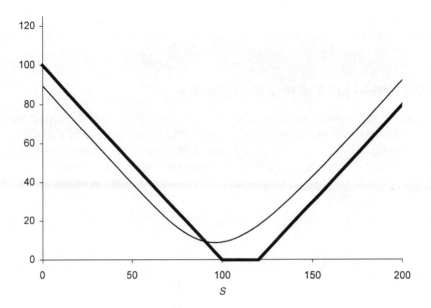

Figure 2.22 Payoff diagram for a strangle.

of volatility. Do you expect the volatility to rise? If so, how can you benefit from this? Until we know more about this relationship, we cannot go into this in more detail.

Straddles and strangles are rarely held until expiry.

A strategy involving options of different types (i.e. both calls and puts) is called a **combination**.

2.16 **RISK REVERSAL**

The **risk reversal** is a combination of a long call, with strike above the current spot, and a short put with a strike below the current spot. Both have the same expiry. The payoff is shown in Figure 2.23.

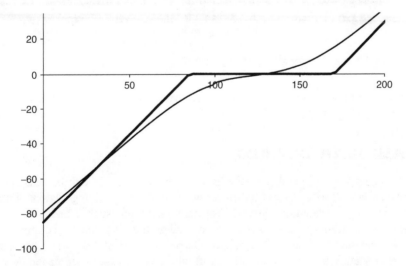

Figure 2.23 Payoff diagram for a risk reversal.

The risk reversal is a very special contract, popular with practitioners. Its value is usually quite small and related to the market's expectations of the behavior of volatility. This is too complex to go into now, but will be explained in Chapter 25.

2.17 BUTTERFLIES AND CONDORS

A more complicated strategy involving the purchase and sale of options with *three* different strikes is a **butterfly spread**. Buying a call with a strike of 90, writing two calls struck at 100 and buying a 110 call gives the payoff in Figure 2.24. This is the kind of position you might enter if you believe that the asset is not going anywhere, either up or down. Because it has no large upside potential (in this case the maximum payoff is 10) the position will be relatively cheap. With options, cheap is good.

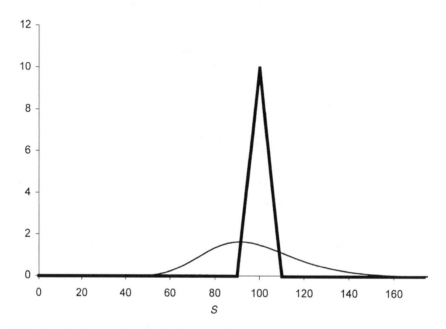

Figure 2.24 Payoff diagram for a butterfly spread.

The **condor** is like a butterfly except that four strikes, and four call options, are used. The payoff is shown in Figure 2.25.

2.18 CALENDAR SPREADS

All of the strategies I have described above have involved buying or writing calls and puts with different strikes *but all with the same expiration*. A strategy involving options with different expiry dates is called a **calendar spread**. You may enter into such a position if you have a precise view on the timing of a market move as well as the direction of the move. As always the motive behind such a strategy is to reduce the payoff at asset values and times which you believe are irrelevant, while increasing the payoff where you think it will matter. Any reduction in payoff will reduce the overall value of the option position.

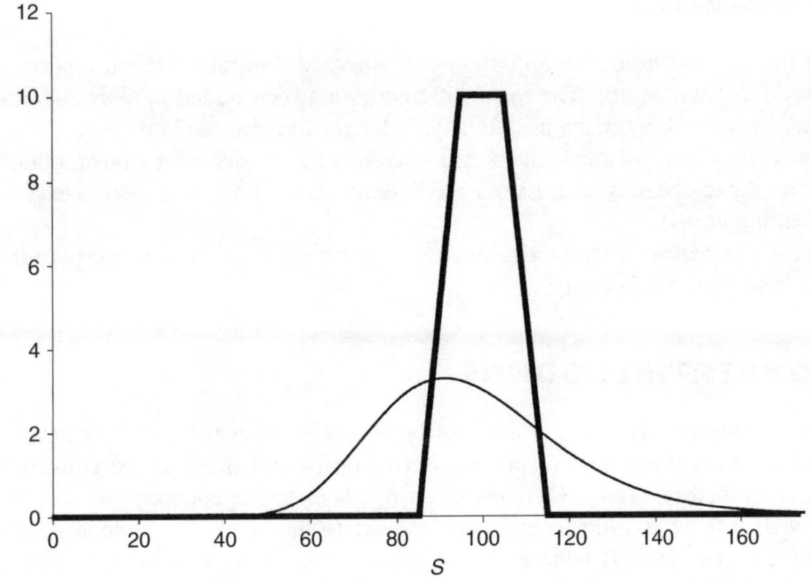

Figure 2.25 Payoff diagram for a condor.

2.19 **LEAPS AND FLEX**

LEAPS or **Long-term equity anticipation securities** are longer-dated exchange-traded calls and puts. They began trading on the CBOE in the late 1980s. They are standardized so that they expire in January each year and are available with expiries up to three years. They come with three strikes, corresponding to at the money and 20% in and out of the money with respect to the underlying asset price when issued.

Figure 2.26 shows LEAPS quoted in the *Wall Street Journal Europe*.

In 1993 the CBOE created **FLEX** or **FLexible EXchange-traded options** on several indices. These allow a degree of customization, in the expiry date (up to five years), the strike price and the exercise style.

LEAPS-LONG TERM

DJ INDUS AVG – CB						
Dec 01	104	p	42	7¾	+ ¾	475
Dec 01	140	p	44	22¼	+ 2½	106
S & P 100 INDEX – CB						
Dec 01	140	c	5	38	+ 6	105
S & P 500 INDEX – CB						
Dec 00	70	p	60	7/16	+ ⅛	8993
Dec 00	90	p	5	1 3/16	+ ¼	15789
Dec 00	100	p	4	1 15/16	+ ½	18242
Dec 00	110	p	85	2¾	+ ¼	15068
Dec 00	112½	p	1	3	+ ⅜	7496
Dec 00	115	p	22	3⅝	+ ⅝	19282
Dec 00	117½	p	10	7⅝	+ 1⅜	839
Dec 00	120	p	52	4⅜	+ ⅝	17787
Dec 00	125	p	6	5⅛	+ ⅝	5809
Dec 00	130	p	18	6¼	+ ⅞	7322
Dec 00	140	p	90	9	+ 1¼	10159
Dec 00	145	p	2	9½	+ ¾	935
Call Volume			**39**	**Open Int 6,469,087**		
Put Volume			**20**	**Open Int 4,671,720**		

Figure 2.26 The *Wall Street Journal Europe* of 5th January 2000, LEAPS. Reproduced by permission of Dow Jones & Company, Inc.

2.20 **WARRANTS**

A contract that is very similar to an option is a **warrant**. Warrants are call options issued by a company on its own equity. The main differences between traded options and warrants are the timescales involved; warrants usually have a longer lifespan, and on exercise the company issues new stock to the warrant holder. On exercise, the holder of a *traded* option receives stock that has already been issued. Exercise is usually allowed any time before expiry, but after an initial waiting period.

The typical lifespan of a warrant is five or more years. Occasionally **perpetual warrants** are issued, these have no maturity.

2.21 **CONVERTIBLE BONDS**

Convertible bonds or **CB**s have features of both bonds and warrants. They pay a stream of coupons with a final repayment of principal at maturity, but they can be converted into the underlying stock before expiry. On conversion rights to future coupons are lost. If the stock price is low then there is little incentive to convert to the stock, the coupon stream is more valuable. In this case the CB behaves like a bond. If the stock price is high then conversion is likely and the CB responds to the movement in the asset. Because the CB can be converted into the asset, its value has to be at least the value of the asset. This makes CBs similar to American options; early exercise and conversion are mathematically the same.

There are other interesting features of convertible bonds, callback, resetting, etc., and the whole of Chapter 43 is devoted to their description and analysis.

2.22 **OVER THE COUNTER OPTIONS**

Not all options are traded on an exchange. Some, known as over the counter or OTC options, are sold privately from one counterparty to another. In Figure 2.27 is the term sheet for an OTC put option, having some special features. A **termsheet** specifies the precise details of an OTC contract. In this OTC put the holder gets a put option on S&P500, but more cheaply than a vanilla put option. This contract is cheap because part of the premium does not have to be paid until and unless the underlying index trades above a specified level. Each time that a new level is reached an extra payment is triggered. This feature means that the contract is not vanilla, and makes the pricing more complicated. We will be discussing special options like the ones in this contract in later chapters. Quantities in square brackets will be set at the time that the deal is struck.

2.23 **SUMMARY**

We now know the basics of options and markets, and a few of the simplest trading strategies. We know some of the jargon and the reasons why people might want to buy an option. We've also seen another example of no arbitrage in put-call parity. This is just the beginning. We don't know how much these instruments are worth, how they are affected by the price of the underlying, or how much risk is involved in the buying or writing of options. And we have only seen the very simplest of contracts; there are many, many more complex products to examine. All of these issues are going to be addressed in later chapters.

Over-the-counter Option linked to the S&P500 Index

Option Type	European put option, with contingent premium feature
Option Seller	XXXX
Option Buyer	[dealing name to be advised]
Notional Amount	USD 20MM
Trade Date	[]
Expiration Date	[]
Underlying Index	S&P500
Settlement	Cash settlement
Cash Settlement Date	5 business days after the Expiration Date
Cash Settlement Amount	Calculated as per the following formula:

$$\text{\#Contracts} * \max[0, \text{S\&Pstrike} - \text{S\&Pfinal}]$$

where #Contracts = Notional Amount / S&Pinitial

This is the same as a conventional put option:
S&Pstrike will be equal **to 95% of the closing price on the Trade Date**
S&Pfinal will be the level of the Underlying Index at the valuation time on the Expiration Date
S&Pinitial is the level of the Underlying Index at the time of execution

Initial Premium Amount	[2%] of Notional Amount
Initial Premium Payment Date	5 business days after Trade Date
Additional Premium Amounts	[1.43%] of Notional Amount per Trigger Level
Additional Premium Payment Dates	The Additional Premium Amounts shall be due only if the Underlying Index at any time from and including the Trade Date and to and including the Expiration Date is equal to or greater than any of the Trigger Levels.
Trigger Levels	103%, 106% and 109% of **S&P500initial**
Documentation	ISDA
Governing Law	New York

This indicative termsheet is neither an offer to buy or sell securities or an OTC derivative product which includes options, swaps, forwards and structured notes having similar features to OTC derivative transactions, nor a solicitation to buy or sell securities or an OTC derivative product. The proposal contained in the foregoing is not a complete description of the terms of a particular transaction and is subject to change without limitation.

Figure 2.27   Termsheet for an OTC 'Put'.

FURTHER READING

- McMillan (1996) and Options Institute (1995) describe many option strategies used in practice.
- Most exchanges have websites. The London International Financial Futures Exchange website contains information about the money markets, bonds, equities, indices and

commodities. See www.liffe.com. For information about options and derivatives generally, see www.cboe.com, the Chicago Board Options Exchange website. The American Stock Exchange is on www.amex.com and the New York Stock Exchange on www.nyse.com.

- Derivatives have often had bad press (and there's probably more to come). See Miller (1997) for a discussion of the pros and cons of derivatives.

- The best books on options are Hull (1997) and Cox & Rubinstein (1985), modesty forbids me from mentioning others.

CHAPTER 3
the random behavior
of assets

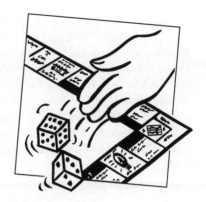

In this Chapter...

- more notation commonly used in mathematical finance
- how to examine time-series data to model returns
- the Wiener process, a mathematical model of randomness
- a simple model for equities, currencies, commodities and indices

3.1 INTRODUCTION

In this chapter I describe a simple continuous-time model for equities and other financial instruments, inspired by our earlier coin-tossing experiment. This takes us into the world of stochastic calculus and Wiener processes. Although there is a great deal of theory behind the ideas I describe, I am going to explain everything in as simple and accessible a manner as possible. We will be modeling the behavior of equities, currencies and commodities, but the ideas are applicable to the fixed-income world as we shall see in Part Four.

3.2 SIMILARITIES BETWEEN EQUITIES, CURRENCIES, COMMODITIES AND INDICES

When you invest in something, whether it is a stock, commodity, work of art or a racehorse, your main concern is that you will make a comfortable return on your investment. By **return** we tend to mean the percentage growth in the value of an asset, together with accumulated dividends, over some period:

$$\text{Return} = \frac{\text{Change in value of the asset} + \text{accumulated cashflows}}{\text{Original value of the asset}}.$$

I want to distinguish here between the percentage or relative growth and the absolute growth. Suppose we could invest in either of two stocks, both of which grow on average by $10 *per annum*. Stock A has a value of $100 and stock B is currently worth $1000. Clearly the former is a better investment, since at the end of the year stock A will probably be worth around $110 (if the past is anything to go by) and stock B $1010. Both have gone up by $10, but A has risen by 10% and B by only 1%. If we have $1000 to invest we would be better off investing in ten of asset A than one of asset B. This illustrates that when we come to model assets, it is the return that we should concentrate on. In this respect, all of equities, currencies, commodities and stock market indices can be treated similarly. What return do we expect to get from them?

Part of the business of estimating returns for each asset is to estimate how much unpredictability there is in the asset value. In the next section I am going to show that randomness plays a large part in financial markets, and start to build up a model for asset returns incorporating this randomness.

3.3 EXAMINING RETURNS

In Figure 3.1 I show the quoted price of Perez Companc, an Argentinian conglomerate, over the period February 1995 to November 1996. This is a typical plot of a financial asset. The asset shows a general upward trend over the period but this is far from guaranteed. If you bought and sold at the wrong times you would lose a lot of money. The unpredictability that is seen in this figure is the main feature of financial modeling. Because there is so much randomness, any mathematical model of a financial asset must acknowledge the randomness and have a probabilistic foundation.

Remembering that the returns are more important to us than the absolute level of the asset price, I show in Figure 3.2 how to calculate returns on a spreadsheet. Denoting the asset value on the ith day by S_i, then the return from day i to day $i+1$ is given by

$$\frac{S_{i+1} - S_i}{S_i} = R_i.$$

Figure 3.1 Perez Companc from February 1995 to November 1996.

Date	Perez	Return					
01-Mar-95	2.11		**Average return**	0.002916			
02-Mar-95	1.90	-0.1	**Standard deviation**	0.024521			
03-Mar-95	2.18	0.149906					
06-Mar-95	2.16	-0.010809					
07-Mar-95	1.91	-0.112583	= AVERAGE(C3:C463)				
08-Mar-95	1.86	-0.029851					
09-Mar-95	1.97	0.061538					
10-Mar-95	2.27	0.15	= STDEVP(C3:C463)				
13-Mar-95	2.49	0.099874					
14-Mar-95	2.76	0.108565					
15-Mar-95	2.61	-0.054264					
16-Mar-95	2.67	0.021858					
17-Mar-95	2.64	-0.010695					
20-Mar-95	2.60	-0.016216	=(B13-B12)/B12				
21-Mar-95	2.59	-0.002747					
22-Mar-95	2.59	-0.002755					
23-Mar-95	2.55	-0.012321					
24-Mar-95	2.73	0.069307					
27-Mar-95	2.91	0.064815					
28-Mar-95	2.92	0.002899					
29-Mar-95	2.92	0					
30-Mar-95	3.12	0.069364					
31-Mar-95	3.14	0.005405					
03-Apr-95	3.13	-0.002688					
04-Apr-95	3.24	0.037736					
05-Apr-95	3.25	0.002597					
06-Apr-95	3.28	0.007772					
07-Apr-95	3.21	-0.020566					
10-Apr-95	3.02	-0.060367					
11-Apr-95	3.08	0.019553					
12-Apr-95	3.19	0.035616					
17-Apr-95	3.21	0.007936					
18-Apr-95	3.17	-0.013123					
19-Apr-95	3.24	0.021277					

IT COULDN'T
BE EASIER

Figure 3.2 Spreadsheet for calculating asset returns.

(I've ignored dividends here, they are easily allowed for, especially since they only get paid two or four times a year typically.) Of course, I didn't need to use data spaced at intervals of a day, I will comment on this later.

In Figure 3.3 I show the daily returns for Perez Companc. This looks very much like 'noise', and that is exactly how we are going to model it. The mean of the returns distribution is

$$\overline{R} = \frac{1}{M} \sum_{i=1}^{M} R_i \qquad (3.1)$$

and the sample standard deviation is

$$\sqrt{\frac{1}{M-1} \sum_{i=1}^{M} (R_i - \overline{R})^2}, \qquad (3.2)$$

where M is the number of returns in the sample (one fewer than the number of asset prices). From the data in this example we find that the mean is 0.002916 and the standard deviation is 0.024521.

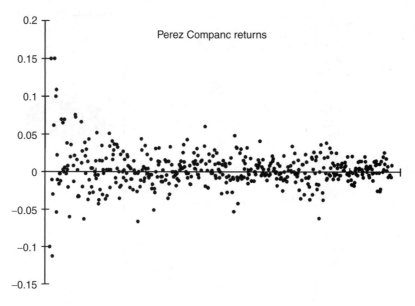

Figure 3.3 Daily returns of Perez Companc.

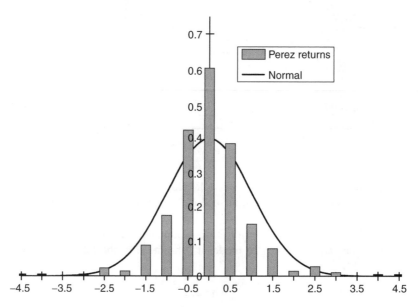

Figure 3.4 Normalized frequency distribution of Perez Companc and the standardized Normal distribution.

The frequency distribution of this time series of daily returns is easily calculated, and very instructive to plot. In Excel use Tools | Data Analysis | Histogram. In Figure 3.4 is shown the frequency distribution of daily returns for Perez Companc. This distribution has been scaled and translated to give it a mean of zero, a standard deviation of one and an area under the curve of one. On the same plot is drawn the probability density function for the standardized

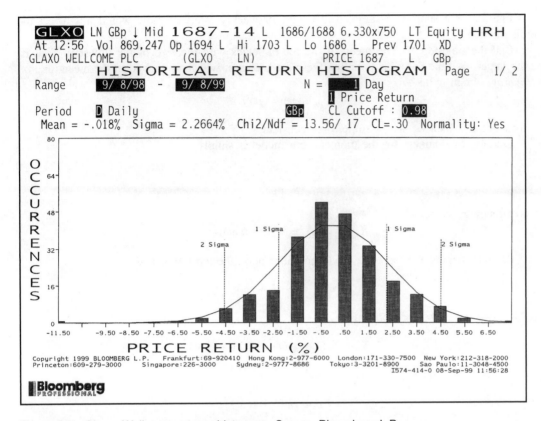

Figure 3.5 Glaxo–Wellcome returns histogram. Source: Bloomberg L.P.

Normal distribution function

$$\frac{1}{\sqrt{2\pi}}e^{-\phi^2/2},$$

where ϕ is a standardized Normal variable. The two curves are not identical but are fairly close.

Supposing that we believe that the empirical returns are close enough to Normal for this to be a good approximation, then we have come a long way towards a model. I am going to write the returns as a random variable, drawn from a Normal distribution with a known, constant, non-zero mean and a known, constant, non-zero standard deviation:

$$R_i = \frac{S_{i+1} - S_i}{S_i} = \text{mean} + \text{standard deviation} \times \phi.$$

Figure 3.5 shows the returns distribution of Glaxo–Wellcome as calculated by Bloomberg. This has not been normalized.

3.4 TIMESCALES

How do the mean and standard deviation of the returns' time series, as estimated by (3.1) and (3.2), scale with the timestep between asset price measurements? In the example the timestep is one day, but suppose I sampled at hourly or weekly intervals, how would this affect the distribution?

Call the timestep δt. The mean of the return scales with the size of the timestep. That is, the larger the time between sampling the more the asset will have moved in the meantime, *on average*. I can write

$$\text{mean} = \mu \, \delta t,$$

for some μ which we will assume to be constant.

Ignoring randomness for the moment, our model is simply

$$\frac{S_{i+1} - S_i}{S_i} = \mu \, \delta t.$$

Rearranging, we get

$$S_{i+1} = S_i(1 + \mu \, \delta t).$$

If the asset begins at S_0 at time $t = 0$ then after one timestep $t = \delta t$ and

$$S_1 = S_0(1 + \mu \, \delta t).$$

After two timesteps $t = 2 \, \delta t$ and

$$S_2 = S_1(1 + \mu \, \delta t) = S_0(1 + \mu \, \delta t)^2,$$

and after M timesteps $t = M$, $\delta t = T$ and

$$S_M = S_0(1 + \mu \, \delta t)^M.$$

This is just

$$S_M = S_0(1 + \mu \, \delta t)^M = S_0 e^{M \log(1 + \mu \, \delta t)} \approx S_0 e^{\mu M \, \delta t} = S_0 e^{\mu T}.$$

In the limit as the timestep tends to zero with the total time T fixed, this approximation becomes exact. This result is important for two reasons.

First, in the absence of any randomness the asset exhibits exponential growth, just like cash in the bank.

Second, the model is meaningful in the limit as the timestep tends to zero. If I had chosen to scale the mean of the returns distribution with any other power of δt it would have resulted in either a trivial model ($S_T = S_0$) or infinite values for the asset.

The second point can guide us in the choice of scaling for the random component of the return. How does the standard deviation of the return scale with the timestep δt? Again, consider what happens after $T/\delta t$ timesteps each of size δt (i.e. after a total time of T). Inside the square root in expression (3.2) there are a large number of terms, $T/\delta t$ of them. In order for the standard deviation to remain finite as we let δt tend to zero, the individual terms in the expression must each be of $O(\delta t)$. Since each term is a square of a return, the standard deviation of the asset return over a timestep δt must be $O(\delta t^{1/2})$:

$$\text{standard deviation} = \sigma \, \delta t^{1/2},$$

where σ is some parameter measuring the amount of randomness. The larger this parameter the more uncertain is the return. For the moment let's assume that it is constant.

Putting these scalings explicitly into our asset return model

$$R_i = \frac{S_{i+1} - S_i}{S_i} = \mu\,\delta t + \sigma\phi\,\delta t^{1/2}. \tag{3.3}$$

I can rewrite Equation (3.3) as

$$S_{i+1} - S_i = \mu S_i\,\delta t + \sigma S_i\phi\,\delta t^{1/2}. \tag{3.4}$$

The left-hand side of this equation is the change in the asset price from timestep i to timestep $i+1$. The right-hand side is the 'model'. We can think of this equation as a model for a **random walk** of the asset price. This is shown schematically in Figure 3.6. We know exactly where the asset price is today but tomorrow's value is unknown. It is distributed about today's value according to (3.4).

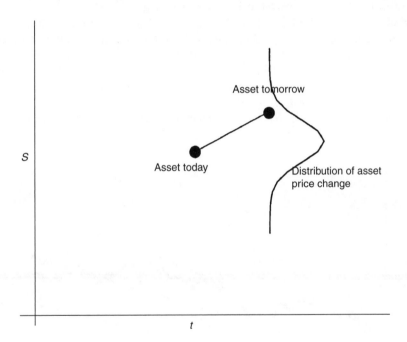

Figure 3.6 A representation of the random walk.

3.4.1 The Drift

The parameter μ is called the **drift rate**, the **expected return** or the **growth rate** of the asset. Statistically it is very hard to measure since the mean scales with the usually small parameter δt. It can be estimated by

$$\mu = \frac{1}{M\,\delta t}\sum_{i=1}^{M} R_i.$$

The unit of time that is usually used is the year, in which case μ is quoted as an *annualized* growth rate.

In the classical option pricing theory the drift plays almost no role. So even though it is hard to measure, this doesn't matter too much.[1]

3.4.2 The Volatility

The parameter σ is called the **volatility** of the asset. It can be estimated by

$$\sqrt{\frac{1}{(M-1)\delta t}\sum_{i=1}^{M}(R_i - \bar{R})^2}.$$

Again, this is almost always quoted in annualized terms.

The volatility is the most important and elusive quantity in the theory of derivatives. I will come back again and again to its estimation and modeling.

Because of their scaling with time, the drift and volatility have different effects on the asset path. The drift is not apparent over short timescales for which the volatility dominates. Over long timescales, for instance decades, the drift becomes important. Figure 3.7 is a realized path of the logarithm of an asset, together with its expected path and a 'confidence interval'. In this example the confidence interval represents one standard deviation. With the assumption of Normality this means that 68% of the time the asset should be within this range. The mean path is growing linearly in time and the confidence interval grows like the square root of time. Thus over short timescales the volatility dominates.[2]

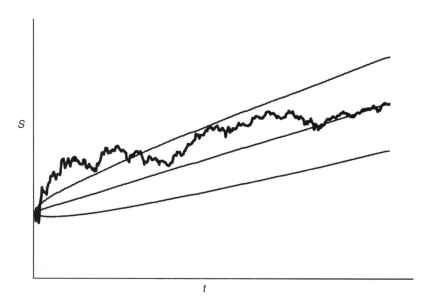

Figure 3.7 Path of the logarithm of an asset, its expected path and one standard deviation above and below.

---

[1] In non-classical theories and in portfolio management, it *does* often matter, very much.

[2] Why did I take the logarithm? Because changes in the logarithm are related to the return on the asset.

3.5 **ESTIMATING VOLATILITY**

The most common estimate of volatility is simply

$$\sqrt{\frac{1}{(M-1)\,\delta t} \sum_{i=1}^{M} (R_i - \overline{R})^2}.$$

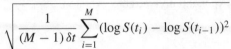

IT LOOKS EASY TO
CALCULATE VOLATILITY...
LATER WE'LL SEE SOME
OF THE PROBLEMS

If δt is sufficiently small the mean return $\overline{R}$ term can be ignored.
For small δt

$$\sqrt{\frac{1}{(M-1)\,\delta t} \sum_{i=1}^{M} (\log S(t_i) - \log S(t_{i-1}))^2}$$

can also be used, where $S(t_i)$ is the closing price on day t_i.

It is highly unlikely that volatility is constant for any given asset. Changing economic circumstances, seasonality etc. will inevitably result in volatility changing with time. If you want to know the volatility today you must use some past data in the calculation. Unfortunately, this means that there is no guarantee that you are actually calculating *today's* volatility.

Typically you would use daily closing prices to work out daily returns and then use the past 10, 30, 100, ... daily returns in the formula above. Or you could use returns over longer or shorter periods. Since all returns are equally weighted, while they are in the estimate of volatility, any large return will stay in the estimate of vol until the 10 (or 30 or 100) days have past. This gives rise to a plateauing of volatility, and is totally spurious.

Because volatility is not directly observable, and because of the plateauing effect in the simple measure of volatility, you might want to use other estimates. Below are a few of the more common ones.

3.5.1 Exponentially Weighted

One can also measure an exponentially-weighted volatility, the square root of a variance that is an exponential moving average of the square of price returns. Such a measure gives decreasing weight to returns further in the past. The plateauing effect is no longer seen, but we get an exponential-decay effect instead. This volatility is estimated according to

$$\sigma_i = \sqrt{\frac{1-\lambda}{\delta t} \sum_{j=-\infty}^{i} \lambda^{i-j} R_j^2},$$

where λ represents the weighting attached to the past volatility versus the present return. In this I have neglected the mean of R, assuming that the time horizon is small. This can be written as

$$\sigma_i^2 = \lambda \sigma_{i-1}^2 + (1-\lambda)\frac{R_i^2}{\delta t}.$$

3.5.2 Using Highs and Lows

If returns are Normally distributed, then we can estimate volatility using the highest price and lowest price reached each day:

$$\sqrt{\frac{1}{(M-1)\,\delta t\,4\log 2}\sum_{i=1}^{M}(\log H(t_i)-\log L(t_i))^2}$$

where $H(t_i)$ is the highest asset price on day t_i and $L(t_i)$ the lowest. This is supposed to be a better estimate of the volatility than the other estimates, requiring fewer days' data for the same sampling error. This is because there is much more information about the distribution in the highs and lows than in the closing prices, the last being at specified times during the day, the first two occurring at any times.

It is possible to use an exponentially-weighted version.

3.5.3 High-Low-Close Estimator

A better estimate that still uses highs and lows but also includes closing prices is

$$\sqrt{\frac{1}{(M-1)\,\delta t}\sum_{i=1}^{M}(0.5(\log H(t_i)-\log L(t_i))^2-0.39(\log S(t_i)-\log S(t_{i-1}))^2)}.$$

This is even more efficient than the estimate using just highs and lows.

It is possible to use an exponentially-weighted version.

Any estimate using highs and lows will be very sensitive to the distribution of the returns. The above results are only true for Normally-distributed returns. There is plenty of doubt about the real nature of returns, which may not be Normal. We'll be discussing this issue later.

3.6 THE RANDOM WALK ON A SPREADSHEET

The random walk (3.4) can be written as a 'recipe' for generating S_{i+1} from S_i:

$$S_{i+1}=S_i(1+\mu\,\delta t+\sigma\phi\,\delta t^{1/2}). \tag{3.5}$$

We can easily simulate the model using a spreadsheet. In this simulation we must input several parameters, a starting value for the asset, a timestep δt, the drift rate μ, the volatility σ and the total number of timesteps. Then, at each timestep, we must choose a random number ϕ from a Normal distribution. I will talk about simulations in depth in Chapter 66, but for the moment let me just say that an approximation to a Normal variable that is fast in a spreadsheet, and quite accurate, is simply to add up twelve random variables drawn from a uniform distribution over zero to one, and subtract six:

$$\left(\sum_{i=1}^{12}\mathrm{RAND}\,(\,)\right)-6.$$

I USE THIS ALL THE TIME, IT'S QUICK AND EASY AND GOOD ENOUGH FOR MOST PURPOSES

The Excel spreadsheet function RAND() gives a uniformly-distributed random variable.

In Figure 3.8 I show the details of a spreadsheet used for simulating the asset price random walk.

	A	B	C	D	E	F	G	H
1	Asset	100		Time	Asset			
2	Drift	0.15		0	100			
3	Volatility	0.25		0.01	101.2378			
4	Timestep	0.01		0.02	103.8329			
5				0.03	106.5909			
6				0.04	110.993			
7		=D4+$B$4		0.05	115.9425			
8				0.06	117.1478			
9				0.07	115.9868			
10				0.08	114.921			
11	=E7*(1+$B$2*$B$4+$B$3*SQRT($B$4)*(RAND()+RAND()+RAND()+RAND()							
12	+RAND()+RAND()+RAND()+RAND()+RAND()+RAND()+RAND()+RAND()-6))							
13				0.11	113.3875			
14				0.12	108.5439			
15				0.13	107.4318			
16				0.14	109.092			
17				0.15	110.8794			
18				0.16	113.5328			
19				0.17	116.099			
20				0.18	116.2446			
21				0.19	119.315			
22				0.2	120.0332			
23				0.21	124.337			
24				0.22	128.3446			
25				0.23	125.5112			
26				0.24	128.2683			
27				0.25	124.0548			
28				0.26	125.9068			
29				0.27	122.4632			
30				0.28	122.4472			
31				0.29	121.3325			
32				0.3	124.593			
33				0.31	121.9263			

Figure 3.8 Simulating the random walk on a spreadsheet.

3.7 THE WIENER PROCESS

So far we have a model that allows the asset to take any value after a timestep. This is a step forward but we have still not reached our goal of continuous time, we still have a discrete timestep. This section is a brief introduction to the continuous-time limit of equations like (3.3). I will start to introduce ideas from the world of stochastic modeling and Wiener processes, delving more deeply in Chapter 4.

I am now going to use the notation $d\cdot$ to mean 'the change in' some quantity. Thus dS is the 'change in the asset price'. But this change will be in *continuous time*. Thus we will go to the limit $\delta t = 0$. The first δt on the right-hand side of (3.4) becomes dt but the second term is more complicated.

I cannot straightforwardly write $dt^{1/2}$ instead of $\delta t^{1/2}$. If I do go to the zero-timestep limit then any random $dt^{1/2}$ term will dominate any deterministic dt term. Yet in our problem the factor in front of $dt^{1/2}$ has a mean of zero, so maybe it does not outweigh the drift after all. Clearly something subtle is happening in the limit.

It turns out, and we will see this in Chapter 4, that because the *variance* of the random term is $O(\delta t)$ we *can* make a sensible continuous-time limit of our discrete-time model. This brings us into the world of Wiener processes.

I am going to write the term $\phi \, \delta t^{1/2}$ as

$$dX.$$

You can think of dX as being a random variable, drawn from a Normal distribution with mean zero and variance dt:

$$E[dX] = 0 \quad \text{and} \quad E[dX^2] = dt.$$

This is not exactly what it is, but it is close enough to give the right idea. This is called a **Wiener process**. The important point is that we can build up a continuous-time theory using Wiener processes instead of Normal distributions and discrete time.

3.8 THE WIDELY ACCEPTED MODEL FOR EQUITIES, CURRENCIES, COMMODITIES AND INDICES

Our asset price model in the continuous-time limit, using the Wiener process notation, can be written as

$$dS = \mu S \, dt + \sigma S \, dX. \tag{3.6}$$

This is our first **stochastic differential equation**. It is a continuous-time model of an asset price. It is the most widely accepted model for equities, currencies, commodities and indices, and the foundation of so much finance theory.

We've now built up a simple model for equities that we are going to be using quite a lot. You could ask, if the stock market is so random how can fund managers justify their fee. Do

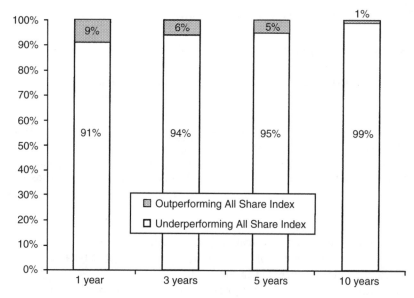

Figure 3.9 Fund performances compared with UK All Share Index. To end December 1998. Data supplied by Virgin Direct.

they manage to outsmart the market? Are they clairvoyant or aren't the markets random? Well, I won't swear that markets are random but I can say with confidence that fund managers don't outperform the market. In Figure 3.9 is shown the percentage of funds that outperform an index of all UK stocks. Whether we look at a one-, three-, five- or 10-year horizon we can see that the vast majority of funds can't even keep up with the market. And statistically speaking, there are bound to be a few that beat the market, but only by chance. Maybe one should invest in a fund that does the opposite of all other funds. Great idea except that the management fee and transaction costs probably mean that that would be a poor investment too. This doesn't prove that markets are random, but it's sufficiently suggestive that most of my personal share exposure is via an index-tracker fund.

3.9 **SUMMARY**

In this chapter I introduced a simple model for the random walk of asset. Initially I built the model up in discrete time, showing what the various terms mean, how they scale with the timestep and showing how to implement the model on a spreadsheet.

Most of this book is about continuous-time models for assets. The continuous-time version of the random walk involves concepts such as stochastic calculus and Wiener processes. I introduced these briefly in this chapter and will now go on to explain the underlying theory of stochastic calculus to give the necessary background for the rest of the book.

FURTHER READING

- Mandelbrot (1963) and Fama (1965) did some of the early work on the analysis of financial data.
- Parkinson (1980) derived the high-low estimator and Garman & Klass (1980) derived the high-low-close estimator.
- For an introduction to random walks and Wiener processes see Øksendal (1992) and Schuss (1980).
- Some high frequency data can be ordered through Olsen Associates, www.olsen.ch. It's not free, but nor is it expensive.
- The famous book by Malkiel (1990) is well worth reading for its insights into the behavior of the stock market. Read what he has to say about chimpanzees, blindfolds and darts. In fact, if you haven't already read Malkiel's book make sure that it is the next book you read after finishing mine.

CHAPTER 4
elementary stochastic calculus

In this Chapter...

- all the stochastic calculus you need to know, and no more
- the meaning of Markov and martingale
- Brownian motion
- stochastic integration
- stochastic differential equations
- Itô's lemma in one and more dimensions

4.1   INTRODUCTION

Stochastic calculus is very important in the mathematical modeling of financial processes. This is because of the underlying random nature of financial markets. Because stochastic calculus is such an important tool I want to ensure that it can be used by everyone. To that end, I am going to try to make this chapter as accessible and intuitive as possible. By the end, I hope that the reader will know what various technical terms mean (and rarely are they very complicated), but, more importantly, will also know how to use the techniques with the minimum of fuss.

Most academic articles in finance have a 'pure' mathematical theme. The mathematical rigor in these works is occasionally justified, but more often than not it only succeeds in obscuring the content. When a subject is young, as is mathematical finance (young*ish*), there is a tendency for technical rigor to feature very prominently in research. This is due to lack of confidence in the methods and results. As the subject ages, researchers will become more cavalier in their attitudes and we will see much more rapid progress.

4.2   A MOTIVATING EXAMPLE

Toss a coin. Every time you throw a head I give you $1, every time you throw a tail you give me $1. Figure 4.1 shows how much money you have after six tosses. In this experiment the sequence was THHTHT, and we finished even.

If I use R_i to mean the random amount, either $1 or −$1, you make on the ith toss then we have

$$E[R_i] = 0, \quad E[R_i^2] = 1 \quad \text{and} \quad E[R_i R_j] = 0.$$

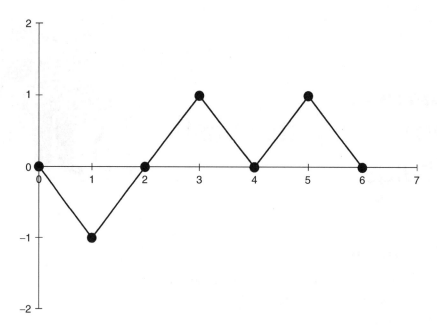

Figure 4.1   The outcome of a coin-tossing experiment.

In this example it doesn't matter whether or not these expectations are conditional on the past. In other words, if I threw five heads in a row it does not affect the outcome of the sixth toss. To the gamblers out there, this property is also shared by a fair die, a balanced roulette wheel, but not by the deck of cards in blackjack. In blackjack the same deck is used for game after game, and the odds during one game depend on what cards were dealt out from the same deck in previous games. That is why you can in the long run beat the house at blackjack but not roulette.

Introduce S_i to mean the total amount of money you have won up to and including the ith toss so that

$$S_i = \sum_{j=1}^{i} R_j.$$

Later on it will be useful if we have $S_0 = 0$, i.e., you start with no money.

If we now calculate expectations of S_i it *does* matter what information we have. If we calculate expectations of future events before the experiment has even begun then

$$E[S_i] = 0 \quad \text{and} \quad E[S_i^2] = E[R_1^2 + 2R_1 R_2 + \cdots] = i.$$

On the other hand, suppose there have been five tosses already, can I use this information and what can we say about expectations for the sixth toss? This is the **conditional expectation**. The expectation of S_6 conditional upon the previous five tosses gives

$$E[S_6 | R_1, \ldots, R_5] = S_5.$$

4.3 **THE MARKOV PROPERTY**

This result is special, the expected value of the random variable S_i conditional upon all of the past events *only depends on the previous value* S_{i-1}. This is the **Markov property**. We say that the random walk has no memory beyond where it is now. Note that it doesn't have to be the case that the expected value of the random variable S_i is the same as the previous value.

MARKOV MEANS NO MEMORY BEYOND THE PRESENT. A PROPERTY OF MOST FINANCE MODELS

This can be generalized to say that, given information about S_j for some values of $1 \leq j < i$, then the only information that is of use to us in estimating S_i is the value of S_j for the largest j for which we have information.

Almost all of the financial models that I will show you have the Markov property. This is of fundamental importance in modeling in finance. I will also show you examples where the system has a small amount of memory, meaning that one or two other pieces of information are important. And I will also give a couple of examples where *all* of the random walk path contains relevant information.

4.4 **THE MARTINGALE PROPERTY**

The coin-tossing experiment possesses another property that can be important in finance. You know how much money you have won after the fifth toss. Your expected winnings after the sixth toss, and indeed after any number of tosses if we keep playing, are just the amount you already hold. That is, the conditional expectation of your winnings at any time in the future is just the amount you already hold:

$$E[S_i | S_j, j < i] = S_j.$$

This is called the **martingale property**.

4.5 **QUADRATIC VARIATION**

I am now going to define the **quadratic variation** of the random walk. This is defined by

$$\sum_{j=1}^{i} (S_j - S_{j-1})^2.$$

Because you either win or lose an amount $1 after each toss, $|S_j - S_{j-1}| = 1$. Thus the quadratic variation is always i:

$$\sum_{j=1}^{i} (S_j - S_{j-1})^2 = i.$$

I want to use the coin-tossing experiment for one more demonstration. And that will lead us to a continuous-time random walk.

4.6 **BROWNIAN MOTION**

I am going to change the rules of my coin-tossing experiment. First of all I am going to restrict the time allowed for the six tosses to a period t, so each toss will take a time $t/6$. Second, the size of the bet will not be \$1 but $\sqrt{t/6}$.

This new experiment clearly still possesses both the Markov and martingale properties, and its quadratic variation measured over the whole experiment is

$$\sum_{j=1}^{6}(S_j - S_{j-1})^2 = 6 \times \left(\sqrt{\frac{t}{6}}\right)^2 = t.$$

I have set up my experiment so that the quadratic variation is just the time taken for the experiment.

I will change the rules again, to speed up the game. We will have n tosses in the allowed time t, with an amount $\sqrt{t/n}$ riding on each throw. Again, the Markov and martingale properties are retained and the quadratic variation is still

$$\sum_{j=1}^{n}(S_j - S_{j-1})^2 = n \times \left(\sqrt{\frac{t}{n}}\right)^2 = t.$$

I am now going to make n larger and larger. All I am doing with my rule changes is to speed up the game, decreasing the time between tosses, with a smaller amount for each bet. But I have chosen my new scalings very carefully, the timestep is decreasing like n^{-1} but the bet size only decreases by $n^{-1/2}$.

In Figure 4.2 I show a series of experiments, each lasting for a time 1, with increasing number of tosses per experiment.

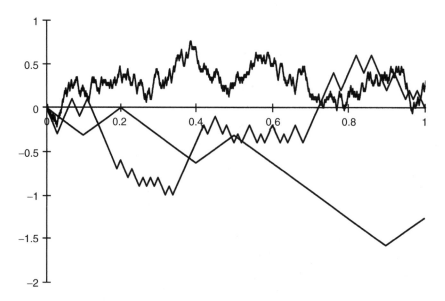

Figure 4.2 A series of coin-tossing experiments, the limit of which is Brownian motion.

As I go to the limit $n = \infty$, the resulting random walk stays finite. It has an expectation, conditional on a starting value of zero, of

$$E[S(t)] = 0$$

and a variance

$$E[S(t)^2] = t.$$

I use $S(t)$ to denote the amount you have won or the value of the random variable after a time t. The limiting process for this random walk as the timesteps go to zero is called **Brownian motion**, and I will denote it by $X(t)$.

The important properties of Brownian motion are as follows:

- *Finiteness*: Any other scaling of the bet size or 'increments' with timestep would have resulted in either a random walk going to infinity in a finite time, or a limit in which there was no motion at all. It is important that the increment scales with the square root of the timestep.
- *Continuity*: The paths are continuous, there are no discontinuities. Brownian motion is the continuous-time limit of our discrete time random walk.
- *Markov*: The conditional distribution of $X(t)$ given information up until $\tau < t$ depends only on $X(\tau)$.
- *Martingale*: Given information up until $\tau < t$ the conditional expectation of $X(t)$ is $X(\tau)$.
- *Quadratic variation*: If we divide up the time 0 to t in a partition with $n + 1$ partition points $t_i = it/n$ then

$$\sum_{j=1}^{n} (X(t_j) - X(t_{j-1}))^2 \to t. \quad \text{(Technically 'almost surely'.)}$$

- *Normality*: Over finite time increments t_{i-1} to t_i, $X(t_i) - X(t_{i-1})$ is Normally distributed with mean zero and variance $t_i - t_{i-1}$.

Having built up the idea and properties of Brownian motion from a series of experiments, we can discard the experiments to leave the Brownian motion that is defined by its properties. These properties will be very important for our financial models.

4.7 STOCHASTIC INTEGRATION

I am going to define a **stochastic integral** by

$$W(t) = \int_0^t f(\tau) \, dX(\tau) = \lim_{n \to \infty} \sum_{j=1}^{n} f(t_{j-1})(X(t_j) - X(t_{j-1}))$$

with

$$t_j = \frac{jt}{n}.$$

Before I manipulate this in any way or discuss its properties, I want to stress that the function $f(t)$ which I am integrating is evaluated in the summation at the *left-hand point* t_{j-1}. It will be crucially important that each function evaluation does not know about the random increment that multiplies it, i.e. the integration is **non-anticipatory**. In financial terms, we will see that we take some action such as choosing a portfolio and only then does the stock price move. This choice of integration is natural in finance, ensuring that we use no information about the future in our current actions.

WE'LL BE USING STOCHASTIC DIFFERENTIAL EQUATIONS ALL THE TIME

4.8 STOCHASTIC DIFFERENTIAL EQUATIONS

Stochastic integrals are important for any theory of stochastic calculus since they can be meaningfully defined. (And in the next section I show how the definition leads to some important properties.) However, it is very common to use a shorthand notation for expressions such as

$$W(t) = \int_0^t f(\tau) \, dX(\tau). \tag{4.1}$$

That shorthand comes from 'differentiating' (4.1) and is

$$dW = f(t) \, dX. \tag{4.2}$$

Think of dX as being an increment in X, i.e. a Normal random variable with mean zero and standard deviation $dt^{1/2}$.

Equations (4.1) and (4.2) are meant to be equivalent. One of the reasons for this shorthand is that Equation (4.2) looks a lot like an ordinary differential equation. We *do not* go the further step of dividing by dt to make it look exactly like an ordinary differential equation because then we would have the difficult task of defining dX/dt.

Pursuing this idea further, imagine what might be meant by

$$dW = g(t) \, dt + f(t) \, dX. \tag{4.3}$$

This is simply shorthand for

$$W(t) = \int_0^t g(\tau) \, d\tau + \int_0^t f(\tau) \, dX(\tau).$$

Equations like (4.3) are called **stochastic differential equations**. Their precise meaning comes, however, from the technically more accurate equivalent stochastic integral. In this book I will use the shorthand versions almost everywhere, so no confusion should arise.

4.9 THE MEAN SQUARE LIMIT

I am going to describe the technical term **mean square limit**. This is useful in the precise definition of stochastic integration. I will explain the idea by way of the simplest example.

Examine the quantity

$$E\left[\left(\sum_{j=1}^n (X(t_j) - X(t_{j-1}))^2 - t \right)^2 \right] \tag{4.4}$$

where

$$t_j = \frac{jt}{n}.$$

This can be expanded as

$$E\left[\sum_{j=1}^{n}(X(t_j) - X(t_{j-1}))^4 + 2\sum_{i=1}^{n}\sum_{j<i}(X(t_i) - X(t_{i-1}))^2(X(t_j) - X(t_{j-1}))^2 \right.$$

$$\left. - 2t\sum_{j=1}^{n}(X(t_j) - X(t_{j-1}))^2 + t^2\right].$$

Since $X(t_j) - X(t_{j-1})$ is Normally distributed with mean zero and variance t/n we have

$$E[(X(t_j) - X(t_{j-1}))^2] = \frac{t}{n}$$

and

$$E[(X(t_j) - X(t_{j-1}))^4] = \frac{3t^2}{n^2}.$$

Thus (4.4) becomes

$$n\frac{3t^2}{n^2} + n(n-1)\frac{t^2}{n^2} - 2tn\frac{t}{n} + t^2 = O\left(\frac{1}{n}\right).$$

As $n \to \infty$ this tends to zero. We therefore say that

$$\sum_{j=1}^{n}(X(t_j) - X(t_{j-1}))^2 = t$$

in the 'mean square limit'. This is often written, for obvious reasons, as

$$\int_0^t (dX)^2 = t.$$

I am not going to use this result, nor will I use the mean square limit technique. However, when I talk about 'equality' in the following 'proof' I mean equality in the mean square sense.

4.10 FUNCTIONS OF STOCHASTIC VARIABLES AND ITÔ'S LEMMA

I am now going to introduce the idea of a function of a stochastic variable. In Figure 4.3 is shown a realization of a Brownian motion $X(t)$ and the function $F(X) = X^2$.

If $F = X^2$ is it true that $dF = 2X\,dX$? No. The ordinary rules of calculus do not generally hold in a stochastic environment. Then what are the rules of calculus?

I am going to 'derive' the most important rule of stochastic calculus, **Itô's lemma**. My derivation is more heuristic than rigorous, but at least it is transparent. I will do this for an arbitrary function $F(X)$.

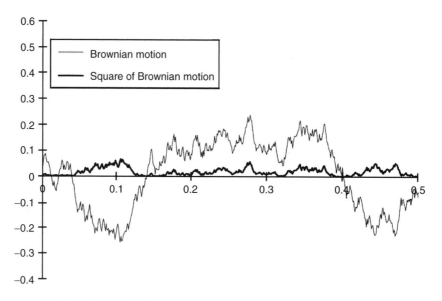

Figure 4.3 A realization of a Brownian motion and its square.

In this derivation I will need to introduce various timescales. The first timescale is very, very small. I will denote it by

$$\frac{\delta t}{n} = h.$$

This timescale is so small that the function $F(X(t+h))$ can be approximated by a Taylor series:

$$F(X(t+h)) - F(X(t)) = (X(t+h) - X(t))\frac{dF}{dX}(X(t)) + \tfrac{1}{2}(X(t+h) - X(t))^2\frac{d^2F}{dX^2}(X(t)) + \cdots.$$

From this it follows that

$$(F(X(t+h)) - F(X(t))) + (F(X(t+2h)) - F(X(t+h))) + \cdots$$
$$+ (F(X(t+nh)) - F(X(t+(n-1)h)))$$
$$= \sum_{j=1}^{n}(X(t+jh) - X(t+(j-1)h))\frac{dF}{dX}(X(t+(j-1)h))$$
$$+ \tfrac{1}{2}\frac{d^2F}{dX^2}(X(t))\sum_{j=1}^{n}(X(t+jh) - X(t+(j-1)h))^2 + \cdots.$$

In this I have used the approximation

$$\frac{d^2F}{dX^2}(X(t+(j-1)h)) = \frac{d^2F}{dX^2}(X(t)).$$

This is consistent with the order of accuracy I require.

The first line in this becomes simply

$$F(X(t+nh)) - F(X(t)) = F(X(t+\delta t)) - F(X(t)).$$

The second is just the definition of

$$\int_{t}^{t+\delta t} \frac{dF}{dX} \, dX$$

and the last is

$$\frac{1}{2} \frac{d^2F}{dX^2}(X(t)) \, \delta t,$$

in the *mean square sense*. Thus we have

$$F(X(t+\delta t)) - F(X(t)) = \int_{t}^{t+\delta t} \frac{dF}{dX}(X(\tau)) \, dX(\tau) + \frac{1}{2} \int_{t}^{t+\delta t} \frac{d^2F}{dX^2}(X(\tau)) \, d\tau.$$

I can now extend this result over longer timescales, from zero up to t, over which F *does* vary substantially to get

$$F(X(t)) = F(X(0)) + \int_{0}^{t} \frac{dF}{dX}(X(\tau)) \, dX(\tau) + \frac{1}{2} \int_{0}^{t} \frac{d^2F}{dX^2}(X(\tau)) \, d\tau.$$

This is the integral version of **Itô's lemma**, which is usually written as

$$dF = \frac{dF}{dX} \, dX + \frac{1}{2} \frac{d^2F}{dX^2} \, dt. \qquad (4.5)$$

READ, MARK, LEARN AND INWARDLY DIGEST. THIS MUST BECOME 2ND NATURE TO YOU

We can now answer the question, if $F = X^2$ what stochastic differential equation does F satisfy? In this example

$$\frac{dF}{dX} = 2X \quad \text{and} \quad \frac{d^2F}{dX^2} = 2.$$

Therefore Itô's lemma tells us that

$$dF = 2X \, dX + dt.$$

This is *not* what we would get if X were a deterministic variable. In integrated form

$$X^2 = F(X) = F(0) + \int_{0}^{t} 2X \, dX + \int_{0}^{t} 1 \, d\tau = \int_{0}^{t} 2X \, dX + t.$$

Therefore

$$\int_{0}^{t} X \, dX = \frac{1}{2}X^2 - \frac{1}{2}t.$$

4.11 **ITÔ AND TAYLOR**

Having derived Itô's lemma, I am going to give some intuition behind the result and then slightly generalize it.

If we were to do a naive Taylor series expansion of F, completely disregarding the nature of X, and treating dX as a small increment in X, we would get

$$F(X + dX) = F(X) + \frac{dF}{dX} dX + \tfrac{1}{2}\frac{d^2F}{dX^2} dX^2,$$

ignoring higher-order terms. We could argue that $F(X + dX) - F(X)$ was just the 'change in' F and so

$$dF = \frac{dF}{dX} dX + \tfrac{1}{2}\frac{d^2F}{dX^2} dX^2.$$

This is very similar to (4.5) (and Taylor series *is* very similar to Itô), with the only difference being that there is a dX^2 instead of a dt. However, since in a sense

$$\int_0^t (dX)^2 = t$$

I could perhaps write

$$dX^2 = dt. \tag{4.6}$$

Although this lacks any rigor (because it's wrong) it does give the correct result. However, on a positive note you can, with little risk of error, use Taylor series with the 'rule of thumb' (4.6) and in practice you will get the right result. Although this is technically incorrect, you almost certainly[1] won't get the wrong result. I will use this rule of thumb almost every time I want to differentiate a function of a random variable. In Chapter 24 I will show when it *is* correct, and better, to use Taylor series.

To end this section I will generalize slightly. Suppose my stochastic differential equation is

$$dS = a(S)\,dt + b(S)\,dX, \tag{4.7}$$

say, for some functions $a(S)$ and $b(S)$. Here dX is the usual Brownian increment. Now if I have a function of S, $V(S)$, what stochastic differential equation does it satisfy? The answer is

$$dV = \frac{dV}{dS} dS + \tfrac{1}{2}b^2\frac{d^2V}{dS^2}\,dt.$$

We could derive this properly or just cheat by using Taylor series with $dX^2 = dt$. I could, if I wanted, substitute for dS from (4.7) to get an equation for dV in terms of the pure Brownian motion X:

$$dV = \left(a(S)\frac{dV}{dS} + \tfrac{1}{2}b(S)^2\frac{d^2V}{dS^2} \right) dt + b(S)\frac{dV}{dS}\,dX.$$

4.12 ITÔ IN HIGHER DIMENSIONS

In financial problems we often have functions of one stochastic variable S and a deterministic variable t, time: $V(S, t)$. If

$$dS = a(S, t)\,dt + b(S, t)\,dX,$$

---

[1] Or should that be 'almost surely'?

then the increment dV is given by

$$dV = \frac{\partial V}{\partial t}\,dt + \frac{\partial V}{\partial S}\,dS + \tfrac{1}{2}b^2 \frac{\partial^2 V}{\partial S^2}\,dt. \tag{4.8}$$

Again, this is shorthand notation for the correct integrated form. This result is obvious, as is the use of partial instead of ordinary derivatives.

Occasionally, we have a function of two, or more, random variables, and time as well: $V(S_1, S_2, t)$. An example would be the value of an option to buy the more valuable out of Nike and Reebok. I will write the behavior of S_1 and S_2 in the general form

$$dS_1 = a_1(S_1, S_2, t)\,dt + b_1(S_1, S_2, t)\,dX_1$$

and

$$dS_2 = a_2(S_1, S_2, t)\,dt + b_2(S_1, S_2, t)\,dX_2.$$

Note that I have *two* Brownian increments dX_1 and dX_2. We can think of these as being Normally distributed with variance dt, but *they are correlated*. The correlation between these two random variables I will call ρ. This can also be a function of S_1, S_2 and t but must satisfy

$$-1 \leq \rho \leq 1.$$

The 'rules of thumb' can readily be imagined:

$$dX_1^2 = dt, \quad dX_2^2 = dt \quad \text{and} \quad dX_1\,dX_2 = \rho\,dt.$$

Itô's lemma becomes

$$dV = \frac{\partial V}{\partial t}\,dt + \frac{\partial V}{\partial S_1}\,dS_1 + \frac{\partial V}{\partial S_2}\,dS_2 + \tfrac{1}{2}b_1^2 \frac{\partial^2 V}{\partial S_1^2}\,dt + \rho b_1 b_2 \frac{\partial^2 V}{\partial S_1 \partial S_2}\,dt + \tfrac{1}{2}b_2^2 \frac{\partial^2 V}{\partial S_2^2}\,dt. \tag{4.9}$$

4.13 SOME PERTINENT EXAMPLES

In this section I am going to introduce a few common random walks and talk about their properties.

4.13.1 Brownian Motion with Drift

The first example is like the simple Brownian motion but with a drift:

$$dS = \mu\,dt + \sigma\,dX.$$

A realization of this is shown in Figure 4.4. The point to note about this realization is that S has gone negative. This random walk would therefore not be a good model for many financial quantities, such as interest rates or equity prices. This stochastic differential equation can be integrated exactly to get

$$S(t) = S(0) + \mu t + \sigma(X(t) - X(0)).$$

4.13.2 The Lognormal Random Walk

My second example is similar to the above but the drift and randomness scale with S:

$$dS = \mu S\,dt + \sigma S\,dX. \tag{4.10}$$

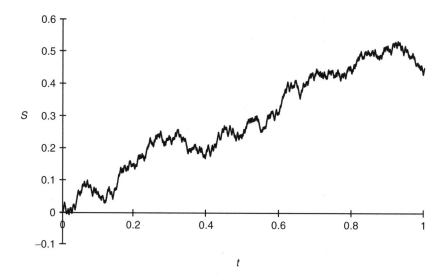

Figure 4.4 A realization of $dS = \mu\, dt + \sigma\, dX$.

A realization of this is shown in Figure 4.5. If S starts out positive it can never go negative; the closer that S gets to zero the smaller the increments dS. For this reason I have had to start the simulation with a non-zero value for S. This property of this random walk is clearly seen if we examine the function $F(S) = \log S$ using Itô's lemma. From Itô we have

$$dF = \frac{dF}{dS}\, dS + \tfrac{1}{2}\sigma^2 S^2 \frac{d^2 F}{dS^2}\, dt = \frac{1}{S}(\mu S\, dt + \sigma S\, dX) - \tfrac{1}{2}\sigma^2\, dt$$

$$= \left(\mu - \tfrac{1}{2}\sigma^2\right) dt + \sigma\, dX.$$

This shows us that $\log S$ can range between minus and plus infinity but cannot reach these limits in a finite time, therefore S cannot reach zero or infinity in a finite time.

How does the time series in Figure 4.5 which was generated on a spreadsheet using random returns compare qualitatively with the time series in Figure 4.6 which is the real series for Glaxo–Wellcome?

The integral form of this stochastic differential equation follows simply from the stochastic differential equation for $\log S$:

$$S(t) = S(0)e^{(\mu - (1/2)\sigma^2)t + \sigma(X(t) - X(0))}.$$

The stochastic differential Equation (4.10) will be particularly important in the modeling of many asset classes. And if we have some function $V(S, t)$ then from Itô it follows that

$$dV = \frac{\partial V}{\partial t}\, dt + \frac{\partial V}{\partial S}\, dS + \tfrac{1}{2}\sigma^2 S^2 \frac{\partial^2 V}{\partial S^2}\, dt. \tag{4.11}$$

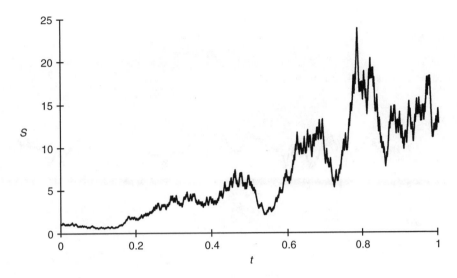

Figure 4.5   A realization of $dS = \mu S\,dt + \sigma S\,dX$.

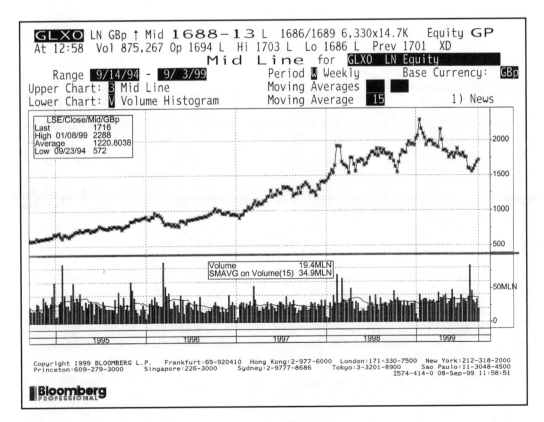

Figure 4.6   Glaxo–Wellcome share price (volume below). Source: Bloomberg L.P.

4.13.3   A Mean-reverting Random Walk

The third example is

$$dS = (v - \mu S)\,dt + \sigma\,dX.$$

A realization of this is shown in Figure 4.7.

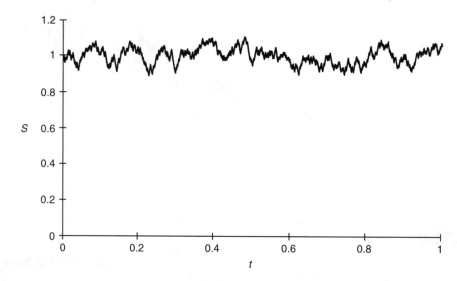

Figure 4.7   A realization of $dS = (v - \mu S)\,dt + \sigma\,dX$.

This random walk is an example of a **mean-reverting** random walk. If S is large, the negative coefficient in front of dt means that S will move down on average, if S is small it rises on average. There is still no incentive for S to stay positive in this random walk. With r instead of S this random walk is the Vasicek model for the short-term interest rate.

Mean-reverting models are used for modeling a random variable that 'isn't going anywhere'. That's why they are often used for interest rates; Figure 4.8 shows the yield on a Japanese Government Bond.

4.13.4   And Another Mean-reverting Random Walk

The final example is similar to the third but I am going to adjust the random term slightly:

$$dS = (v - \mu S)\,dt + \sigma S^{1/2}\,dX.$$

Now if S ever gets close to zero the randomness decreases, perhaps this will stop S from going negative? Let's play around with this example for a while. And we'll see Itô in practice.

Write $F = S^{1/2}$. What stochastic differential equation does F satisfy? Since

$$\frac{dF}{dS} = \tfrac{1}{2}S^{-1/2} \quad \text{and} \quad \frac{d^2F}{dS^2} = -\tfrac{1}{4}S^{-3/2}$$

we have

$$dF = \left(\frac{4v - \sigma^2}{8F} - \tfrac{1}{2}\mu F\right) dt + \tfrac{1}{2}\sigma\,dX.$$

Figure 4.8 Time series of the yield on a JGB. Source: Bloomberg L.P.

I have just turned the original stochastic differential equation with a variable coefficient in front of the random term into a stochastic differential equation with a constant random term. In so doing I have made the drift term nastier. In particular, the drift is now singular at $F = S = 0$. Something special is happening at $S = 0$.

Instead of examining $F(S) = S^{1/2}$, can I find a function $F(S)$ such that its stochastic differential equation has a zero drift term? For this I will need

$$(\nu - \mu S)\frac{dF}{dS} + \tfrac{1}{2}\sigma^2 S \frac{d^2 F}{dS^2} = 0.$$

This is easily integrated once to give

$$\frac{dF}{dS} = AS^{-(2\nu/\sigma^2)}e^{2\mu S/\sigma^2} \tag{4.12}$$

for any constant A. I won't take this any further but just make one observation. If

$$\frac{2\nu}{\sigma^2} \geq 1$$

we cannot integrate (4.12) at $S = 0$. This makes the origin **non-attainable**. In other words, if the parameter ν is sufficiently large it forces the random walk to stay away from zero.

This particular stochastic differential equation for S will be important later on; it is the Cox, Ingersoll & Ross model for the short-term interest rate.

These are just four of the many random walks we will be seeing.

4.14 SUMMARY

This chapter introduced the most important tool of the trade, Itô's lemma. Itô's lemma allows us to manipulate functions of a random variable. If we think of S as the value of an asset for which we have a stochastic differential equation, a 'model', then we can handle functions of the asset, and ultimately value contracts such as options.

If we use Itô as a tool we do not need to know why or how it works, only how to use it. Essentially all we require to successfully use the lemma is a rule of thumb, as explained in the text. Unless we are using Itô in highly unusual situations, then we are unlikely to make any errors.

FURTHER READING

- Neftci (1996) is the only readable book on stochastic calculus for beginners. It does not assume any knowledge about anything. It takes the reader very slowly through the basics as applied to finance.

- Once you have got beyond the basics, move on to Øksendal (1992) and Schuss (1980).

CHAPTER 5
the Black–Scholes model

In this Chapter...

- the foundations of derivatives theory: delta hedging and no arbitrage
- the derivation of the Black–Scholes partial differential equation
- the assumptions that go into the Black–Scholes equation
- how to modify the equation for commodity and currency options

5.1 INTRODUCTION

This is, without doubt, the most important chapter in the book. In it I describe and explain the basic building blocks of derivatives theory. These building blocks are delta hedging and no arbitrage. They form a moderately sturdy foundation to the subject and have performed well since 1973 when the ideas became public.

In this chapter I begin with the stochastic differential equation model for equities and exploit the correlation between this asset and an option on this asset to make a perfectly risk-free portfolio. I then appeal to no arbitrage to equate returns on all risk-free portfolios to the risk-free interest rate, the so-called 'no free lunch' argument.

The arguments are trivially modified to incorporate dividends on the underlying and also to price commodity and currency options and options on futures.

This chapter is theoretical, yet all of the ideas contained here are regularly used in practice. Even though all of the assumptions can be shown to be wrong to a greater or lesser extent, the Black–Scholes model is profoundly important both in theory and in practice.

5.2 A VERY SPECIAL PORTFOLIO

In Chapter 2 I described some of the characteristics of options and options markets. I introduced the idea of call and put options, amongst others. The value of a call option is clearly going to be a function of various parameters in the contract, such as the strike price E and the time to expiry $T - t$, T is the date of expiry, and t is the current time. The value will also depend on properties of the asset itself, such as its price, its drift and its volatility, as well as the risk-free rate of interest.[1] We can write the option value as

$$V(S, t; \sigma, \mu; E, T; r).$$

Notice that the semi-colons separate different types of variables and parameters:

- S and t are variables;
- σ and μ are parameters associated with the asset price;
- E and T are parameters associated with the details of the particular contract;
- r is a parameter associated with the currency in which the asset is quoted.

I'm not going to carry all the parameters around, except when it is important. For the moment I'll just use $V(S, t)$ to denote the option value.

One simple observation is that a call option will rise in value if the underlying asset rises, and will fall if the asset falls. This is clear since a call has a larger payoff the greater the value of the underlying at expiry. This is an example of **correlation** between two financial instruments, and in this case the correlation is positive. A put and the underlying have a negative correlation. We can exploit these correlations to construct a very special portfolio.

Use Π to denote the value of a portfolio of one long option position and a short position in some quantity Δ, **delta**, of the underlying:

$$\Pi = V(S, t) - \Delta S. \tag{5.1}$$

The first term on the right is the option and the second term is the short asset position. Notice the minus sign in front of the second term. The quantity Δ will for the moment be some constant quantity of our choosing. We will assume that the underlying follows a lognormal random walk

$$dS = \mu S \, dt + \sigma S \, dX.$$

It is natural to ask how the value of the portfolio changes from time t to $t + dt$. The change in the portfolio value is due partly to the change in the option value and partly to the change in the underlying:

$$d\Pi = dV - \Delta \, dS.$$

Notice that Δ has not changed during the timestep; we have not anticipated the change in S. From Itô we have

$$dV = \frac{\partial V}{\partial t} \, dt + \frac{\partial V}{\partial S} \, dS + \frac{1}{2}\sigma^2 S^2 \frac{\partial^2 V}{\partial S^2} \, dt.$$

Thus the portfolio changes by

$$d\Pi = \frac{\partial V}{\partial t} \, dt + \frac{\partial V}{\partial S} \, dS + \frac{1}{2}\sigma^2 S^2 \frac{\partial^2 V}{\partial S^2} \, dt - \Delta \, dS. \tag{5.2}$$

5.3 ELIMINATION OF RISK: DELTA HEDGING

The right-hand side of (5.2) contains two types of terms, the deterministic and the random. The deterministic terms are those with the dt, and the random terms are those with the dS.

---

[1] Actually, I'm lying. One of these parameters does not affect the option value.

Pretending for the moment that we know V and its derivatives then we know everything about the right-hand side of (5.2) *except for the value of dS*. And this quantity we can never know in advance.

These random terms are the risk in our portfolio. Is there any way to reduce or even eliminate this risk? This can be done in theory (and *almost* in practice) by carefully choosing Δ. The random terms in (5.2) are

$$\left(\frac{\partial V}{\partial S} - \Delta\right) dS.$$

If we choose

$$\Delta = \frac{\partial V}{\partial S} \tag{5.3}$$

then the randomness is reduced to zero.

THIS IS NOT JUST A THEORETICAL CONCEPT, IT IS USED IN REAL LIFE AS WELL

Any reduction in randomness is generally termed **hedging**, whether that randomness is due to fluctuations in the stock market or the outcome of a horse race. The perfect elimination of risk, by exploiting correlation between two instruments (in this case an option and its underlying) is generally called **delta hedging**.

Delta hedging is an example of a **dynamic hedging** strategy. From one timestep to the next the quantity $\partial V/\partial S$ changes since it is, like V, a function of the ever-changing variables S and t. This means that the perfect hedge must be continually rebalanced. In later chapters we will see examples of static hedging, where a hedging position is *not* changed as the variables evolve.

Delta hedging was effectively first described by Thorp & Kassouf (1967) but they missed the crucial (Nobel prize winning) next step. (We will see more of Thorp when we look at casino blackjack as an investment in Chapter 51.)

5.4 NO ARBITRAGE

After choosing the quantity Δ as suggested above, we hold a portfolio whose value changes by the amount

$$d\Pi = \left(\frac{\partial V}{\partial t} + \frac{1}{2}\sigma^2 S^2 \frac{\partial^2 V}{\partial S^2}\right) dt. \tag{5.4}$$

This change is completely *riskless*. If we have a completely risk-free change $d\Pi$ in the portfolio value Π then it must be the same as the growth we would get if we put the equivalent amount of cash in a risk-free interest-bearing account:

$$d\Pi = r\Pi \, dt. \tag{5.5}$$

This is an example of the **no arbitrage** principle.

To see why this should be so, consider in turn what might happen if the return on the portfolio were, first, greater and, second, less than the risk-free rate. If we were guaranteed to get a return of greater than r from the delta-hedged portfolio then what we could do is borrow from the bank, paying interest at the rate r, invest in the risk-free option/stock portfolio and make a profit. If on the other hand the return were less than the risk-free rate we should go short the option, delta hedge it, and invest the cash in the bank. Either way, we make a riskless profit in excess of the risk-free rate of interest. At this point we say that, all things being equal, the action of investors buying and selling to exploit the arbitrage opportunity will cause the market price of the option to move in the direction that eliminates the arbitrage.

5.5 **THE BLACK–SCHOLES EQUATION**

Substituting (5.1), (5.3) and (5.4) into (5.5) we find that

$$\left(\frac{\partial V}{\partial t} + \tfrac{1}{2}\sigma^2 S^2 \frac{\partial^2 V}{\partial S^2}\right) dt = r\left(V - S\frac{\partial V}{\partial S}\right) dt.$$

On dividing by dt and rearranging we get

THE FAMOUS,
NOBEL PRIZE –
WINNING EQUATION
THAT GOT DERIVATIVES
THEORY STARTED

$$\frac{\partial V}{\partial t} + \tfrac{1}{2}\sigma^2 S^2 \frac{\partial^2 V}{\partial S^2} + rS\frac{\partial V}{\partial S} - rV = 0. \qquad (5.6)$$

This is the **Black–Scholes equation**. The equation was first written down in 1969, but a few years passed, with Fischer Black and Myron Scholes justifying the model, before it was published. The derivation of the equation was finally published in 1973, although the call and put formulae had been published a year earlier.

The Black–Scholes equation is a **linear parabolic partial differential equation**. In fact, almost all partial differential equations in finance are of a similar form. They are almost always linear, meaning that if you have two solutions of the equation then the sum of these is itself also a solution. Or at least they tended to be linear until recently. In Part Three I will show you some examples of recent models which lead to nonlinear equations. Financial equations are also usually parabolic, meaning that they are related to the heat or diffusion equation of mechanics. One of the good things about this is that such equations are relatively easy to solve numerically.

The Black–Scholes equation contains all the obvious variables and parameters such as the underlying, time, and volatility, but there is no mention of the drift rate μ. Why is this? Any dependence on the drift dropped out at the same time as we eliminated the dS component of the portfolio. The economic argument for this is that since we can perfectly hedge the option with the underlying we should not be rewarded for taking unnecessary risk; only the risk-free rate of return is in the equation. This means that if you and I agree on the volatility of an asset we will agree on the value of its derivatives *even if we have differing estimates of the drift*.

Another way of looking at the hedging argument is to ask what happens if we hold a portfolio consisting of just the stock, in a quantity Δ, and cash. If Δ is the partial derivative of some option value then such a portfolio will yield an amount at expiry that is simply that option's payoff. In other words, we can use the same Black–Scholes argument to **replicate** an option just by buying and selling the underlying asset. This leads to the idea of a **complete market**. In a complete market an option can be replicated with the underlying, thus making options redundant. Why buy an option when you can get the same payoff by trading in the asset? Many things conspire to make markets incomplete and we will discuss some of these, such as transaction costs and stochastic volatility, in later chapters.

Since the Black–Scholes equation is of such importance I'm going to spend a moment relating the equation to ideas and models in other parts of the book.

For the rest of this part of the book I'm going to explain what the Black–Scholes equation means as a differential equation and show how to solve it in a few special cases. I will generalize the model slightly but not too much. I will also show what happens to the equation when early exercise is allowed. Superficially, early exercise does not make much difference, but on a closer inspection it changes the whole nature of the problem.

The second part of the book is devoted to more serious generalizations to accommodate the pricing of exotic derivative contracts. Still the extensions are what might be termed 'classical', there is nothing too outrageous here and there is nothing outside the 'Black–Scholes world'.

Part Three is devoted to more serious generalizations and some advanced topics. Some of the generalizations are minor modifications to the Black–Scholes world and can still be thought of as classical. These are just attempts to broaden the Black–Scholes world, to relax some of the assumptions. Other models take us a long way from the Black–Scholes world into fairly new and uncharted waters. Delta hedging and no arbitrage are seen to be not quite as straightforward as I had led you to understand. You are free to believe or disbelieve any of these models, but I do ask that you appreciate them.

Many of the Black–Scholes ideas, plus a few new ones, will be found in Part Four on the pricing of interest rate products. There are some technical reasons that make the fixed-income world harder than the equity, currency, commodity world. Nevertheless delta hedging and no arbitrage play the same role.

Risk management is the subject of Part Five. Delta hedging is supposed to eliminate risk. Not only is this not quite true but also there are times when you might not want to hedge at all. We also discuss credit risk issues here.

In Part Six I show how to solve the Black–Scholes equation, and related equations, numerically. I describe the two main numerical methods: finite-difference methods and Monte Carlo simulations. I include Visual Basic code to illustrate many of the methods. None of these methods are hard; we are lucky in this subject that the governing equation, usually being parabolic, is simple to solve numerically.

5.6 THE BLACK–SCHOLES ASSUMPTIONS

What are the 'assumptions' that I've just been referring to? Here is a partial list, together with some discussion.

- *The underlying follows a lognormal random walk:* This is not entirely necessary. To find explicit solutions we will need the random term in the stochastic differential equation for S to be proportional to S. The 'factor' σ does not need to be constant to find solutions, but it must only be time dependent, see Chapter 7. As far as the validity of the equation is concerned it doesn't matter if the volatility is also asset-price dependent, but then the equation will either have very messy explicit solutions, if it has any at all, or have to be solved numerically. Then there is the question of the drift term μS. Do we need this term to take this form, after all it doesn't even appear in the equation? There is a technicality here that whatever the stochastic differential equation for the asset S, the domain over which the asset can range must be zero to infinity. This is a technicality I am not going into, but it amounts to another elimination of arbitrage. It is possible to choose the drift so that the asset is restricted to lie within a range, such a drift would not be allowed.

- *The risk-free interest rate is a known function of time:* This restriction is just to help us find explicit solutions again. If r were constant this job would be even easier. In practice, the interest rate is often taken to be time-dependent but known in advance. Explicit formulae still exist for the prices of simple contracts and I discuss this issue in Chapter 8. In reality the rate r is not known in advance and is itself stochastic, or so it seems from data. I will discuss stochastic interest rates in Part Four. We've also assumed that lending and borrowing rates are the same. It is not difficult to relax this assumption, and it is related to ideas in Chapter 27.

- *There are no dividends on the underlying:* I will drop this restriction in a moment, and discuss the subject more generally in Chapter 8.

- *Delta hedging is done continuously:* This is definitely impossible. Hedging must be done in discrete time. Often the time between rehedges will depend on the level of transaction costs in the market for the underlying; the lower the costs, the more frequent the rehedging. This subject is covered in depth in Chapter 23.

- *There are no transaction costs on the underlying:* The dynamic business of delta hedging is in reality expensive since there is a bid-offer spread on most underlyings. In some markets this matters and in some it doesn't. Chapter 24 is devoted to a discussion of these issues.

- *There are no arbitrage opportunities:* This is a beauty. Of course there are arbitrage opportunities, a lot of people make a lot of money finding them.[2] It is extremely important to stress that we are ruling out model-dependent arbitrage. This is highly dubious since it depends on us having the correct model in the first place, and that is unlikely. I am happier ruling out model-independent arbitrage i.e. arbitrage arising when two identical cashflows have different values. But even that can be criticized.

There are many more assumptions but the above are the most important. In other parts of the book I will drop these assumptions or, if I don't drop them, I will at least loosen them a bit.

5.7 FINAL CONDITIONS

The Black–Scholes Equation (30.7) knows nothing about what kind of option we are valuing, whether it is a call or a put, nor what is the strike and the expiry. These points are dealt with by the **final condition**. We must specify the option value V as a function of the underlying at the expiry date T. That is, we must prescribe $V(S, T)$, the payoff.

For example, if we have a call option then we know that

$$V(S, T) = \max(S - E, 0).$$

For a put we have

$$V(S, T) = \max(E - S, 0),$$

for a binary call

$$V(S, T) = \mathcal{H}(S - E)$$

and for a binary put

$$V(S, T) = \mathcal{H}(E - S),$$

where $\mathcal{H}(\cdot)$ is the **Heaviside function**, which is zero when its argument is negative and one when it is positive.

The imposition of the final condition will be explained in Chapters 6 and 7, and implemented numerically in Part Six.

---

[2] Life, and everything in it, is based on arbitrage opportunities and their exploitation.

As an aside, observe that both the asset, S, and 'money in the bank', e^{rt} satisfy the Black–Scholes equation.

5.8   OPTIONS ON DIVIDEND-PAYING EQUITIES

The first generalization we discuss is how to value options on stocks paying dividends. This is just about the simplest generalization of the Black–Scholes model. To keep things simple (I will complicate matters in Chapter 8) let's assume that the asset receives a continuous and constant dividend yield, D. Thus in a time dt each asset receives an amount $DS\,dt$. This must be factored into the derivation of the Black–Scholes equation. I take up the Black–Scholes argument at the point where we are looking at the change in the value of the portfolio:

$$d\Pi = \frac{\partial V}{\partial t}\,dt + \frac{\partial V}{\partial S}\,dS + \tfrac{1}{2}\sigma^2 S^2 \frac{\partial^2 V}{\partial S^2}\,dt - \Delta\,dS - D\Delta S\,dt.$$

The last term on the right-hand side is simply the amount of the dividend per asset, $DS\,dt$, multiplied by the number of the asset held, $-\Delta$. The Δ is still given by the rate of change of the option value with respect to the underlying, but after some simple substitutions we now get

$$\frac{\partial V}{\partial t} + \tfrac{1}{2}\sigma^2 S^2 \frac{\partial^2 V}{\partial S^2} + (r - D)S\frac{\partial V}{\partial S} - rV = 0. \tag{5.7}$$

5.9   CURRENCY OPTIONS

Options on currencies are handled in exactly the same way. In holding the foreign currency we receive interest at the foreign rate of interest r_f. This is just like receiving a continuous dividend. I will skip the derivation but we readily find that

$$\frac{\partial V}{\partial t} + \tfrac{1}{2}\sigma^2 S^2 \frac{\partial^2 V}{\partial S^2} + (r - r_f)S\frac{\partial V}{\partial S} - rV = 0. \tag{5.8}$$

5.10   COMMODITY OPTIONS

The relevant feature of commodities requiring that we adjust the Black–Scholes equation is that they have a **cost of carry**. That is, the storage of commodities is not without cost. Let us introduce q as the fraction of the value of a commodity that goes towards paying the cost of carry. This means that just holding the commodity will result in a gradual loss of wealth even if the commodity price remains fixed. To be precise, for each unit of the commodity held, an amount $qS\,dt$ will be required during short time dt to finance the holding. This is just like having a negative dividend and so we get

$$\frac{\partial V}{\partial t} + \tfrac{1}{2}\sigma^2 S^2 \frac{\partial^2 V}{\partial S^2} + (r + q)S\frac{\partial V}{\partial S} - rV = 0. \tag{5.9}$$

5.11 **OPTIONS ON FUTURES**

The final modification to the Black–Scholes model in this chapter is to value options on futures. Recall that the future price of a non-dividend paying equity F is related to the spot price by

$$F = e^{r(T_F - t)}S$$

where T_F is the maturity date of the futures contract. We can easily change variables, and look for a solution $V(S, t) = \mathcal{V}(F, t)$. We find that

$$\frac{\partial \mathcal{V}}{\partial t} + \tfrac{1}{2}\sigma^2 F^2 \frac{\partial^2 \mathcal{V}}{\partial F^2} - r\mathcal{V} = 0. \qquad (5.10)$$

The equation for an option on a future is actually simpler than the Black–Scholes equation.

5.12 **SOME OTHER WAYS OF DERIVING THE BLACK–SCHOLES EQUATION**

The derivation of the Black–Scholes equation above is the classical one, and similar to the original Black & Scholes derivation. There are other ways of getting to the same result. Here are a few, without any of the details. The details, and more examples, are contained in the final reference in the Further reading.

5.12.1 The Martingale Approach

The value of an option can be shown to be an expectation, not a real expectation but a special, risk-neutral one. We'll be seeing lots of this subject later. This is a useful result, since it forms the basis for pricing by simulation, see Chapter 66.

The partial differential equation can be derived from the expectation, see Chapter 10. The concepts of hedging and no arbitrage are obviously still used in this derivation.

5.12.2 The Binomial Model

The binomial model is a discrete time, discrete asset price model for underlyings and again uses hedging and no arbitrage to derive a pricing algorithm for options. We shall see this in detail in Chapter 12. In taking the limit as the timestep shrinks to zero we get the continuous-time Black–Scholes equation.

5.12.3 CAPM/Utility

Again, we'll be seeing the Capital Asset Pricing Model later. For the moment you just need to know that it is a model for the behavior of risky assets and a principle and algorithm for defining and finding optimal ways to allocate wealth among the assets. Portfolios are described in terms of their risk (standard deviation of returns) and reward (expected growth). If you include options in this framework then the possible combinations of risk and reward are not increased. This is because options are, in a sense, just functions of their underlyings. This is market completeness. The risk and reward on an option and on its underlying are related and the Black–Scholes equation follows.

5.13 **SUMMARY**

This was an important but not too difficult chapter. In it I introduced some very powerful and beautiful concepts such as delta hedging and no arbitrage. These two fundamental principles led to the Black–Scholes option pricing equation. Everything from this point on is based on, or is inspired by, these ideas.

FURTHER READING

- The history of option theory leading up to Black–Scholes is described in Briys, Mai, Bellalah & de Varenne (1998).
- The story of the derivation of the Black–Scholes equation, written by Bob Whaley, can be found in the 10th anniversary issue of *Risk* magazine, published in December 1997.
- Of course, you must read the original work, Black & Scholes (1973) and Merton (1973).
- See Black (1976) for the details of the pricing of options on futures, and Garman & Kohlhagen (1983) for the pricing of FX options.
- For details of other ways to derive the Black–Scholes equation see Andreasen, Jensen & Poulson (1998).

CHAPTER 6
partial differential equations

In this Chapter...

- properties of the parabolic partial differential equation
- the meaning of terms in the Black–Scholes equation
- some solution techniques

6.1 INTRODUCTION

The analysis and solution of partial differential equations is a BIG subject. We can only skim the surface in this book. If you don't feel comfortable with the subject, then the list of books at the end should be of help. However, to understand finance, and even to solve partial differential equations numerically, does not require any great depth of understanding. The aim of this chapter is to give just enough background to the subject to permit any reasonably numerate person to follow the rest of the book; I want to keep the entry requirements to the subject as low as possible.

6.2 PUTTING THE BLACK–SCHOLES EQUATION INTO HISTORICAL PERSPECTIVE

The Black–Scholes partial differential equation is in two dimensions, S and t. It is a parabolic equation, meaning that it has a second derivative with respect to one variable, S, and a first derivative with respect to the other, t. Equations of this form are more colloquially known as **heat** or **diffusion equations**.

The equation, in its simplest form, goes back to almost the beginning of the 19th century. Diffusion equations have been successfully used to model

- diffusion of one material within another, smoke particles in air
- flow of heat from one part of an object to another
- chemical reactions, such as the Belousov–Zhabotinsky reaction which exhibits fascinating wave structure
- electrical activity in the membranes of living organisms, the Hodgkin–Huxley model

- dispersion of populations, individuals move both randomly and to avoid overcrowding
- pursuit and evasion in predator–prey systems
- pattern formation in animal coats, the formation of zebra stripes
- dispersion of pollutants in a running stream

In most of these cases the resulting equations are more complicated than the Black–Scholes equation.

The simplest heat equation for the temperature in a bar is usually written in the form

$$\frac{\partial u}{\partial t} = \frac{\partial^2 u}{\partial x^2}$$

where u is the temperature, x is a spatial co-ordinate and t is time. This equation comes from a heat balance. Consider the flow into and out of a small section of the bar. The flow of heat along the bar is proportional to the spatial gradient of the temperature

$$\frac{\partial u}{\partial x}$$

and thus the derivative of this, the *second* derivative of the temperature, is the heat retained by the small section. This retained heat is seen as a rise in the temperature, represented mathematically by

$$\frac{\partial u}{\partial t}.$$

The balance of the second x-derivative and the first t-derivative results in the heat equation. (There would be a coefficient in the equation, depending on the properties of the bar, but I have set this to one.)

6.3 THE MEANING OF THE TERMS IN THE BLACK–SCHOLES EQUATION

The Black–Scholes equation can be accurately interpreted as a reaction-convection-diffusion equation. The basic diffusion equation is a balance of a first-order t derivative and a second-order S derivative:

$$\frac{\partial V}{\partial t} + \frac{1}{2}\sigma^2 S^2 \frac{\partial^2 V}{\partial S^2}.$$

If these were the only terms in the Black–Scholes equation it would still exhibit the smoothing-out effect, that any discontinuities in the payoff would be instantly diffused away. The only difference between these terms and the terms as they appear in the basic heat or diffusion equation, is that the diffusion coefficient is a function of one of the variables S. Thus we really have diffusion in a non-homogeneous medium.

The first-order S-derivative term

$$rS\frac{\partial V}{\partial S}$$

can be thought of as a convection term. If this equation represented some physical system, such as the diffusion of smoke particles in the atmosphere, then the convective term would be due to a breeze, blowing the smoke in a preferred direction.

The final term

$$-rV$$

is a reaction term. Balancing this term and the time derivative would give a model for decay of a radioactive body, with the half-life being related to r.

Putting these terms together we get a reaction-convection-diffusion equation. An almost identical equation would be arrived at for the dispersion of pollutant along a flowing river with absorption by the sand. In this, the dispersion is the diffusion, the flow is the convection, and the absorption is the reaction.

6.4 BOUNDARY AND INITIAL/FINAL CONDITIONS

To uniquely specify a problem we must prescribe **boundary conditions** and an **initial** or **final condition**. Boundary conditions tell us how the solution must behave for all times at certain values of the asset. In financial problems we usually specify the behavior of the solution at $S = 0$ and as $S \to \infty$. We must also tell the problem how the solution begins. The Black–Scholes equation is a backward equation, meaning that the signs of the t derivative and the second S derivative in the equation are the same when written on the same side of the equals sign. We therefore have to impose a final condition. This is usually the payoff function at expiry.

The Black–Scholes equation in its basic form is linear and satisfies the superposition principle; add together two solutions of the equation and you will get a third. This is not true of nonlinear equations. Linear diffusion equations have some very nice properties. Even if we start out with a discontinuity in the final data, due to a discontinuity in the payoff, this *immediately* gets smoothed out, due to the diffusive nature of the equation. Another nice property is the uniqueness of the solution. Provided that the solution is not allowed to grow too fast as S tends to infinity the solution will be unique. This precise definition of 'too fast' need not worry us, we will not have to worry about uniqueness for any problems we encounter.

6.5 SOME SOLUTION METHODS

We are not going to spend much time on the exact solution of the Black–Scholes equation. Such solution is important, but current market practice is such that models have features which preclude the exact solution. The few explicit, closed-form solutions that are used by practitioners will be covered in the next two chapters.

6.5.1 Transformation to Constant Coefficient Diffusion Equation

It can sometimes be useful to transform the basic Black–Scholes equation into something a little bit simpler by a change of variables. If we write

$$V(S, t) = e^{\alpha x + \beta \tau} U(x, \tau),$$

where

$$\alpha = -\tfrac{1}{2}\left(\frac{2r}{\sigma^2} - 1\right), \quad \beta = -\tfrac{1}{4}\left(\frac{2r}{\sigma^2} + 1\right)^2, \quad S = e^x \text{ and } t = T - \frac{2\tau}{\sigma^2},$$

then $U(x, \tau)$ satisfies the basic diffusion equation

$$\frac{\partial U}{\partial \tau} = \frac{\partial^2 U}{\partial x^2}. \tag{6.1}$$

This simpler equation is easier to handle than the Black–Scholes equation. Sometimes that can be important, for example when seeking closed-form solutions, or in some simple numerical schemes. We shall not pursue this any further.

6.5.2 Green's Functions

One solution of the Black–Scholes equation is

$$V'(S, t) = \frac{e^{-r(T-t)}}{\sigma S' \sqrt{2\pi(T-t)}} e^{-\left(\log(S/S') + (r - (1/2)\sigma^2)(T-t)\right)^2 / 2\sigma^2(T-t)} \tag{6.2}$$

for any S'. (You can verify this by substituting back into the equation, but we'll also be seeing it derived in the next chapter.) This solution is special because as $t \to T$ it becomes zero everywhere, except at $S = S'$. In this limit the function becomes what is known as a **Dirac delta function**. Think of this as a function that is zero everywhere except at one point where it is infinite, in such a way that its integral is one. How is this of help to us?

Expression (6.2) is a solution of the Black–Scholes equation for any S'. Because of the linearity of the equation we can multiply (6.2) by any constant, and we get another solution. But then we can also get another solution by adding together expressions of the form (6.2) but with different values for S'. Putting this together, and thinking of an integral as just a way of adding together many solutions, we find that

$$\frac{e^{-r(T-t)}}{\sigma \sqrt{2\pi(T-t)}} \int_0^\infty e^{-\left(\log(S/S') + (r - (1/2)\sigma^2)(T-t)\right)^2 / 2\sigma^2(T-t)} f(S') \frac{dS'}{S'}$$

is also a solution of the Black–Scholes equation for any function $f(S')$. (If you don't believe me, substitute it into the Black–Scholes equation.)

Because of the nature of the integrand as $t \to T$ (i.e. that it is zero everywhere except at S' and has integral one), if we choose the arbitrary function $f(S')$ to be the payoff function then this expression becomes the solution of the problem:

$$V(S, t) = \frac{e^{-r(T-t)}}{\sigma \sqrt{2\pi(T-t)}} \int_0^\infty e^{-\left(\log(S/S') + (r - (1/2)\sigma^2)(T-t)\right)^2 / 2\sigma^2(T-t)} \text{Payoff}(S') \frac{dS'}{S'}$$

The function $V'(S, t)$ given by (6.2) is called the **Green's function**.

6.5.3 Series Solution

Sometimes we have boundary conditions at two finite (and non-zero) values of S, S_u and S_d, say (we see examples in Chapter 16). For this type of problem, we postulate that the required solution of the Black–Scholes equation can be written as an infinite sum of special functions. First of all, transform to the nicer basic diffusion equation in x and τ. Now write the solution as

$$e^{\alpha x + \beta \tau} \sum_{i=0}^\infty a_i(\tau) \sin(i\omega x) + b_i(\tau) \cos(i\omega x),$$

for some ω and some functions a and b to be found. The linearity of the equation suggests that a sum of solutions might be appropriate. If this is to satisfy the Black–Scholes equation then we must have

$$\frac{da_i}{d\tau} = -i^2 \omega^2 a_i(\tau) \quad \text{and} \quad \frac{db_i}{d\tau} = -i^2 \omega^2 b_i(\tau).$$

You can easily show this by substitution. The solutions are thus

$$a_i(\tau) = A_i e^{-i^2\omega^2\tau} \quad \text{and} \quad b_i(\tau) = B_i e^{-i^2\omega^2\tau}.$$

The solution of the Black–Scholes equation is therefore

$$e^{\alpha x + \beta \tau} \sum_{i=0}^{\infty} e^{-i^2\omega^2\tau} (A_i \sin(i\omega x) + B_i \cos(i\omega x)). \tag{6.3}$$

We have solved the equation; all that we need to do now to satisfy boundary and initial conditions.

Consider the example where the payoff at time $\tau = 0$ is $f(x)$ (although it would be expressed in the original variables, of course) but the contract becomes worthless if ever $x = x_d$ or $x = x_u$.[1]

Rewrite the term in brackets in (6.3) as

$$C_i \sin\left(i\omega' \frac{x - x_d}{x_u - x_d}\right) + D_i \cos\left(i\omega' \frac{x - x_d}{x_u - x_d}\right).$$

To ensure that the option is worthless on these two x values, choose $D_i = 0$ and $\omega' = \pi$. The boundary conditions are thereby satisfied. All that remains is to choose the C_i to satisfy the final condition:

$$e^{\alpha x} \sum_{i=0}^{\infty} C_i \sin\left(i\omega' \frac{x - x_d}{x_u - x_d}\right) = f(x).$$

This also is simple. Multiplying both sides by

$$\sin\left(j\omega' \frac{x - x_d}{x_u - x_d}\right),$$

and integrating between x_d and x_u we find that

$$C_j = \frac{2}{x_u - x_d} \int_{x_d}^{x_u} f(x) e^{-\alpha x} \sin\left(j\omega' \frac{x - x_d}{x_u - x_d}\right) dx.$$

This technique, which can be generalized, is the **Fourier series method**. There are some problems with the method if you are trying to represent a discontinuous function with a sum of trigonometrical functions. The oscillatory nature of an approximate solution with a finite number of terms is known as **Gibbs phenomenon**.

6.6 SIMILARITY REDUCTIONS

Apart from the Green's function, we're not going to use any of the above techniques in this book; rarely will we even find explicit solutions. But one technique that we will find useful is the **similarity reduction**. I will demonstrate the idea using the simple diffusion equation, but we will later use it in many other, more complicated problems.

The basic diffusion equation

$$\frac{\partial u}{\partial t} = \frac{\partial^2 u}{\partial x^2} \tag{6.4}$$

---

[1] This is an example of a double knock-out option, see Chapter 16.

is an equation for the function u which depends on the two variables x and t. Sometimes, in very, very special cases we can write the solution as a function of just one variable. Let me give an example. Verify that the function

$$u(x, t) = \int_0^{x/t^{1/2}} e^{-(1/4)\xi^2} d\xi$$

satisfies (6.4). But in this function x and t only appear in the combination

$$\frac{x}{t^{1/2}}.$$

Thus, in a sense, u is a function of only one variable.

A slight generalization, but also demonstrating the idea of similarity solutions, is to look for a solution of the form

$$u = t^{-1/2} f(\xi) \tag{6.5}$$

where

$$\xi = \frac{x}{t^{1/2}}.$$

Substitute (6.5) into (6.4) to find that a solution for f is

$$f = e^{-(1/4)\xi^2},$$

so that

$$t^{-1/2} e^{-(1/4)(x^2/t)}$$

is also a special solution of the diffusion equation.

Be warned, though. You can't always find similarity solutions; not only must the equation have a particularly nice structure but also the similarity form must be consistent with any initial condition or boundary conditions.

6.7 OTHER ANALYTICAL TECHNIQUES

The other two main solution techniques for linear partial differential equations are Fourier and Laplace transforms. These are such large and highly technical subjects that I really cannot begin to give an idea of how they work, space is far too short. But be reassured that it is probably not worth your while learning the techniques, since in finance they can be used to solve only a very small number of problems. If you want to learn something useful then move on to the next section.

6.8 NUMERICAL SOLUTION

Even though there are several techniques that we can use for finding solutions, in the vast majority of cases we must solve the Black–Scholes equation numerically. But we are lucky. Parabolic differential equations are just about the easiest equations to solve numerically. Obviously, there are any number of really sophisticated techniques, but if you stick with the simplest then you can't go far wrong. In Chapters 12, 63 and 64 we discuss these methods in detail. I want to stress that I am going to derive many partial differential equations from now on, and I am going to assume you trust me that we will at the end of the book see how to solve them.

6.9 **SUMMARY**

This short chapter is only intended as a primer on partial differential equations. If you want to study this subject in depth, see the books and articles mentioned below.

FURTHER READING

- Grindrod (1991) is all about reaction-diffusion equations, where they come from and their analysis. The book includes many of the physical models described above.

- Murray (1989) also contains a great deal on reaction-diffusion equations, but concentrates on models of biological systems.

- Wilmott & Wilmott (1990) describe the diffusion of pollutant along a river with convection and absorption by the river bed.

- The classical reference works for diffusion equations are Crank (1989) and Carslaw & Jaeger (1989). But also see the book on partial differential equations by Sneddon (1957) and the book on general applied mathematical methods by Strang (1986).

CHAPTER 7
the Black–Scholes formulae and the 'greeks'

In this Chapter...

- the derivation of the Black–Scholes formulae for calls, puts and simple digitals
- the meaning and importance of the 'greeks', delta, gamma, theta, vega and rho
- the difference between differentiation with respect to variables and to parameters
- formulae for the greeks for calls, puts and simple digitals

7.1 INTRODUCTION

The Black–Scholes equation has simple solutions for calls, puts and some other contracts. In this chapter I'm going to walk you through the derivation of these formulae step by step. This is one of the few places in the book where I do derive formulae. The reason that I don't often derive formulae is that the majority of contracts do not have explicit solutions for their theoretical value. Instead much of my emphasis will be placed on finding numerical solutions of the Black–Scholes equation.

We've seen how the quantity 'delta', the first derivative of the option value with respect to the underlying, occurs as an important quantity in the derivation of the Black–Scholes equation. In this chapter I describe the importance of other derivatives of the option price, with respect to the variables (the underlying asset and time) and with respect to some of the parameters. These derivatives are important in the hedging of an option position, playing key roles in risk management. It can be argued that it is more important to get the hedging correct than to be precise in the pricing of a contract. The reason for this is that if you are accurate in your hedging you will have reduced or eliminated future uncertainty. This leaves you with a profit (or loss) that is set the moment that you buy or sell the contract. But if your hedging is inaccurate, then it doesn't matter, within reason, what you sold the contract for initially, as future uncertainty could easily dominate any initial profit. Of course, life is not so simple, in reality we are exposed to model error, which can make a mockery of anything we do. However, this illustrates the importance of good hedging, and that's where the 'greeks' come in.

7.2 DERIVATION OF THE FORMULAE FOR CALLS, PUTS AND SIMPLE DIGITALS

The Black–Scholes equation is

$$\frac{\partial V}{\partial t} + \tfrac{1}{2}\sigma^2 S^2 \frac{\partial^2 V}{\partial S^2} + rS\frac{\partial V}{\partial S} - rV = 0. \tag{7.1}$$

This equation must be solved with final condition depending on the payoff: each contract will have a different functional form prescribed at expiry $t = T$, depending on whether it is a call, a put or something more fancy. This is the final condition that must be imposed to make the solution unique. We'll worry about final conditions later, for the moment concentrate on manipulating (7.1) into something we can easily solve.

The first step in the manipulation is to change from present value to future value terms. Recalling that the payoff is received at time T but that we are valuing the option at time t this suggests that we write

$$V(S, t) = e^{-r(T-t)} U(S, t).$$

This takes our differential equation to

$$\frac{\partial U}{\partial t} + \tfrac{1}{2}\sigma^2 S^2 \frac{\partial^2 U}{\partial S^2} + rS\frac{\partial U}{\partial S} = 0.$$

The second step is really trivial. Because we are solving a backward equation, discussed in Chapter 6, we'll write

$$\tau = T - t.$$

This now takes our equation to

$$\frac{\partial U}{\partial \tau} = \tfrac{1}{2}\sigma^2 S^2 \frac{\partial^2 U}{\partial S^2} + rS\frac{\partial U}{\partial S}.$$

When we first started modeling equity prices we used intuition about the asset price *return* to build up the stochastic differential equation model. Let's go back to examine the return and write

$$\xi = \log S.$$

With this as the new variable, we find that

$$\frac{\partial}{\partial S} = e^{-\xi}\frac{\partial}{\partial \xi} \quad \text{and} \quad \frac{\partial^2}{\partial S^2} = e^{-2\xi}\frac{\partial^2}{\partial \xi^2} - e^{-2\xi}\frac{\partial}{\partial \xi}.$$

Now the Black–Scholes equation becomes

$$\frac{\partial U}{\partial \tau} = \tfrac{1}{2}\sigma^2 \frac{\partial^2 U}{\partial \xi^2} + \left(r - \tfrac{1}{2}\sigma^2\right)\frac{\partial U}{\partial \xi}.$$

What has this done for us? It has taken the problem defined for $0 \le S < \infty$ to one defined for $-\infty < \xi < \infty$. But more importantly, the coefficients in the equation are now all constant, independent of the underlying. This is a big step forward, made possible by the lognormality of the underlying asset. We are nearly there.

The last step is simple, but the motivation is not so obvious. Write

$$x = \xi + \left(r - \tfrac{1}{2}\sigma^2\right)\tau,$$

and $U = W(x, \tau)$. This is just a 'translation' of the co-ordinate system. It's a bit like using the forward price of the asset instead of the spot price as a variable. After this change of variables the Black–Scholes becomes the simpler

$$\frac{\partial W}{\partial \tau} = \tfrac{1}{2}\sigma^2\frac{\partial^2 W}{\partial x^2}. \tag{7.2}$$

To summarize,

$$V(S, t) = e^{-r(T-t)}U(S, t) = e^{-r\tau}U(S, T - \tau) = e^{-r\tau}U(e^{\xi}, T - \tau)$$

$$= e^{-r\tau}U(e^{x-(r-(1/2)\sigma^2)\tau}, T - \tau) = e^{-r\tau}W(x, \tau).$$

To those of you who already know the Black–Scholes formulae for calls and puts the variable x will ring a bell:

$$x = \xi + \left(r - \tfrac{1}{2}\sigma^2\right)\tau = \log S + \left(r - \tfrac{1}{2}\sigma^2\right)(T - t).$$

Having turned the original Black–Scholes equation into something much simpler, let's take a break for a moment while I explain where we are headed.

I'm going to derive an expression for the value of any option whose payoff is a known function of the asset price at expiry. This includes calls, puts and digitals. This expression will be in the form of an integral. For special cases, I'll show how to rewrite this integral in terms of the cumulative distribution function for the Normal distribution. This is particularly useful since the function can be found on spreadsheets, calculators and in the backs of books. But there are two steps before I can write down this integral.

The first step is to find a special solution of (7.2), called the fundamental solution. This solution has useful properties. The second step is to use the linearity of the equation and the useful properties of the special solution to find the *general solution* of the equation. Here we go.

I'm going to look for a special solution of (7.2) of the following form

$$W(x, \tau) = \tau^{\alpha}f\left(\frac{(x - x')}{\tau^{\beta}}\right), \tag{7.3}$$

where x' is an arbitrary constant. And I'll call this special solution $W_f(x, \tau; x')$. Note that the unknown function depends on only *one* variable $(x - x')/\tau^{\beta}$. As well as finding the function f we must find the constant parameters α and β. We can expect that if this approach works, the equation for f will be an ordinary differential equation since the function only has one variable. This reduction of dimension is an example of a similarity reduction, discussed in Chapter 6.

Substituting expression (7.3) into (7.2) we get

$$\tau^{\alpha-1}\left(\alpha f - \beta\eta\frac{df}{d\eta}\right) = \tfrac{1}{2}\sigma^2\tau^{\alpha-2\beta}\frac{d^2f}{d\eta^2}, \tag{7.4}$$

where

$$\eta = \frac{x - x'}{\tau^{\beta}}.$$

Examining the dependence of the two terms in (7.4) on both τ and η we see that we can only have a solution if

$$\alpha - 1 = \alpha - 2\beta \quad \text{i.e.} \quad \beta = \tfrac{1}{2}.$$

I want to ensure that my 'special solution' has the property that its integral over all ξ is independent of τ, for reasons that will become apparent. To ensure this, I require

$$\int_{-\infty}^{\infty} \tau^\alpha f\left(\frac{x - x'}{\tau^\beta}\right) dx$$

to be constant. I can write this as

$$\int_{-\infty}^{\infty} \tau^{\alpha+\beta} f(\eta)\, d\eta$$

and so I need

$$\alpha = -\beta = -\tfrac{1}{2}.$$

The function f now satisfies

$$-f - \eta \frac{df}{d\eta} = \sigma^2 \frac{d^2 f}{d\eta^2}.$$

This can be written

$$\sigma^2 \frac{d^2 f}{d\eta^2} + \frac{d(\eta f)}{d\eta} = 0,$$

which can be integrated once to give

$$\sigma^2 \frac{df}{d\eta} + \eta f = a,$$

where a is a constant. For my special solution I'm going to choose $a = 0$. This equation can be integrated again to give

$$f(\eta) = be^{-\eta^2/(2\sigma^2)}.$$

I will choose the constant b such that the integral of f from minus infinity to plus infinity is one:

$$f(\eta) = \frac{1}{\sqrt{2\pi}\sigma} e^{-\eta^2/(2\sigma^2)}.$$

This is the special solution I have been seeking:[1]

$$W(x, \tau) = \frac{1}{\sqrt{2\pi\tau}\sigma} e^{-((x-x')^2/2\sigma^2\tau)}.$$

Now I will explain why it is useful in our quest for the Black–Scholes formulae.

In Figure 7.1 W is plotted as a function of x' for several values of τ. Observe how the function rises in the middle but decays at the sides. As $\tau \to 0$ this becomes more pronounced. The 'middle' is the point $x' = x$. At this point the function grows unboundedly and away from this point the function decays to zero as $\tau \to 0$. Although the function is increasingly confined

---

[1] It is just the probability density function for a Normal random variable with mean zero and standard deviation σ.

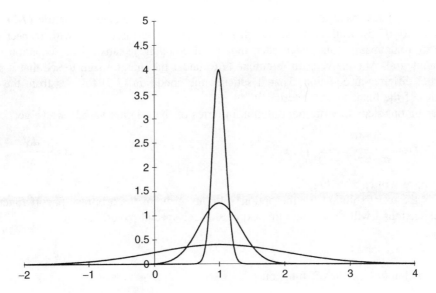

Figure 7.1   The fundamental solution.

to a narrower and narrower region its area remains fixed at one. These properties of decay away from one point, unbounded growth at that point and constant area, result in a **Dirac delta function** $\delta(x' - x)$ as $\tau \to 0$. The delta function has one important property, namely

$$\int \delta(x' - x)g(x')\,dx' = g(x)$$

where the integration is from any point below x to any point above x. Thus the delta function 'picks out' the value of g at the point where the delta function is singular i.e. at $x' = x$. In the limit as $\tau \to 0$ the function W becomes a delta function at $x = x'$. This means that

$$\lim_{\tau \to 0} \frac{1}{\sigma\sqrt{2\pi\tau}} \int_{-\infty}^{\infty} e^{-((x'-x)^2/2\sigma^2\tau)}g(x')\,dx' = g(x).$$

This property of the special solution, together with the linearity of the Black–Scholes equation, are all that are needed to find some explicit solutions.

Now is the time to consider the payoff. Let's call it

$$\text{Payoff}(S).$$

This is the value of the option at time $t = T$. It is the final condition for the function V, satisfying the Black–Scholes equation:

$$V(S, T) = \text{Payoff}(S).$$

With our new variables, this final condition is

$$W(x, 0) = \text{Payoff}(e^x). \tag{7.5}$$

I claim that the solution of this for $\tau > 0$ is

$$W(x, \tau) = \int_{-\infty}^{\infty} W_f(x, \tau; x')\,\text{Payoff}(e^{x'})\,dx'. \tag{7.6}$$

To show this, I just have to demonstrate that the expression satisfies Equation (7.2) and the final condition (7.5). Both of these are straightforward. The integration with respect to x' is similar to a summation, and since each individual component satisfies the equation so does the sum/integral. Alternatively, differentiate (7.6) under the integral sign to see that it satisfies the partial differential equation. That it satisfies the condition (7.5) follows from the special properties of the fundamental solution W_f.

Retracing our steps to write our solution in terms of the original variables, we get

$$V(S,t) = \frac{e^{-r(T-t)}}{\sigma\sqrt{2\pi(T-t)}} \int_0^\infty e^{-\left(\log(S/S')+(r-(1/2)\sigma^2)(T-t)\right)^2/2\sigma^2(T-t)} \, \text{Payoff}\,(S')\frac{dS'}{S'}, \qquad (7.7)$$

where I have written $x' = \log S'$.

This is the exact solution for the option value in terms of the arbitrary payoff function. In the next sections I will manipulate this expression for special payoff functions.

7.2.1 Formula for a Call

The call option has the payoff function

$$\text{Payoff}(S) = \max(S - E, 0).$$

Expression (7.7) can then be written as

$$\frac{e^{-r(T-t)}}{\sigma\sqrt{2\pi(T-t)}} \int_E^\infty e^{-\left(\log(S/S')+(r-(1/2)\sigma^2)(T-t)\right)^2/2\sigma^2(T-t)}(S' - E)\frac{dS'}{S'}.$$

Return to the variable $x' = \log S'$, to write this as

$$\frac{e^{-r(T-t)}}{\sigma\sqrt{2\pi(T-t)}} \int_{\log E}^\infty e^{-\left(-x'+\log S+(r-(1/2)\sigma^2)(T-t)\right)^2/2\sigma^2(T-t)}(e^{x'} - E)\,dx'$$

$$= \frac{e^{-r(T-t)}}{\sigma\sqrt{2\pi(T-t)}} \int_{\log E}^\infty e^{-\left(-x'+\log S+(r-(1/2)\sigma^2)(T-t)\right)^2/2\sigma^2(T-t)}e^{x'}\,dx'$$

$$- E\frac{e^{-r(T-t)}}{\sigma\sqrt{2\pi(T-t)}} \int_{\log E}^\infty e^{-\left(-x'+\log S+(r-(1/2)\sigma^2)(T-t)\right)^2/2\sigma^2(T-t)}\,dx'.$$

Both integrals in this expression can be written in the form

$$\int_d^\infty e^{-(1/2)x'^2}\,dx'$$

for some d (the second is just about in this form already, and the first just needs a completion of the square).

Apart from a couple of minor differences, this integral is just like the cumulative distribution function for the standardized Normal distribution[2] defined by

$$N(x) = \frac{1}{\sqrt{2\pi}} \int_{-\infty}^x e^{-(1/2)\phi^2}\,d\phi.$$

---

[2] I.e. having zero mean and unit standard deviation.

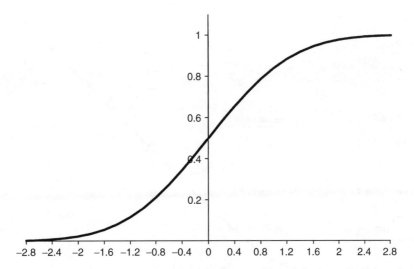

Figure 7.2 The cumulative distribution function for a standardized Normal random variable, $N(x)$.

This function, plotted in Figure 7.2, is the probability that a Normally distributed variable is less than x.

Thus the option price can be written as two separate terms involving the cumulative distribution function for a Normal distribution:

$$\text{Call option value} = SN(d_1) - Ee^{-r(T-t)}N(d_2)$$

where

$$d_1 = \frac{\log\left(\dfrac{S}{E}\right) + \left(r + \frac{1}{2}\sigma^2\right)(T - t)}{\sigma\sqrt{T - t}}$$

and

$$d_2 = \frac{\log\left(\dfrac{S}{E}\right) + \left(r - \frac{1}{2}\sigma^2\right)(T - t)}{\sigma\sqrt{T - t}}.$$

When there is continuous dividend yield on the underlying, or it is a currency, then

Call option value

$$Se^{-D(T-t)}N(d_1) - Ee^{-r(T-t)}N(d_2)$$

$$d_1 = \frac{\log\left(\dfrac{S}{E}\right) + \left(r - D + \frac{1}{2}\sigma^2\right)(T - t)}{\sigma\sqrt{T - t}}$$

$$d_2 = \frac{\log\left(\dfrac{S}{E}\right) + \left(r - D - \frac{1}{2}\sigma^2\right)(T - t)}{\sigma\sqrt{T - t}}$$

$$= d_1 - \sigma\sqrt{T - t}$$

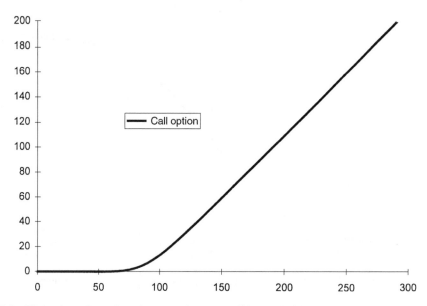

Figure 7.3 The value of a call option as a function of the underlying asset price at a fixed time to expiry.

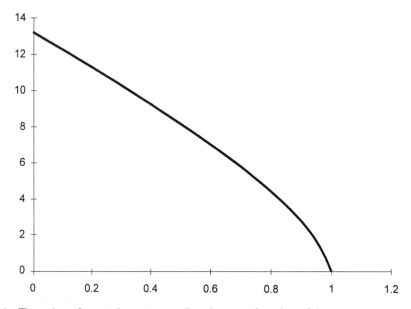

Figure 7.4 The value of an at-the-money call option as a function of time.

The option value is shown in Figure 7.3 as a function of the underlying asset at a fixed time to expiry. In Figure 7.4 the value of the at-the-money option is shown as a function of time, and expiry is $t = 1$. In Figure 7.5 the call value is shown as a function of both the underlying and time.

When the asset is 'at-the-money forward', i.e. $S = E^{-(r-D)(T-t)}$, then there is a simple approximation for the call value (Brenner & Subrahmanyam, 1994):

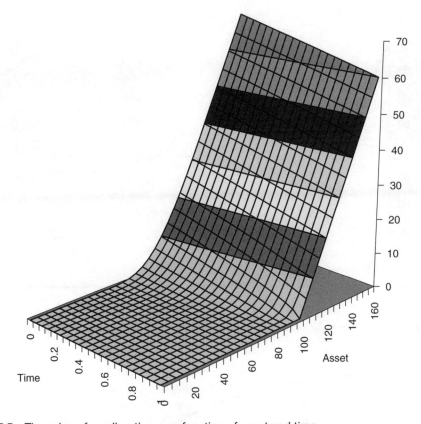

Figure 7.5 The value of a call option as a function of asset and time.

$$\text{Call} \approx 0.4 \ Se^{-D(T-t)}\sigma\sqrt{T-t}.$$

7.2.2 Formula for a Put

The put option has payoff

$$\text{Payoff}(S) = \max(E - S, 0).$$

The value of a put option can be found in the same way as above, or using put-call parity

$$\text{Put option value} = -SN(-d_1) + Ee^{-r(T-t)}N(-d_2),$$

with the same d_1 and d_2.

When there is continuous dividend yield on the underlying, or it is a currency, then

Put option value

$$-Se^{-D(T-t)}N(-d_1) + Ee^{-r(T-t)}N(-d_2)$$

YOU MEMORIZED THE CALL VALUE, I HOPE, NOW MEMORIZE THE PUT

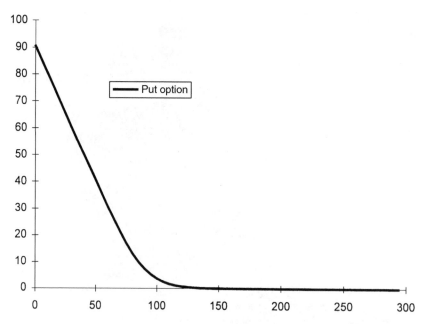

Figure 7.6 The value of a put option as a function of the underlying asset at a fixed time to expiry.

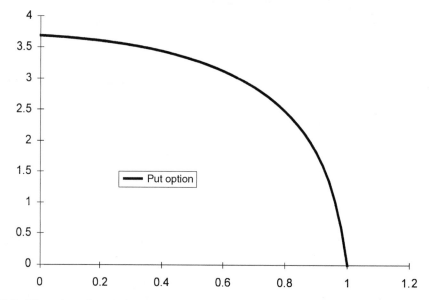

Figure 7.7 The value of an at-the money put option as a function of time.

The option value is shown in Figure 7.6 against the underlying asset and in Figure 7.7 against time. In Figure 7.8 the option value is shown as a function of both the underlying asset and time.

When the asset is at-the-money forward the simple approximation for the put value (Brenner & Subrahmanyam, 1994) is

$$\text{Put} \approx 0.4 \ S e^{-D(T-t)} \sigma \sqrt{T - t}.$$

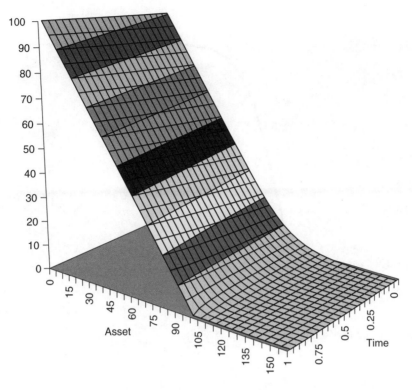

Figure 7.8 The value of a put option as a function of asset and time.

7.2.3 Formula for a Binary Call

The binary call has payoff

$$\text{Payoff}(S) = \mathcal{H}(S - E),$$

where $\mathcal{H}$ is the Heaviside function taking the value one when its argument is positive and zero otherwise.

Incorporating a dividend yield, we can write the option value as

$$\frac{e^{-r(T-t)}}{\sigma\sqrt{2\pi(T-t)}} \int_{\log E}^{\infty} e^{-\left(x'-\log S-(r-D-(1/2)\sigma^2)(T-t)\right)^2/2\sigma^2(T-t)} \, dx'.$$

This term is just like the second term in the call option equation and so

> **Binary call option value**
>
> $$e^{-r(T-t)}N(d_2)$$

The option value is shown in Figure 7.9.

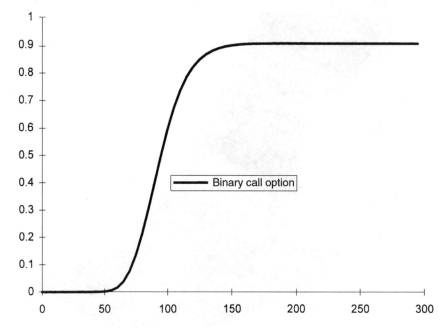

Figure 7.9 The value of a binary call option.

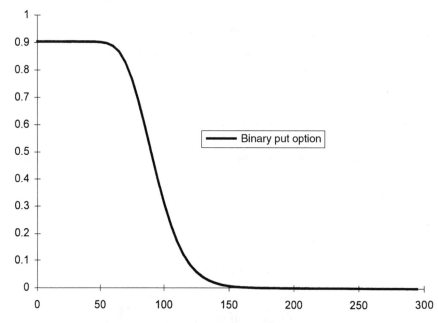

Figure 7.10 The value of a binary put option.

7.2.4   Formula for a Binary Put

The binary put has a payoff of one if $S < E$ at expiry. It has a value of

> **Binary put option value**
>
> $$e^{-r(T-t)}(1 - N(d_2))$$

since a binary call and a binary put must add up to the present value of \$1 received at time T. The option value is shown in Figure 7.10.

7.3   **DELTA**

The **delta** of an option or a portfolio of options is the sensitivity of the option or portfolio to the underlying. It is the rate of change of value with respect to the asset:

$$\Delta = \frac{\partial V}{\partial S}$$

THE DELTA IS OUR FIRST, AND MOST IMPORTANT, 'GREEK'

Here V can be the value of a single contract or of a whole portfolio of contracts. The delta of a portfolio of options is just the sum of the deltas of all the individual positions.

The theoretical device of delta hedging, introduced in Chapter 5, for eliminating risk is far more than that, it is a very important practical technique.

Roughly speaking, the financial world is divided up into speculators and hedgers. The speculators take a view on the direction of some quantity such as the asset price (or a more abstract quantity such as volatility) and implement a strategy to take advantage of their view. Such people may not hedge at all.

Then there are the hedgers. There are two kinds of hedger: the ones who hold a position already and want to eliminate some very specific risk (usually using options) and the ones selling (or buying) the options because they believe they have a better price and can make money by hedging away *all* risk. It is the latter type of hedger who is delta hedging. They can only guarantee to make a profit by selling a contract for a high value if they can eliminate all of the risk due to the random fluctuation in the underlying.

Delta hedging means holding one of the option and short a quantity Δ of the underlying. Delta can be expressed as a function of S and t, I'll give some formulae later in this section. This function varies as S and t vary. This means that the number of assets held must be continuously changed to maintain a **delta neutral** position; this procedure is called **dynamic hedging**. Changing the number of assets held requires the continual purchase and/or sale of the stock. This is called **rehedging** or **rebalancing** the portfolio.

This delta hedging may take place very frequently in highly liquid markets where it is relatively cheap to buy and sell. Thus the Black–Scholes assumption of continuous hedging may be quite accurate. In less liquid markets, you lose a lot on bid-offer spread and will therefore hedge less frequently. Moreover, you may not even be able to buy or sell in the quantities you want. Even in the absence of costs, you cannot be sure that your model for the underlying is accurate. There will certainly be some risk associated with the model. These issues make delta hedging less than perfect and in practice the risk in the underlying cannot be hedged away perfectly. Issues of discrete hedging and transaction costs are covered in depth in Chapters 23 and 24.

Some contracts (see especially Chapter 16) have a delta that becomes very large at special times or asset values. The size of the delta makes delta hedging impossible; what can you do if you find yourself with a theoretical delta requiring you to buy more stock than exists? In such a situation the basic foundation of the Black–Scholes world has collapsed and you would be right to question the validity of any pricing formula. This happens at expiry close to the strike for binary options. Although I've given a formula for their price above and a formula for their delta below, I'd be careful using them if I were you.

Here are some formulae for the deltas of common contracts (all formulae assume that the underlying pays dividends or is a currency):

Deltas of common contracts

Call $\qquad e^{-D(T-t)}N(d_1)$

Put $\qquad e^{-D(T-t)}(N(d_1) - 1)$

Binary call $\dfrac{e^{-r(T-t)}N'(d_2)}{\sigma S\sqrt{T-t}}$

Binary put $\quad -\dfrac{e^{-r(T-t)}N'(d_2)}{\sigma S\sqrt{T-t}}$

$$N'(x) = \frac{1}{\sqrt{2\pi}}e^{-(1/2)x^2}$$

Examples of these functions are plotted in Figure 7.11, with some scaling of the binaries.

7.4 GAMMA

The **gamma**, Γ, of an option or a portfolio of options is the second derivative of the position with respect to the underlying:

$$\Gamma = \frac{\partial^2 V}{\partial S^2}$$

THE GAMMA IS IMPORTANT WHEN EXAMINING THE HIGHER-ORDER BEHAVIOR OF A CONTRACT

Since gamma is the sensitivity of the delta to the underlying it is a measure of by how much or how often a position must be rehedged in order to maintain a delta-neutral position. Although

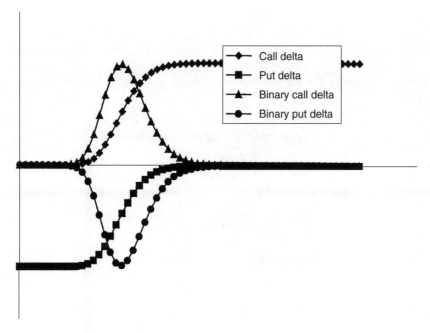

Figure 7.11 The deltas of a call, a put, a binary call and a binary put option. (Binary values scaled to a maximum value of one.)

the delta also varies with time this effect is dominated by the Brownian nature of the movement in the underlying.

In a delta-neutral position the gamma is partly responsible for making the return on the portfolio equal to the risk-free rate, the no-arbitrage condition of Chapter 5. The rest of this task falls to the time-derivative of the option value, discussed below. Actually, the situation is far more complicated than this because of the necessary discreteness in the hedging, there is a finite time between rehedges. In any delta-hedged position you make money on some hedges and lose some on others. In a long gamma position ($\Gamma > 0$) you make money on the large moves in the underlying and lose it on the small moves. To be precise, you make money 32% of the time and lose it 68%. But when you make it, you make more. The net effect is to get the risk-free rate of return on the portfolio. You won't have a clue where this fact came from, but all will be made clear in Chapter 23.

Gamma also plays an important role when there is a mismatch between the market's view of volatility and the actual volatility of the underlying; again this is discussed in Chapter 23.

Because costs can be large and because one wants to reduce exposure to model error it is natural to try to minimize the need to rebalance the portfolio too frequently. Since gamma is a measure of sensitivity of the hedge ratio Δ to the movement in the underlying, the hedging requirement can be decreased by a gamma-neutral strategy. This means buying or selling more *options*, not just the underlying. Because the gamma of the underlying (its second derivative) is zero, we cannot add gamma to our position just with the underlying. We can have as many options in our position as we want, we choose the quantities of each such that both delta and gamma are zero. The minimal requirement is to hold two different types of option and the underlying. In practice, the option position is not readjusted too often because, if the cost of transacting in the underlying is large, then the cost of transacting in its derivatives is even larger.

Here are some formulae for the gammas of common contracts:

Gammas of common contracts

Call $\dfrac{e^{-D(T-t)}N'(d_1)}{\sigma S\sqrt{T-t}}$

Put $\dfrac{e^{-D(T-t)}N'(d_1)}{\sigma S\sqrt{T-t}}$

Binary call $-\dfrac{e^{-r(T-t)}d_1 N'(d_2)}{\sigma^2 S^2(T-t)}$

Binary put $\dfrac{e^{-r(T-t)}d_1 N'(d_2)}{\sigma^2 S^2(T-t)}$

Examples of these functions are plotted in Figure 7.12, with some scaling for the binaries.

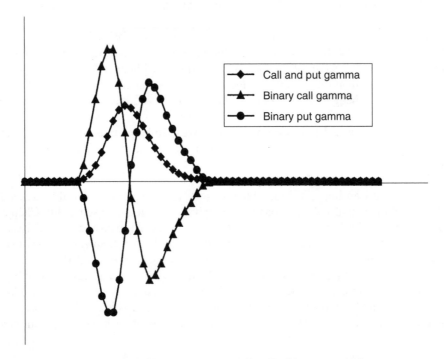

Call and put gamma

Binary call gamma

Binary put gamma

Figure 7.12   The gammas of a call, a put, a binary call and a binary put option.

7.5 **THETA**

Theta, Θ, is the rate of change of the option price with time.

$$\Theta = \frac{\partial V}{\partial t}$$

IF THE ASSET DOESN'T MOVE, THE OPTION WILL CHANGE BY THE THETA WITH TIME

The theta is related to the option value, the delta and the gamma by the Black–Scholes equation. In a delta-hedged portfolio the theta contributes to ensuring that the portfolio earns the risk-free rate. But it contributes in a completely certain way, unlike the gamma which contributes the right amount *on average*.

Here are some formulae for the thetas of common contracts:

Thetas of common contracts

Call $\quad -\dfrac{\sigma S e^{-D(T-t)} N'(d_1)}{2\sqrt{T-t}} + DSN(d_1)e^{-D(T-t)} - rEe^{-r(T-t)}N(d_2)$

Put $\quad -\dfrac{\sigma S e^{-D(T-t)} N'(-d_1)}{2\sqrt{T-t}} - DSN(-d_1)e^{-D(T-t)} + rEe^{-r(T-t)}N(-d_2)$

Binary call $\quad re^{-r(T-t)}N(d_2) + e^{-r(T-t)}N'(d_2)\left(\dfrac{d_1}{2(T-t)} - \dfrac{r-D}{\sigma\sqrt{T-t}}\right)$

Binary put $\quad re^{-r(T-t)}(1 - N(d_2)) - e^{-r(T-t)}N'(d_2)\left(\dfrac{d_1}{2(T-t)} - \dfrac{r-D}{\sigma\sqrt{T-t}}\right)$

These functions are plotted in Figure 7.13.

7.6 **VEGA**

Vega, a.k.a. zeta and kappa, is a very important but confusing quantity. It is the sensitivity of the option price to volatility.

$$\text{Vega} = \frac{\partial V}{\partial \sigma}$$

WE DON'T KNOW THE VOLATILITY PRECISELY. VEGA MEASURES SENSITIVITY OF VALUE TO VOL

This is completely different from the other greeks[3] since it is a derivative with respect to a parameter and not a variable. This makes something of a difference when we come to finding numerical solutions for such quantities.

---

[3] It's not even Greek. Among other things it is an American car, a star (Alpha Lyrae), the real name of Zorro: there are a couple of 16th century Spanish authors called Vega, an Op art painting by Vasarely and a character in the computer game 'Street Fighter'. And who could forget Vincent and his brother?

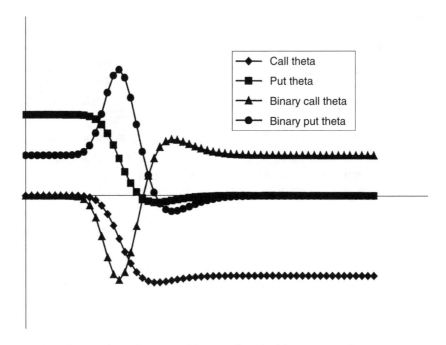

Figure 7.13   The thetas of a call, a put, a binary call and a binary put option.

In practice, the volatility of the underlying is not known with certainty. Not only is it very difficult to measure at any time, it is even harder to predict what it will do in the future. Suppose that we put a volatility of 20% into an option pricing formula, how sensitive is the price to that number? That's the vega.

As with gamma hedging, one can vega hedge to reduce sensitivity to the volatility. This is a major step towards eliminating some model risk, since it reduces dependence on a quantity that, to be honest, is not known very accurately.

There is a downside to the measurement of vega. It is only really meaningful for options having single-signed gamma everywhere. For example it makes sense to measure vega for calls and puts but not binary calls and binary puts. I have included the formulae for the vega of such contracts (see page 117), but they should be used with care, if at all. The reason for this is that call and put values (and options with single-signed gamma) have values that are monotonic in the volatility: increase volatility in a call and its value increases everywhere. Contracts with a gamma that changes sign may have a vega measured at zero because as we increase the volatility the price may rise somewhere and fall somewhere else. Such a contract is very exposed to volatility risk but that risk is not measured by the vega. See Chapter 27 for more details.

The second derivative with respect to σ has been called 'vomma' and the second-order derivative with respect to the asset and the volatility has been called 'kabanga'. I doubt that they represent what their fans think they represent, and I'm going to make no further mention of them.

Here are formulae for the vegas of common contracts:

Vegas of common contracts

Call $S\sqrt{T-t}e^{-D(T-t)}N'(d_1)$

Put $S\sqrt{T-t}e^{-D(T-t)}N'(d_1)$

Binary call $-e^{-r(T-t)}N'(d_2)\left(\sqrt{T-t}+\dfrac{d_2}{\sigma}\right)$

Binary put $e^{-r(T-t)}N'(d_2)\left(\sqrt{T-t}+\dfrac{d_2}{\sigma}\right)$

In Figure 7.14 is shown the value of an at-the-money call option as a function of the volatility. There is one year to expiry, the strike is 100, the interest rate is 10% and there are no dividends. No matter how far in or out of the money this curve is always monotonically increasing for call options and put options; uncertainty adds value to the contract. The slope of this curve is the vega.

In Figure 7.15 is shown the value of an out-of-the-money binary call option as a function of the volatility. There is one year to expiry, the asset value is 88, strike is 100, the interest rate is 10% and there are no dividends. Observe that there is maximum at a volatility of about 24%. The value of the option is not monotonic in the volatility. We will see later why this makes the meaning of vega somewhat suspect.

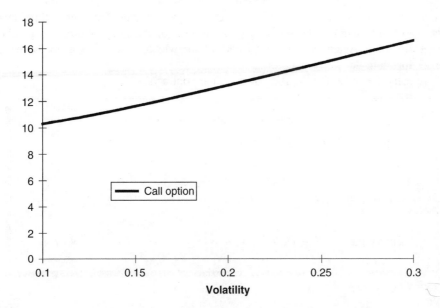

Figure 7.14   The value of an at-the-money call option as a function of volatility.

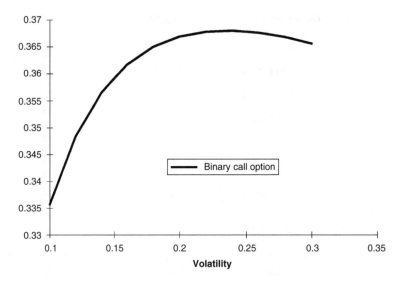

Figure 7.15   The value of an out-of-the-money binary call option as a function of volatility.

7.7 **RHO**

Rho, ρ, is the sensitivity of the option value to the interest rate used in the Black–Scholes formulae:

$$\rho = \frac{\partial V}{\partial r}$$

In practice one often uses a whole term structure of interest rates, meaning a time-dependent rate $r(t)$. Rho would then be the sensitivity to the level of the rates assuming a parallel shift in rates at all times. Again, you must be careful for which contracts you measure rho; see Chapter 27 for more details.

Here are some formulae for the rhos of common contracts:

Rhos of common contracts

Call $E(T-t)e^{-r(T-t)}N(d_2)$

Put $-E(T-t)e^{-r(T-t)}N(-d_2)$

Binary call $-(T-t)e^{-r(T-t)}N(d_2) + \dfrac{\sqrt{T-t}}{\sigma}e^{-r(T-t)}N'(d_2)$

Binary put $-(T-t)e^{-r(T-t)}(1-N(d_2)) - \dfrac{\sqrt{T-t}}{\sigma}e^{-r(T-t)}N'(d_2)$

The sensitivities of common contracts to the dividend yield or foreign interest rate are given by the following formulae:

Sensitivity to dividend for common contracts

Call $\qquad -(T-t)Se^{-D(T-t)}N(d_1)$

Put $\qquad (T-t)Se^{-D(T-t)}N(-d_1)$

Binary call $\quad -\dfrac{\sqrt{T-t}}{\sigma}e^{-r(T-t)}N'(d_2)$

Binary put $\quad \dfrac{\sqrt{T-t}}{\sigma}e^{-r(T-t)}N'(d_2)$

7.8 IMPLIED VOLATILITY

The Black–Scholes formula for a call option takes as input the expiry, the strike, the underlying and the interest rate *together with the volatility* to output the price. All but the volatility are easily measured. How do we know what volatility to put into the formulae? A trader can see on his screen that a certain call option with four months until expiry and a strike of 100 is trading at 6.51 with the underlying at 101.5 and a short-term interest rate of 8%. Can we use this information in some way?

Turn the relationship between volatility and an option price on its head. If we can see the price at which the option is trading, we can ask 'What volatility must I use to get the correct market price?' This is called the **implied volatility**. The implied volatility is the volatility of the underlying which when substituted into the Black–Scholes formula gives a theoretical price equal to the market price. In a sense it is the market's view of volatility over the life of the option. Assuming that we are using call prices to estimate the implied volatility then provided the option price is less than the asset and greater than zero then we can find a unique value for the implied volatility. (If the option price is outside these bounds then there's a very extreme arbitrage opportunity.) Because there is no simple formula for the implied volatility as a function of the option value we must solve the equation

$$V_{BS}(S_0, t_0; \sigma, r; E, T) = \text{known value}$$

for σ, where V_{BS} is the Black–Scholes formula. Today's asset price is S_0, the date is t_0 and everything is known in this equation except for σ. Below is an algorithm for finding the implied volatility from the market price of a call option to any required degree of accuracy. The method used is **Newton–Raphson** which uses the derivative of the option price with respect to the volatility (the vega) in the calculation. This method is particularly good for a well-behaved function such as a call value.

A SIMPLE VB FUNCTION FOR CALCULATING IMPLIED VOL FROM CALL PRICES

```
Function ImpVolCall(MktPrice As Double, Strike As
        Double, Expiry As Double, _ Asset As Double,
        IntRate As Double, Error As Double)
Volatility = 0.2
dv = error + 1
While Abs(dv) > error
    d1 = Log(Asset / Strike) + (IntRate + 0.5 *
        Volatility * Volatility) * Expiry
    d1 = d1 / (Volatility * Sqr(Expiry))
    d2 = d1 - Volatility * Sqr(Expiry)
    PriceError = Asset * cdf(d1) - Strike *
                Exp(-IntRate * Expiry) _
                * cdf(d2) - MktPrice
    Vega = Asset * Sqr(Expiry / 3.1415926 / 2) *
            Exp(-0.5 * d1 * d1)
    dv = PriceError / Vega
    Volatility = Volatility - dv
Wend
ImpVolCall = Volatility
End Function
```

In this we need the cumulative distribution function for the Normal distribution. The following is a simple algorithm which gives an accurate, and fast, approximation to the cumulative distribution function of the standardized Normal:

$$\text{For} \quad x \geq 0 \quad N(x) \approx 1 - \frac{1}{\sqrt{2\pi}} e^{-(1/2)x^2} \left(a_1 d + a_2 d^2 + a_3 d^3 + a_4 d^4 + a_5 d^5 \right)$$

where

$$d = \frac{1}{1 + 0.2316419x}$$

and

$a1 = 0.31938153, \quad a2 = -0.356563782, \quad a3 = 1.781477937, \quad a4 = -1.821255978$ and $a5 = 1.330274429.$

For $x < 0$ use the fact that $N(x) + N(-x) = 1.$

```
Function cdf(x As Double) As Double
Dim d As Double
Dim temp as Double
Dim a1 As Double
Dim a2 As Double
Dim a3 As Double
Dim a4 As Double
Dim a5 As Double
d = 1 / (1 + 0.2316419 * Abs(x))
a1 = 0.31938153
a2 = -0.356563782
a3 = 1.781477937
a4 = -1.821255978
a5 = 1.330274429
temp = a5
temp = a4 + d * temp
```

```
temp = a3 + d * temp
temp = a2 + d * temp
temp = a1 + d * temp
temp = d * temp
cdf = 1 - 1 / Sqr(2 * 3.1415926) * Exp(-0.5 * x * x) * temp
If x < 0 Then cdf = 1 - cdf
End Function
```

In practice if we calculate the implied volatility for many different strikes and expiries on the same underlying then we find that *the volatility is not constant*. A typical result is that of Figure 7.16 which shows the implied volatilities for the S&P500 on 9th September 1999 for options expiring later in the month. The implied volatilities for the calls and puts should be identical, because of put-call parity. The differences seen here could be due to bid-offer spread or calculations performed at slightly different times.

This shape is commonly referred to as the **smile**, but it could also be in the shape of a **frown**. In this example it's a rather lopsided wry grin. Whatever the shape, it tends to persist with time, with certain shapes being characteristic of certain markets.

The dependence of the implied volatility on strike and expiry can be interpreted in many ways. The easiest interpretation is that it represents the market's view of future volatility in some complex way. This issue is covered in depth in Chapter 25. Another possibility is that it reflects the uncertainty in volatility, perhaps volatility is also a stochastic variable; see Chapter 26.

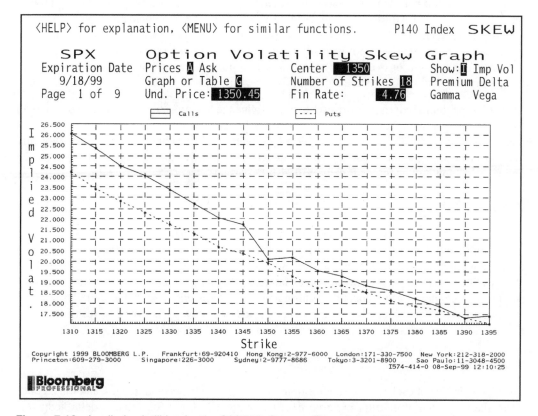

Figure 7.16 Implied volatilities for the S&P500. Source: Bloomberg L.P.

HEDGING MEANS MANY THINGS. HERE ARE A FEW EXAMPLES

7.9 A CLASSIFICATION OF HEDGING TYPES

7.9.1 Why Hedge?

'Hedging' in its broadest sense means the reduction of risk by exploiting relationships or correlation between various risky investments (or bets). The concept is used widely in horse racing, other sports betting and, of course, high finance. The reason for hedging is that it can lead to an improved risk/return. In the classical Modern Portfolio Theory framework (Chapter 51), for example, it is usually possible to construct many portfolios having the same expected return but with different variance of returns ('risk'). Clearly, if you have two portfolios with the same expected return the one with the lower risk is the better investment.

7.9.2 The Two Main Classifications

Probably the most important distinction between types of hedging is between model-independent and model-dependent hedging strategies.

Model-Independent Hedging: An example of such hedging is Put-call Parity. There is a simple relationship between calls and puts on an asset (when they are both European and with the same strikes and expiries), the underlying stock and a zero-coupon bond with the same maturity. This relationship is completely independent of how the underlying asset changes in value. Another example is Spot-forward Parity. In neither case do we have to specify the dynamics of the asset, or even its volatility, to find a possible hedge. Such model-independent hedges are few and far between.

Model-Dependent Hedging: Most sophisticated finance hedging strategies depend on a model for the underlying asset. The obvious example is the hedging used in the Black–Scholes analysis that leads to a whole theory for the value of derivatives. In pricing derivatives we typically need to at least know the volatility of the underlying asset. If the model is wrong then the option value and any hedging strategy will also be wrong.

7.9.3 Delta Hedging

One of the building blocks of derivatives theory is **delta hedging**. This is the theoretically perfect elimination of all risk by using a very clever hedge between the option and its underlying. Delta hedging exploits the perfect correlation between the changes in the option value and the changes in the stock price. This is an example of 'dynamic' hedging; the hedge must be continually monitored and frequently adjusted by the sale or purchase of the underlying asset. Because of the frequent rehedging, any dynamic hedging strategy is going to result in losses due to transaction costs. In some markets this can be very important.

7.9.4 Gamma Hedging

To reduce the size of each rehedge and/or to increase the time between rehedges, and thus reduce costs, the technique of **gamma hedging** is often employed. A portfolio that is delta hedged is insensitive to movements in the underlying as long as those movements are quite small. There is a small error in this due to the convexity of the portfolio with respect to the underlying. Gamma hedging is a more accurate form of hedging that theoretically eliminates these second-order effects. Typically, one hedges one, exotic, say, contract with a vanilla contract and the underlying. The quantities of the vanilla and the underlying are chosen so as to make both the portfolio delta and the portfolio gamma instantaneously zero.

7.9.5 Vega Hedging

As I said above, the prices and hedging strategies are only as good as the model for the underlying. The key parameter that determines the value of a contract is the volatility of the underlying asset. Unfortunately, this is a very difficult parameter to measure or even estimate. Nor is it usually a constant as assumed in the simple theories. Obviously, the value of a contract depends on this parameter, and so to ensure that our portfolio value is insensitive to this parameter we can **vega hedge.** This means that we hedge one option with both the underlying and another option in such a way that both the delta and the vega, the sensitivity of the portfolio value to volatility, are zero. This is often quite satisfactory in practice but is usually theoretically inconsistent; we should not use a constant volatility (basic Black–Scholes) model to calculate sensitivities to parameters that are assumed not to vary. The distinction between variables (underlying asset price and time) and parameters (volatility, dividend yield, interest rate) is extremely important here. It is justifiable to rely on sensitivities of prices to variables, but usually not on sensitivity to parameters. To get around this problem it is possible to independently model volatility etc. as variables themselves. In such a way it is possible to build up a consistent theory.

7.9.6 Static Hedging

There are quite a few problems with delta hedging, on both the practical and the theoretical side. In practice, hedging must be done at discrete times and is costly. Sometimes one has to buy or sell a prohibitively large number of the underlying in order to follow the theory. This is a problem with barrier options and options with discontinuous payoff. On the theoretical side, the model for the underlying is not perfect, since at the very least we do not know parameter values accurately. Delta hedging alone leaves us very exposed to the model; this is model risk. Many of these problems can be reduced or eliminated if we follow a strategy of **static hedging** as well as delta hedging: buy or sell more liquid traded contracts to reduce the cashflows in the original contract. The static hedge is put into place now, and left until expiry. In an extreme case where an exotic contract has all of its cashflows matched by cashflows from traded options then its value is given by the cost of setting up the static hedge; a model is not needed.

7.9.7 Margin Hedging

Often what causes banks, and other institutions, to suffer during volatile markets is not the change in the paper value of their assets but the requirement to suddenly come up with a large amount of cash to cover an unexpected margin call. Recent examples where margin has caused significant damage are Metallgesellschaft and Long Term Capital Management. Writing options is very risky. The downside of buying an option is just the initial premium, while the upside may be unlimited. The upside of writing an option is limited, but the downside could be huge. For this reason, to cover the risk of default in the event of an unfavorable outcome, the clearing houses that register and settle options insist on the deposit of a margin by the writers of options.

Margin comes in two forms, the initial margin and the maintenance margin. The initial margin is the amount deposited at the initiation of the contract. The total amount held as margin must stay above a prescribed maintenance margin. If it ever falls below this level then more money (or equivalent in bonds, stocks etc.) must be deposited. The amount of margin that must be deposited depends on the particular contract. A dramatic market move could result in a sudden large margin call that may be difficult to meet. To prevent this situation it is possible to **margin hedge**; that is, set up a portfolio such that a margin calls on one part of the portfolio are balanced by refunds from other parts. Usually over-the-counter contracts have no associated margin requirements and so won't appear in the calculation.

7.9.8 Crash (Platinum) Hedging

The final variety of hedging that we discuss is specific to extreme markets. Market crashes have at least two obvious effects on our hedging. First of all, the moves are so large and rapid that they cannot be traditionally delta hedged. The convexity effect is not small. Second, normal market correlations become meaningless. Typically all correlations become one (or minus one). **Crash** or **Platinum hedging** exploits the latter effect in such a way as to minimize the worst possible outcome for the portfolio. The method, called CrashMetrics (Chapter 58), does not rely on difficult to measure parameters such as volatilities and so is a very robust hedge. Platinum hedging comes in two types: hedging the paper value of the portfolio and hedging the margin calls.

7.10 **SUMMARY**

In this chapter we went through the derivation of some of the most important formulae. We also saw the definitions and descriptions of the hedge ratios. Trading in derivatives would be no more than gambling if you took away the ability to hedge. Hedging is all about managing risk and reducing uncertainty.

FURTHER READING

- See Taleb (1997) for a lot of detailed analysis of vega.

- See Press *et al.* (1992) for more routines for finding roots, i.e. for finding implied volatilities.

- There are many 'virtual' option pricers on the Internet. See, for example, www.cboe.com.

- I'm not going to spend much time on deriving or even presenting formulae. There are 1001 books that contain option formulae, there is even one book with 1001 formulae (Haug, 1997).

	Call	Put	Binary Call	Binary Put
Value V Black–Scholes value	$Se^{-D(T-t)}N(d_1)$ $-Ee^{-r(T-t)}N(d_2)$	$-Se^{-D(T-t)}N(-d_1)$ $+Ee^{-r(T-t)}N(-d_2)$	$e^{-r(T-t)}N(d_2)$	$e^{-r(T-t)}(1-N(d_2))$
Delta $\dfrac{\partial V}{\partial S}$ Sensitivity to underlying	$e^{-D(T-t)}N(d_1)$	$e^{-D(T-t)}(N(d_1)-1)$	$\dfrac{e^{-r(T-t)}N'(d_2)}{\sigma S\sqrt{T-t}}$	$-\dfrac{e^{-r(T-t)}N'(d_2)}{\sigma S\sqrt{T-t}}$
Gamma $\dfrac{\partial^2 V}{\partial S^2}$ Sensitivity of delta to underlying	$\dfrac{e^{-D(T-t)}N'(d_1)}{\sigma S\sqrt{T-t}}$	$\dfrac{e^{-D(T-t)}N'(d_1)}{\sigma S\sqrt{T-t}}$	$-\dfrac{e^{-r(T-t)}d_1N'(d_2)}{\sigma^2 S^2(T-t)}$	$\dfrac{e^{-r(T-t)}d_1N'(d_2)}{\sigma^2 S^2(T-t)}$
Theta $\dfrac{\partial V}{\partial t}$ Sensitivity to time	$-\dfrac{\sigma Se^{-D(T-t)}N'(d_1)}{2\sqrt{T-t}}$ $+DSN(d_1)e^{-D(T-t)}$ $-rEe^{-r(T-t)}N(d_2)$	$-\dfrac{\sigma Se^{-D(T-t)}N'(-d_1)}{2\sqrt{T-t}}$ $-DSN(-d_1)e^{-D(T-t)}$ $+rEe^{-r(T-t)}N(-d_2)$	$re^{-r(T-t)}N(d_2)$ $+e^{-r(T-t)}N'(d_2)$ $\times\left(\dfrac{d_1}{2(T-t)}-\dfrac{r-D}{\sigma\sqrt{T-t}}\right)$	$re^{-r(T-t)}(1-N(d_2))$ $-e^{-r(T-t)}N'(d_2)$ $\times\left(\dfrac{d_1}{2(T-t)}-\dfrac{r-D}{\sigma\sqrt{T-t}}\right)$
Vega $\dfrac{\partial V}{\partial \sigma}$ Sensitivity to volatility	$S\sqrt{T-t}\,e^{-D(T-t)}N'(d_1)$	$S\sqrt{T-t}\,e^{-D(T-t)}N'(d_1)$	$-e^{-r(T-t)}N'(d_2)$ $\times\left(\sqrt{T-t}+\dfrac{d_2}{\sigma}\right)$	$e^{-r(T-t)}N'(d_2)$ $\times\left(\sqrt{T-t}+\dfrac{d_2}{\sigma}\right)$
Rho (r) $\dfrac{\partial V}{\partial r}$ Sensitivity to interest rate	$E(T-t)e^{-r(T-t)}N(d_2)$	$-E(T-t)e^{-r(T-t)}N(-d_2)$	$-(T-t)e^{-r(T-t)}N(d_2)$ $+\dfrac{\sqrt{T-t}}{\sigma}e^{-r(T-t)}N'(d_2)$	$-(T-t)e^{-r(T-t)}(1-N(d_2))$ $-\dfrac{\sqrt{T-t}}{\sigma}e^{-r(T-t)}N'(d_2)$
Rho (D) $\dfrac{\partial V}{\partial D}$ Sensitivity to dividend yield	$-(T-t)Se^{-D(T-t)}N(d_1)$	$(T-t)Se^{-D(T-t)}N(-d_1)$	$-\dfrac{\sqrt{T-t}}{\sigma}e^{-r(T-t)}N'(d_2)$	$\dfrac{\sqrt{T-t}}{\sigma}e^{-r(T-t)}N'(d_2)$

$$d_1 = \frac{\log\left(\dfrac{S}{E}\right)+\left(r-D+\tfrac{1}{2}\sigma^2\right)(T-t)}{\sigma\sqrt{T-t}}, \quad d_2 = \frac{\log\left(\dfrac{S}{E}\right)+\left(r-D-\tfrac{1}{2}\sigma^2\right)(T-t)}{\sigma\sqrt{T-t}} = d_1-\sigma\sqrt{T-t}, \quad N(x)=\frac{1}{\sqrt{2\pi}}\int_{-\infty}^{x}e^{-(1/2)\xi^2}\,d\xi \quad\text{and}\quad N'(x)=\frac{1}{\sqrt{2\pi}}e^{-(1/2)x^2}$$

CHAPTER 8
simple generalizations of the Black–Scholes world

In this Chapter...

- complex dividend structures
- jump conditions
- time-dependent volatility, interest rates and dividend yield

8.1 INTRODUCTION

This chapter is an introduction to some of the possible generalizations of the 'Black–Scholes world'. In particular, I will discuss the effect of dividend payments on the underlying asset and how to incorporate time-dependent parameters into the framework. These subjects lead to some interesting and important mathematical and financial conclusions.

The generalizations are very straightforward. However, later, in Part Three, I describe other models of the financial world that take us a long way from Black–Scholes.

8.2 DIVIDENDS, FOREIGN INTEREST AND COST OF CARRY

In Chapter 5 I showed how to incorporate certain types of dividend structures into the Black–Scholes option pricing framework, and then in Chapter 7 I gave some formulae for the values of some common vanilla contracts, again with dividends on the underlying. The dividend structure that I dealt with was the very simplest from a mathematical point of view. I assumed that an amount was paid to the holder of the asset that was proportional to the value of the asset and that it was paid continuously. In other words, the owner of one asset received a dividend of $DS\,dt$ in a timestep dt. This dividend structure is realistic if the underlying is an index on a large number of individual assets each receiving a lump sum dividend but with all these dividends spread out through the year. It is also a good model if the underlying is a currency in which case we simply take the 'dividend yield' to be the foreign interest rate. Similarly, if the underlying is a commodity with a cost of carry that is proportional to its value, then the 'dividend yield' is just the cost of carry (with a minus sign, we benefit from dividends but must pay out the cost of carry).

To recap, if the underlying receives a dividend of $DS\,dt$ in a timestep dt when the asset price is S then

$$\frac{\partial V}{\partial t} + \tfrac{1}{2}\sigma^2 S^2 \frac{\partial^2 V}{\partial S^2} + (r - D)S\frac{\partial V}{\partial S} - rV = 0.$$

However, if the underlying is a stock, then the assumption of constant and continuously-paid dividend yield is not a good one.

8.3 DIVIDEND STRUCTURES

Typically, dividends are paid out quarterly in the US and semi-annually or quarterly in the UK. The dividend is set by the board of directors of a company some time before it is paid out and the amount of the payment is made public. The amount is often chosen to be similar to previous payments, but will obviously reflect the success or otherwise of the company. The amount specified is a dollar amount, it is *not* a percentage of the stock price on the day that the payment is made. So reality differs from the above simple model in three respects:

- the amount of the dividend is not known until shortly before it is paid
- the payment is a given dollar amount, independent of the stock price
- the dividend is paid discretely, and not continuously throughout the year.

In what follows I am going to make some assumptions about the dividend. I will assume that

- the amount of the dividend is a known amount, possibly with some functional dependence on the asset value *at the payment date*
- the dividend is paid discretely on a known date.

Other assumptions that I could, but won't, make because of the subsequent complexity of the modeling are that the dividend amount and/or date are random, that the dividend amount is a function of the stock price on the day that the dividend is set, that the dividend depends on how well the stock has done in the previous quarter...

8.4 DIVIDEND PAYMENTS AND NO ARBITRAGE

How does the stock react to the payment of a dividend? To put the question another way, if you have a choice whether to buy a stock just before or just after it goes ex-dividend, which should you choose?

Let me introduce some notation. The dates of dividends will be t_i and the amount of the dividend paid on that day will be D_i. This may be a function of the underlying asset, but it then must be a deterministic function. The moment just before the stock goes ex-dividend will be denoted by t_i^- and the moment just after will be t_i^+.

The person who buys the stock on or before t_i^- will also get the rights to the dividend. The person who buys it at t_i^+ or later will not receive the dividend. It looks like there is an advantage in buying the stock just before the dividend date. Of course, this advantage is balanced by *a fall in the stock price as it goes ex-dividend*. Across a dividend date the stock falls by the amount

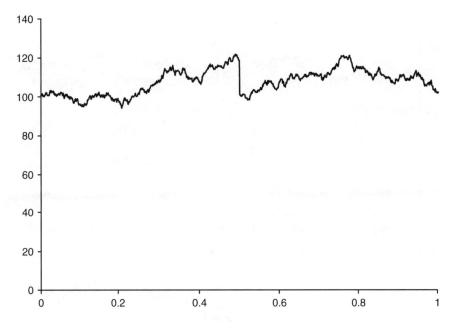

Figure 8.1 A stock price path across a dividend date.

of the dividend. If it did not, then there would be arbitrage opportunities. We can write

$$S(t_i^+) = S(t_i^-) - D_i. \tag{8.1}$$

In Figure 8.1 is shown an asset price path showing the fall in the asset price as it goes ex-dividend; the drop has been exaggerated.

This jump in the stock price will presumably have some effect on the value of an option. We will discuss this next.

8.5 THE BEHAVIOR OF AN OPTION VALUE ACROSS A DIVIDEND DATE

We have just seen how the underlying asset jumps in value, in a completely predictable way, across a dividend date.

Jump conditions tell us about the value of a dependent variable, an option price, when there is a discontinuous change in one of the independent variables. In the present case, there is a discontinuous change in the asset price due to the payment of a dividend but how does this affect the option price? Does the option price also jump? The jump condition relates the values of the option across the jump, from times t_i^- to t_i^+. The jump condition will be derived by a simple no-arbitrage argument.

To see what the jump condition should be, ask the question, 'By how much do I profit or lose when the stock price jumps?' If you hold the option then you do not see any of the dividend; that goes to the holder of the stock, not you, the holder of the option. If the dividend amount and date are known in advance then there is no surprise in the fall in the stock price. The conclusion must be that the option does not change in value across the dividend date, its path

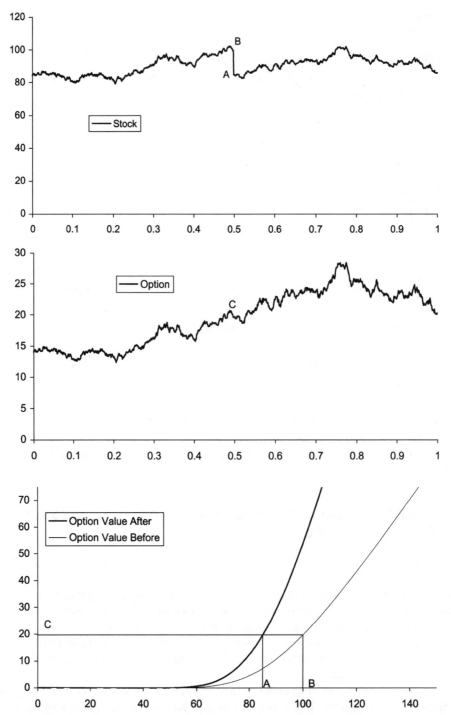

Figure 8.2   Top picture, a realization of the stock price showing a fall across the dividend date. Middle picture, the corresponding realization of the option price (in this example a call). Bottom picture, the option value as a function of the stock price just before and just after the dividend date.

is continuous. Continuity of the option value across a dividend date can be written as

$$V(S(t_i^-), t_i^-) = V(S(t_i^+), t_i^+) \qquad (8.2)$$

or, in terms of the amount of the dividend,

$$V(S, t_i^-) = V(S - D_i, t_i^+). \qquad (8.3)$$

OUR FIRST JUMP CONDITION, ARISING FROM THE NO-ARB PRINCIPLE

The jump condition and its effect on the option value can be explained by reference to Figure 8.2. In this figure, the top picture shows a realization of the stock price with a fall across the dividend date. The middle picture shows the corresponding realization of a call option price. The bottom picture shows the option value as a function of the stock price just before and just after the dividend date. Observe the points 'A' and 'B' on these pictures. 'A' is the stock price after the dividend has been paid and 'B' is the price before. On the bottom picture we see the values of the option associated with these before and after asset prices. *These option values are the same* and are denoted by 'C'. Even though there is a fall in the asset value, the option value is unchanged because the whole *V versus S* plot changes. The relationship between the before and after values of the option are related by (8.3). I will give two examples.

Suppose that the dividend paid out is proportional to the asset value, $D_i = DS$. In this case

$$S(t_i^+) = (1 - D)S(t_i^-).$$

Equation (8.3) is then just

$$V(S, t_i^-) = V((1 - D)S, t_i^+).$$

The two option price curves are identical if one stretches the after curve by a factor of $(1 - D)^{-1}$ in the horizontal direction. Thus, even though the option value is continuous across a dividend date, the delta changes discontinuously.

If the dividend is independent of the stock price then

$$S(t_i^+) = S(t_i^-) - D_i,$$

where D_i is independent of the asset value. The before curve is then identical to the after curve, but shifted by an amount D_i.

8.6 TIME-DEPENDENT PARAMETERS

The next generalization concerns the term structure of parameters. In this section I show how to derive formulae for options when the interest rate, volatility and dividend yield/foreign interest rate are time-dependent.

The Black–Scholes partial differential equation is valid as long as the parameters r, D and σ are known functions of time; in practice one often has a view on the future behavior of these parameters. For instance, you may want to incorporate the market's view on the direction of interest rates. Assume that you want to price options knowing $r(t)$, $D(t)$ and $\sigma(t)$. Note that when I write '$D(t)$' I am specifically assuming a time-dependent dividend yield, that is, the amount of the dividend is $D(t)S\,dt$ in a timestep dt.

The equation that we must solve is now

$$\frac{\partial V}{\partial t} + \tfrac{1}{2}\sigma^2(t)S^2\frac{\partial^2 V}{\partial S^2} + (r(t) - D(t))S\frac{\partial V}{\partial S} - r(t)V = 0, \tag{8.4}$$

where the dependence on t is shown explicitly.

Introduce new variables as follows.

$$\overline{S} = Se^{\alpha(t)}, \quad \overline{V} = Ve^{\beta(t)}, \quad \overline{t} = \gamma(t).$$

We are free to choose the functions α, β and γ and so we will choose them so as to eliminate all time-dependent coefficients from (8.4). After changing variables (8.4) becomes

$$\dot{\gamma}(t)\frac{\partial \overline{V}}{\partial \overline{t}} + \tfrac{1}{2}\sigma(t)^2\overline{S}^2\frac{\partial^2 \overline{V}}{\partial \overline{S}^2} + (r(t) - D(t) + \dot{\alpha}(t))\overline{S}\frac{\partial \overline{V}}{\partial \overline{S}} - (r(t) + \dot{\beta}(t))\overline{V} = 0, \tag{8.5}$$

where $\dot{} = d/dt$. By choosing

$$\beta(t) = \int_t^T r(\tau)\,d\tau$$

we make the coefficient of $\overline{V}$ zero and then by choosing

$$\alpha(t) = \int_t^T (r(\tau) - D(\tau))\,d\tau,$$

we make the coefficient of $\partial\overline{V}/\partial\overline{S}$ also zero. Finally, the remaining time dependence, in the volatility term, can be eliminated by choosing

$$\gamma(t) = \int_t^T \sigma^2(\tau)\,d\tau.$$

Now (8.5) becomes the much simpler equation

$$\frac{\partial \overline{V}}{\partial \overline{t}} = \tfrac{1}{2}\overline{S}^2\frac{\partial^2 \overline{V}}{\partial \overline{S}^2}. \tag{8.6}$$

The important point about this equation is that it has coefficients which are *independent of time*, and there is no mention of r, D or σ. If we use $\overline{V}(\overline{S}, \overline{t})$ to denote any solution of (8.6), then the corresponding solution of (8.5), in the original variables, is

$$V = e^{-\beta(t)}\overline{V}(Se^{\alpha(t)}, \gamma(t)). \tag{8.7}$$

Now use V_{BS} to mean any solution of the Black–Scholes equation for *constant* interest rate r_c, dividend yield D_c and volatility σ_c. This solution can be written in the form

$$V_{BS} = e^{-r_c(T-t)}\overline{V}_{BS}(Se^{-(r_c-D_c)(T-t)}, \sigma_c^2(T - t)) \tag{8.8}$$

for some function $\overline{V}_{BS}$. By comparing (8.7) and (8.8) it follows that the solution of the time-dependent parameter problem is the same as the solution of the constant parameter problem if we use the following substitutions:

$$
r_c = \frac{1}{T-t} \int_t^T r(\tau)\, d\tau
$$

$$
D_c = \frac{1}{T-t} \int_t^T D(\tau)\, d\tau
$$

$$
\sigma_c^2 = \frac{1}{T-t} \int_t^T \sigma^2(\tau)\, d\tau
$$

These formulae give the average, over the remaining lifetime of the option, of the interest rate, the dividend yield and the squared volatility.

Just to make things absolutely clear, here is the formula for a European call option with time-dependent parameters:

$$
Se^{-\int_t^T D(\tau)\, d\tau} N(d_1) - Ee^{-\int_t^T r(\tau)\, d\tau} N(d_2)
$$

where

$$
d_1 = \frac{\log\left(\dfrac{S}{E}\right) + \displaystyle\int_t^T (r(\tau) - D(\tau))\, d\tau + \tfrac{1}{2} \int_t^T \sigma^2(\tau)\, d\tau}{\sqrt{\displaystyle\int_t^T \sigma^2(\tau)\, d\tau}}
$$

and

$$
d_2 = \frac{\log\left(\dfrac{S}{E}\right) + \displaystyle\int_t^T (r(\tau) - D(\tau))\, d\tau - \tfrac{1}{2} \int_t^T \sigma^2(\tau)\, d\tau}{\sqrt{\displaystyle\int_t^T \sigma^2(\tau)\, d\tau}}.
$$

There are some conditions that I must attach to the use of these formulae. They are generally not correct if there is early exercise, or for certain types of exotic option. The question to ask to decide whether they are correct is 'Are all the conditions, final and boundary, preserved by the transformations?'

8.7 **FORMULAE FOR POWER OPTIONS**

An option with a payoff that depends on the asset price at expiry raised to some power is called a **power option**. Suppose that it has a payoff

$$
\text{Payoff } (S^\alpha)
$$

we can find a simple formula for the value of the option if we have a simple formula for an option with payoff given by

$$
\text{Payoff } (S). \tag{8.9}
$$

This is because of the lognormality of the underlying asset.

Writing

$$S = S^\alpha$$

the Black–Scholes equation becomes, in the new variable S,

$$\frac{\partial V}{\partial t} + \tfrac{1}{2}\alpha^2\sigma^2 S^2 \frac{\partial^2 V}{\partial S^2} + \alpha\left(\tfrac{1}{2}\sigma^2(\alpha - 1) + r\right) S\frac{\partial V}{\partial S} - rV = 0.$$

Thus whatever the formula for the option value with simple payoff (8.9), the formula for the power version has S^α instead of S and adjustment made to σ, r and D.

8.8 THE log CONTRACT

The **log contract** has the payoff

$$\log\left(\frac{S}{E}\right).$$

The theoretical fair value for this contract is of the form

$$a(t) + b(t)\log\left(\frac{S}{E}\right).$$

Substituting this expression into the Black–Scholes equation results in

$$\dot{a} + \dot{b}\log\left(\frac{S}{E}\right) - \tfrac{1}{2}\sigma^2 b + (r - D)b - ra - rb\log\left(\frac{S}{E}\right) = 0,$$

where ˙ denotes d/dt. Equating terms in $\log(S/E)$ and those independent of S results in

$$b(t) = e^{-r(T-t)} \quad \text{and} \quad a(t) = \left(r - D - \tfrac{1}{2}\sigma^2\right)(T - t)e^{-r(T-t)}.$$

The two arbitrary constants of integration have been chosen to match the solution with the payoff at expiry.

This value is rather special in that the dependence of the option price on the underlying asset, S, and the volatility, σ, uncouples. One term contains S and no σ and the other contains σ and no S. We briefly saw in Chapter 7 the concept of vega hedging to eliminate volatility risk. It is conceivable, even though not entirely justifiably, that the simplicity of the log contract value makes it a useful weapon for hedging other contracts against fluctuations in volatility. Having said that, it's not exactly a highly liquid contract.

The log contract payoff can be positive or negative depending on whether $S > E$ or $S < E$. If we modify the payoff to be

$$\max\left(\log\left(\frac{S}{E}\right), 0\right)$$

then we have a genuine 'option' which may or may not be exercised. The value of this option is

$$e^{-r(T-t)}\sigma\sqrt{T - t}\,N'(d_2) + e^{-r(T-t)}\left(\log\left(\frac{S}{E}\right) + \left(r - D - \tfrac{1}{2}\sigma^2\right)(T - t)\right)N'(d_2).$$

8.9 SUMMARY

In this chapter I made some very simple generalizations to the Black–Scholes world. I showed the effect of discretely-paid dividends on the value of an option, deriving a jump condition by a no-arbitrage argument. Generally, this condition would be applied numerically and its implementation is discussed in Chapter 64. I also showed how time-dependent parameters can be incorporated into the pricing of simple vanilla options.

FURTHER READING

- See Merton (1973) for the original derivation of the Black–Scholes formulae with time-dependent parameters.
- For a model with stochastic dividends see Geske (1978).
- The practical implications of discrete dividend payments are discussed by Gemmill (1992).
- See Neuberger (1994) for further info on the log contract.

CHAPTER 9
early exercise and American options

In this Chapter...

- the meaning of 'early exercise'
- the difference between European, American and Bermudan options
- how to value American options in the partial differential equation framework
- how to decide when to exercise early
- early exercise and dividends

9.1 INTRODUCTION

American options are contracts that may be exercised early, *prior* to expiry. For example, if the option is a call, we may hand over the exercise price and receive the asset whenever we wish. These options contrast with European options for which exercise is only permitted *at* expiry. Most traded stock and futures options are American style, but most index options are European.

The right to exercise at any time at will is clearly valuable. The value of an American option cannot be less than an equivalent European option. But as well as giving the holder more rights, they also give him more headaches; when should he exercise? Part of the valuation problem is deciding when is the best time to exercise. This is what makes American options much more interesting than their European cousins. Moreover, the issues I am about to raise have repercussions in many other financial problems.

9.2 THE PERPETUAL AMERICAN PUT

There is a very simple example of an American option that we can examine for the insight that it gives us in the general case. This simple example is the **perpetual American put**. This contract can be exercised for a put payoff at *any* time. There is no expiry, that's why it is called a 'perpetual' option. So we can, at any time of *our* choosing, sell the underlying and receive an amount E. That is, the payoff is

$$\max(E - S, 0).$$

We want to find the value of this option before exercise.

The first point to note is that the solution is independent of time, $V(S)$. It depends only on the level of the underlying. This is a property of perpetual options when the contract details are time-homogeneous, provided that there is a finite solution. When we come to the general, non-perpetual, American option, we unfortunately lose this property. ('Unfortunately', since it makes it easy for us to find the solution in this special case.)

The second point to note, which is important for all American options, is that the option value can never go below the early-exercise payoff. In the case under consideration

$$V \geq \max(E - S, 0). \tag{9.1}$$

Consider what would happen if this 'constraint' were violated. Suppose that the option value is less than $\max(E - S, 0)$, I could buy the option for V, immediately exercise it by handing over the asset (worth S) and receive an amount E. I thus make

$$-\text{cost of put} - \text{cost of asset} + \text{strike price} = -V - S + E > 0.$$

This is a riskless profit. If we believe that there are no arbitrage opportunities then we must believe (9.1).

While the option value is strictly greater than the payoff, it must satisfy the Black–Scholes equation; I return to this point in the next section. Recalling that the option is perpetual and therefore that the value is independent of t, it must satisfy

$$\tfrac{1}{2}\sigma^2 S^2 \frac{d^2 V}{dS^2} + rS \frac{dV}{dS} - rV = 0.$$

This is the ordinary differential equation you get when the option value is a function of S only. The general solution of this second-order ordinary differential equation is

$$V(S) = AS + BS^{-2r/\sigma^2},$$

where A and B are arbitrary constants.

The first part of this solution (that with coefficient A) is simply the asset: the asset itself satisfies the Black–Scholes equation. If we can find A and B we have found the solution for the perpetual American put.

Clearly, for the perpetual American put the coefficient A must be zero; as $S \to \infty$ the value of the option must tend to zero. What about B?

Let us postulate that while the asset value is 'high' we won't exercise the option. But if it falls too low we immediately exercise the option, receiving $E - S$. (Common sense tells us we don't exercise when $S > E$.) Suppose that we decide that $S = S^*$ is the value at which we exercise, i.e. as soon as S reaches this value from above we exercise. How do we choose S^*?

When $S = S^*$ the option value must be the same as the exercise payoff:

$$V(S^*) = E - S^*.$$

It cannot be less, which would result in an arbitrage opportunity, and it cannot be more or we wouldn't exercise. Continuity of the option value with the payoff gives us one equation:

$$V(S^*) = B(S^*)^{-2r/\sigma^2} = E - S^*.$$

But since both B and S^* are unknown, we need one more equation. Let's look at the value of the option as a function of S^*, eliminating B using the above. We find that for $S > S^*$

$$V(S) = (E - S^*) \left(\frac{S}{S^*} \right)^{-2r/\sigma^2}. \tag{9.2}$$

We are going to choose S^* to *maximize the option's value at any time before exercise*. In other words, what choice of S^* makes V given by (9.2) as large as possible? The reason for this is obvious; if we can exercise whenever we like then we do so in such a way to maximize our worth. We find this value by differentiating (9.2) with respect to S^* and setting the resulting expression equal to zero:

$$\frac{\partial}{\partial S^*} (E - S^*) \left(\frac{S}{S^*} \right)^{-2r/\sigma^2} = \frac{1}{S^*} \left(\frac{S}{S^*} \right)^{-2r/\sigma^2} \left(-S^* + \frac{2r}{\sigma^2}(E - S^*) \right) = 0.$$

We find that

$$S^* = \frac{E}{1 + \dfrac{\sigma^2}{2r}}.$$

This choice maximizes $V(S)$ for *all* $S \geq S^*$. The solution with this choice for S^* and with the corresponding B given by

$$\frac{\sigma^2}{2r} \left(\frac{E}{1 + \dfrac{\sigma^2}{2r}} \right)^{1 + 2r/\sigma^2}$$

is shown in Figure 9.1.

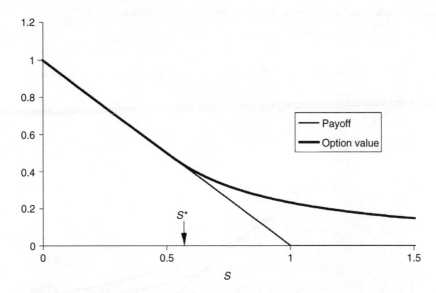

Figure 9.1   The solution for the perpetual American put.

The observant reader will notice something special about this function: the slope of the option value and the slope of the payoff function are the same at $S = S^*$. To see that this follows from the choice of S^* let us examine the difference between the option value and the payoff function:

$$(E - S^*) \left(\frac{S}{S^*} \right)^{-2r/\sigma^2} - (E - S).$$

Differentiate this with respect to S and you will find that the expression is zero at $S = S^*$.

This demonstrates, in a completely non-rigorous way, that if we want to maximize our option's value by a careful choice of exercise strategy, then this is equivalent to solving the Black–Scholes equation with continuity of option *value* and option *delta*, the slope. This is called the **high-contact** or **smooth-pasting condition**.

> The American option value is maximized by
> an exercise strategy that makes the
> option value and option delta continuous

We exercise the option as soon as the asset price reaches the level at which the option price and the payoff meet. This position, S^*, is called the **optimal exercise point**.

Another way of looking at the condition of continuity of delta is to consider what happens if the delta is not continuous at the exercise point. The two possibilities are shown in Figure 9.2. In this figure the curve (a) corresponds to exercise that is not optimal because it is premature, the option value is lower than it could be. In case (b) there is clearly an arbitrage opportunity. If we take case (a) but progressively delay exercise by lowering the exercise point, we will maximize the option value everywhere when the delta is continuous.

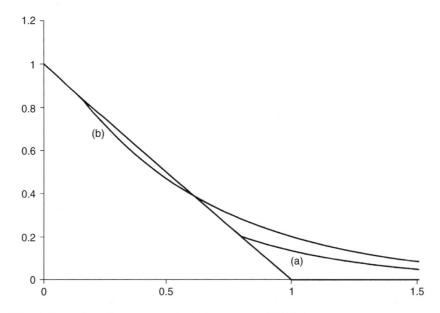

Figure 9.2 Option price when exercise is (a) too soon or (b) too late.

When there is a continuously paid and constant dividend yield on the asset, or the asset is a foreign currency, the relevant ordinary differential equation for the perpetual option is

$$\tfrac{1}{2}\sigma^2 S^2 \frac{d^2V}{dS^2} + (r - D)S\frac{dV}{dS} - rV = 0.$$

The general solution is now

$$AS^{\alpha^+} + BS^{\alpha^-},$$

where

$$\alpha^{\pm} = \frac{1}{\sigma^2}\left(-\left(r - D - \tfrac{1}{2}\sigma^2\right) \pm \sqrt{\left(r - D - \tfrac{1}{2}\sigma^2\right)^2 + 2r\sigma^2}\right),$$

with $\alpha^- < 0 < \alpha^+$. The perpetual American put now has value

$$BS^{\alpha^-},$$

where

$$B = -\frac{1}{\alpha^-}\left(\frac{E}{1 - \dfrac{1}{\alpha^-}}\right)^{1-\alpha^-}.$$

It is optimal to exercise when S reaches the value

$$\frac{E}{1 - \dfrac{1}{\alpha^-}}.$$

Before considering the formulation of the general American option problem, we consider one more special case.

9.3 PERPETUAL AMERICAN CALL WITH DIVIDENDS

The solution for the American perpetual call is

$$AS^{\alpha^+},$$

where

$$A = \frac{1}{\alpha^+}\left(\frac{E}{1 - \dfrac{1}{\alpha^+}}\right)^{1-\alpha^+}$$

and it is optimal to exercise as soon as S reaches

$$S^* = \frac{E}{1 - \dfrac{1}{\alpha^+}}$$

from below.

An interesting special case is when $D = 0$. Then the solution is $V = S$ and S^* becomes infinite. Thus it is never optimal to exercise the American perpetual call when there are no dividends on the underlying; its value is the same as the underlying. As we see in a moment, this result also holds for the ordinary non-perpetual American call in the absence of dividends.

9.4 MATHEMATICAL FORMULATION FOR GENERAL PAYOFF

Now I build up the theory for American-style contracts with arbitrary payoff and expiry, following the standard Black–Scholes argument with minor modifications. The contract will no longer be perpetual and the option value will now be a function of both S and t.

Construct a portfolio of one American option with value $V(S, t)$ and short a number Δ of the underlying:

$$\Pi = V - \Delta S.$$

The change in value of this portfolio in excess of the risk-free rate is given by

$$d\Pi - r\Pi \, dt = \left(\frac{\partial V}{\partial t} + \tfrac{1}{2}\sigma^2 S^2 \frac{\partial^2 V}{\partial S^2} - r(V - \Delta S) \right) dt + \left(\frac{\partial V}{\partial S} - \Delta \right) dS.$$

With the choice

$$\Delta = \frac{\partial V}{\partial S}$$

this becomes

$$\left(\frac{\partial V}{\partial t} + \tfrac{1}{2}\sigma^2 S^2 \frac{\partial^2 V}{\partial S^2} + rS \frac{\partial V}{\partial S} - rV \right) dt.$$

In the Black–Scholes argument for European options we set this expression equal to zero, since this precludes arbitrage. But it precludes arbitrage whether we are buying or selling the contract. When the contract is American the long/short relationship is asymmetrical; it is the holder of the exercise rights who controls the early-exercise feature. The writer of the option can do no more than sit back and enjoy the view. If V is the value of a long position in an American option then all we can say is that we can earn *no more* than the risk-free rate on our portfolio. Thus we arrive at the *inequality*

$$\frac{\partial V}{\partial t} + \tfrac{1}{2}\sigma^2 S^2 \frac{\partial^2 V}{\partial S^2} + rS \frac{\partial V}{\partial S} - rV \leq 0. \tag{9.3}$$

The writer of the American option *can* make more than the risk-free rate if the holder does not exercise *optimally*. He also makes more profit if the holder of the option has a poor estimate of the volatility of the underlying and exercises in accordance with that estimate.

Equation (9.3) can easily be modified to accommodate dividends on the underlying.

If the payoff for early exercise is $P(S, t)$, possibly time-dependent, then the no-arbitrage constraint

$$V(S, t) \geq P(S, t), \tag{9.4}$$

must apply everywhere. At expiry we have the final condition

$$V(S, T) = P(S, T).$$

> The option value is maximized if the owner of the option exercises such that
>
> $$\Delta = \frac{\partial V}{\partial S} \text{ is continuous}$$

AMERICAN OPTION PRICING IS AN OPTIMIZATION PROBLEM

The American option valuation problem consists of (9.3), (9.4), (9.4) and (9.4).

If we substitute the Black–Scholes European call solution, in the absence of dividends, into the inequality (9.3) then it is clearly satisfied; it actually satisfies the *equality*. If we substitute the expression into the constraint (9.4) with $P(S, t) = \max(S - E, 0)$ then this too is satisfied. The conclusion is that the value of an American call option is the same as the value of a European call option when the underlying pays no dividends. Compare this with our above result that the perpetual call option should not be exercised — the American call option with a finite time to expiry should also not be exercised before expiry. To exercise before expiry would be 'sub-optimal'.

None of this is true if there are dividends on the underlying. Again, to see this, simply substitute the expressions from Chapter 8 into the constraint. Since the call option has a value which approaches $Se^{-D(T-t)}$ as $S \to \infty$ there is clearly a point at which the European value fails to satisfy the constraint (9.4). *If the constraint is not satisfied somewhere then the problem has not been solved anywhere.* This is very important: our 'solution' must satisfy the inequalities everywhere or the 'solution' is invalid. This is due to the diffusive nature of the differential equation; an error in the solution at any point is immediately propagated *everywhere*.

The problem for the American option is what is known as a **free boundary problem**. In the European option problem we know that we must solve for all values of S from zero to infinity. When the option is American we do not know *a priori* where the Black–Scholes equation is to be satisfied; this must be found as part of the solution. This means that we do not know the position of the early exercise boundary. Moreover, except in special and trivial cases, this position is time-dependent. For example, we should exercise the American put if the asset value falls below $S^*(t)$, but how do we find $S^*(t)$?

Not only is this problem much harder than the fixed boundary problem (for example, where we know that we solve for S between zero and infinity), but this also makes the problem nonlinear. That is, if we have two solutions of the problem we do not get another solution if we add them together. This is easily shown by considering the perpetual American straddle on a dividend-paying stock. If this is defined as a *single* contract that may at any time be exercised for an amount $\max(S - E, 0) + \max(E - S, 0) = |S - E|$ then its value is not the same as the sum of a perpetual American put and a perpetual American call. Its solution is again of the form

$$V(S) = AS^{\alpha^+} + BS^{\alpha^-},$$

and I suggest that the reader find the solution for himself. The reason that this contract is not the sum of two other American options is that there is only one exercise opportunity, and the two-option contract has one exercise opportunity per contract. If the contracts were both European then the sum of the two separate solutions would give the correct answer; the European valuation problem is linear. This contract can also be used to demonstrate that there can easily be more than one optimal exercise boundary. With the perpetual American straddle,

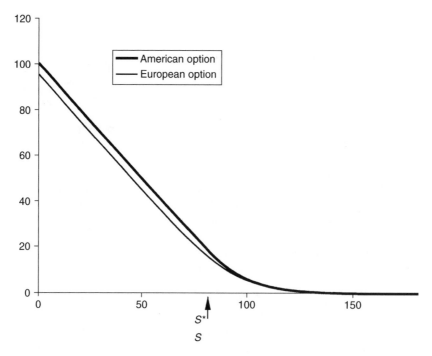

Figure 9.3 Values of a European and an American put, see text for parameter values. The optimal exercise point is marked.

as defined here, one should exercise either if the asset gets too low or too high. The exact positions of the boundaries can be determined by making the option and its delta everywhere continuous. One can imagine that if a contract has a really strange payoff, there could be any number of free boundaries. Since we don't know *a priori* how many free boundaries there are going to be (although common sense gives us a clue) it is useful to have a numerical method that can find these boundaries without having to be told how many to look for. I discuss these issues in Chapter 64.

In Figure 9.3 are shown the values of a European and American put with strike 100, volatility 20%, interest rate 5% and with one year to expiry. The position of the free boundary, the optimal exercise point, is marked. Remember that this point moves in time.

In Figure 9.4 are shown the values of a European and American call with strike 100, volatility 20%, interest rate 5% and with one year to expiry. There is a constant dividend yield of 5% on the underlying. (If there were no dividend payment then the two curves would be identical.) The position of the free boundary, the optimal exercise point, is marked.

Figure 9.5 shows the Bloomberg option valuation calculator, applied to an American option.

9.5 **LOCAL SOLUTION FOR CALL WITH CONSTANT DIVIDEND YIELD**

If we cannot find full solutions to non-trivial problems, we can at least find local solutions, solutions that are good approximations for some values of the asset at some times. We have

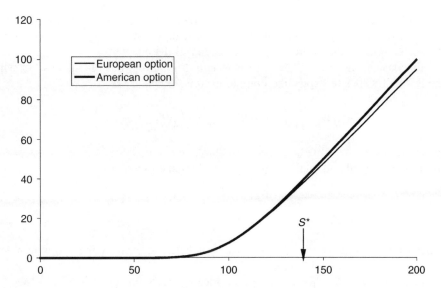

Figure 9.4 Values of a European and an American call, see text for parameter values. The optimal exercise point is marked.

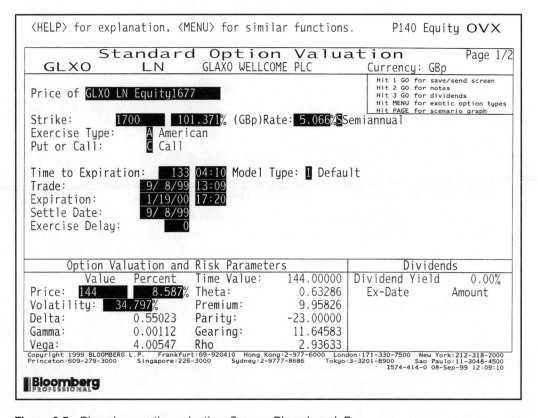

Figure 9.5 Bloomberg option valuation. Source: Bloomberg L.P.

seen the solution for the American call with dividends when there is a long time to expiry, but what about close to expiry? I will state the results without any proof. The proofs are simple but tedious, the relevant literature is cited at the end of the chapter.

First let's consider the case $r > D$. This is usually true for options on equities, for which the dividend is small. Close to expiry the optimal exercise boundary is

$$S^*(t) \sim \frac{rE}{D}\left(1 + 0.9034\ldots\sigma\sqrt{\tfrac{1}{2}(T-t)} + \cdots\right).$$

The call should be exercised if the asset rises above this value. Note that as $T - t \to -\infty$ we have from the perpetual call analysis that the free boundary tends to

$$\frac{E}{1 - \dfrac{1}{\alpha^+}}.$$

If the asset value rises above the free boundary it is better to exercise the option to receive the dividends than to continue holding it.

Near the point $t = T$, $S = Er/D$ the option price is approximately

$$V \sim S - E + E(T-t)^{3/2} f\left(\frac{\log\left(\dfrac{SD}{Er}\right)}{\sqrt{T-t}}\right),$$

where

$$f(x) = -\frac{2r}{\sigma^2}x + 0.075\ldots\left((x^2+4)e^{-(1/4)x^2} + \tfrac{1}{2}(x^2+6x)\int_{-\infty}^{x}e^{-(1/4)s^2}\,ds\right).$$

When $D = 0$ there is no free boundary, it is never optimal to exercise early.

When $r < D$ the free boundary 'starts' from $S = E$ at time $t = T$. The local analysis is more subtle. (There is a nasty $\sqrt{(T-t)\log(T-t)}$ term.)

9.6 OTHER DIVIDEND STRUCTURES

Other dividend structures present no difficulties. The discretely-paid dividend does have an interesting effect on the option value, and on the jump condition across the dividend date.

In Chapter 8 I showed that if there is a discretely paid dividend then the asset falls by the amount of the dividend:

$$S(t_d^+) = S(t_d^-) - D.$$

If the dividend takes the asset value from $S(t_d^-)$ to $S(t_d^+)$ then we must apply the jump condition

$$V(S(t_d^-)a, t_d^-) = V(S(t_d^+), t_d^+)$$

across the dividend date T_d. Thus

$$V(S, t_d^-) = V(S - D, t_d^+).$$

This ensures that the realized option value is continuous. When the option is American it is possible that such a jump condition takes the option value below the payoff just before the dividend date. This is not allowed. If we find that this happens, we must impose the no-arbitrage constraint that the option value is at least the payoff function. Thus the jump condition becomes

$$V(S, t_d^-) = \max(V(S - D, t_d^+), P(S, t_d^-))$$

But this means that the realized option value may no longer be continuous: is this correct? Yes, this does not matter because continuity is only lost *if one should have already optimally exercised before the dividend is paid.*

9.7 ONE-TOUCH OPTIONS

We saw the European binary option in Chapters 2 and 7. The payoff for that option is $1 if the asset is above, for a binary call, a specified level at expiry. The **one-touch option** is an American version of this. This contract can be exercised at any time for a fixed amount, $1 if the asset is above some specified level. There is no benefit in holding the option once the level has been reached therefore it should be exercised immediately the level is reached for the first time, hence the name 'one touch'. These contracts fall into the class of 'once exotic now vanilla', due to their popularity. They are particularly useful for hedging other contracts that also have a payoff that depends on whether or not the specified level is reached.

Since they are American-style options we must decide as part of the solution when to optimally exercise. As I have said, they would clearly be exercised as soon as the level is reached. This makes an otherwise complicated free boundary problem into a rather simple *fixed* boundary problem. For a one-touch call, we must solve the Black–Scholes equation with $V(S_u, t) = 1$, where S_u is the strike price of the contract, and $V(S, T) = 0$. We only need solve for S less than S_u. The solution of this problem is

$$V(S, t) = \left(\frac{S_u}{S}\right)^{2r/\sigma^2} N(d_5) + \frac{S}{S_u} N(d_1),$$

with the usual d_1, and

$$d_5 = \frac{\log\left(\frac{S}{S_u}\right) - \left(r + \frac{1}{2}\sigma^2\right)(T - t)}{\sigma\sqrt{T - t}}.$$

(The subscript '5' is to make the notation consistent with that in a later chapter.) The option value is shown in Figure 9.6 and the delta in Figure 9.7.

The problem for the one-touch put is obvious and the solution is

$$V(S, t) = \left(\frac{S_l}{S}\right)^{2r/\sigma^2} N(-d_5) + \frac{S}{S_l} N(-d_1),$$

with

$$d_5 = \frac{\log\left(\frac{S}{S_l}\right) - \left(r + \frac{1}{2}\sigma^2\right)(T - t)}{\sigma\sqrt{T - t}}.$$

The double one-touch option has both an upper and a lower level on which the payoff of $1 is received. Thus $V(S_l, t) = V(S_u, t) = 1$. The solution, shown in Figure 9.8, can be found

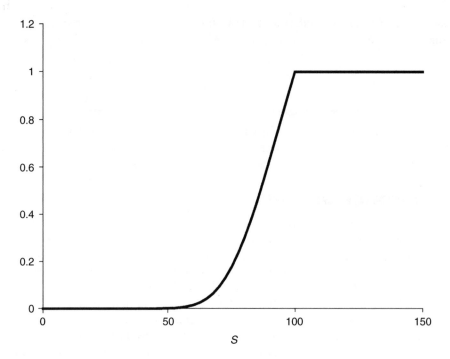

Figure 9.6 The value of a one-touch call option.

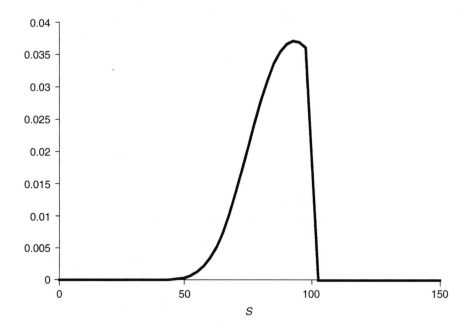

Figure 9.7 The delta of a one-touch call option.

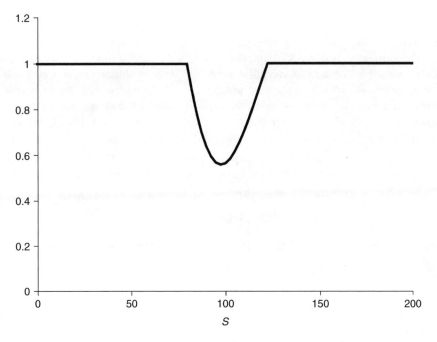

Figure 9.8 The value of a double one-touch option.

by Fourier series (see Chapter 6). Note that the value is not the sum of a one-touch call and a one-touch put.

9.8 OTHER FEATURES IN AMERICAN-STYLE CONTRACTS

The American option can be made even more interesting in many ways. Two possibilities are described in this section. These are the intermittent exercise opportunities of the Bermudan option, and the make-your-mind-up feature where the decision to exercise and the actual exercise occur at different times.

9.8.1 Bermudan Options

It is common for contracts that allow early exercise to permit the exercise only at certain specified times, and not at *all* times before expiry. For example, exercise may only be allowed on Thursdays between certain times. An option with such intermittent exercise opportunities is called a **Bermudan option**. All that this means mathematically is that the constraint (9.4) is only 'switched on' at these early exercise dates. The pricing of such a contract numerically is, as we shall see, no harder than the pricing of American options where exercise is permitted at all times.

This situation can be made more complicated by the dependence of the exercise dates on a second asset. For example, early exercise is permitted only when a second asset is above a certain level. This makes the contract a multi-asset contract; see Chapter 11.

9.8.2 Make Your Mind Up

In some contracts the decision to exercise must be made before exercise takes place. For example, we must give two weeks' warning before we exercise, and we cannot change our mind. This contract is not hard to value theoretically. Suppose that we must give a warning of time τ. If at time t we decide to exercise at time $t + \tau$ then on exercise we receive a certain deterministic amount. To make the analysis easier to explain, assume that there is no time dependence in this payoff, so that on exercise we receive $P(S)$. The value of this payoff at a time τ earlier is $V^\tau(S, \tau)$ where $V^\tau(S, t)$ is the solution of

$$\frac{\partial V^\tau}{\partial t} + \tfrac{1}{2}\sigma^2 S^2 \frac{\partial^2 V^\tau}{\partial S^2} + rS\frac{\partial V^\tau}{\partial S} - rV^\tau = 0$$

with

$$V^\tau(S, 0) = P(S).$$

This would have to be modified if the problem were time-inhomogeneous.

Obviously, we have

$$\frac{\partial V}{\partial t} + \tfrac{1}{2}\sigma^2 S^2 \frac{\partial^2 V}{\partial S^2} + rS\frac{\partial V}{\partial S} - rV \leq 0.$$

Because $V^\tau(S, \tau)$ is the value of the contract at decision time if we have decided to exercise, then our early-exercise constraint becomes

$$V(S, t) \geq V^\tau(S, \tau).$$

As an example, suppose that we get a payoff of $S - E$, this is $P(S)$. Note that there is no $\max(\cdot)$ function in this; we have said we will exercise and exercise we must, even if the asset is out of the money. The function $V^\tau(S, \tau)$ is clearly $S - Ee^{-r\tau}$ so that our **make-your-mind-up option** satisfies the constraint

$$V(S, t) \geq S - Ee^{-r\tau}.$$

A further complication is to allow one change of mind. That is, we say we will exercise in two weeks' time, but when that date comes we change our mind, and do not exercise. But the next time we say we will exercise, we must. This is also not too difficult to price theoretically.

The trick is to introduce two functions for the option value, $V_0(S, t)$ and $V_1(S, t)$. The former is the value before making the first decision to exercise, the latter is the value having made that decision but having then changed your mind. We also need $V_0^\tau(S, t', t)$ and $V_1^\tau(S, t)$. The latter is simply the earlier V^τ. The former is slightly more complicated. In $V_0^\tau(S, t', t)$ the t' represents the time at which the option will be exercised or exercise is declined. The t represents the time before that date.

The problem for V_1 is exactly the same as for the basic make-your-mind-up option i.e.

$$\frac{\partial V_1}{\partial t} + \tfrac{1}{2}\sigma^2 S^2 \frac{\partial^2 V_1}{\partial S^2} + rS\frac{\partial V_1}{\partial S} - rV_1 \leq 0$$

with

$$V_1(S, t) \geq V_1^\tau(S, \tau)$$

where

$$\frac{\partial V_0^\tau}{\partial t} + \tfrac{1}{2}\sigma^2 S^2 \frac{\partial^2 V_0^\tau}{\partial S^2} + rS\frac{\partial V_0^\tau}{\partial S} - rV_0^\tau = 0$$

with

$$V_1^\tau(S, 0) = P(S).$$

The function $V_0^\tau(S, t', t)$ satisfies

$$\frac{\partial V_0^\tau}{\partial t} + \tfrac{1}{2}\sigma^2 S^2 \frac{\partial^2 V_0^\tau}{\partial S^2} + rS \frac{\partial V_0^\tau}{\partial S} - rV_0^\tau = 0$$

(with time derivatives with respect to t and *not* t') with

$$V_0^\tau(S, t', 0) = \max(P(S), V_1(S, t')).$$

Then we have

$$\frac{\partial V_0}{\partial t} + \tfrac{1}{2}\sigma^2 S^2 \frac{\partial^2 V_0}{\partial S^2} + rS \frac{\partial V_0}{\partial S} - rV_0 \leq 0$$

with the optimality constraint

$$V_0(S, t) \geq V_0^\tau(S, t + \tau, \tau).$$

Obviously, we can introduce more levels if we are permitted to change our minds a specified number of times.

In Part Three we will see many problems where we must introduce more than one function to value a single contract.

9.9 **OTHER ISSUES**

The pricing of American options and all the issues that this raises are important for many reasons. Some of these we describe here, but we will come back to these ideas again and again.

9.9.1 Nonlinearity

The pricing of American options is a nonlinear problem because of the free boundary. There are other nonlinear problems in finance, some nonlinear because of the free boundary and some because the governing differential equation is itself nonlinear. Nonlinearity can be important for several reasons. Most obviously, nonlinear problems are harder to solve than linear problems, usually requiring numerical solution.

Nonlinear governing equations are found in Chapter 24 for models of pricing with transaction costs, Chapter 27 for uncertain parameter models, Chapter 30 for models of market crashes, and Chapter 31 for models of options used for speculative purposes.

9.9.2 Free Boundary Problems

Free boundary problems, in other contexts, will be found scattered throughout the book. Again, the solution must almost always be found numerically. As an example of a free boundary problem that is not quite an American option (but is similar), consider the **installment option**. In this contract the owner must keep paying a premium, on prescribed dates, to keep the contract alive. If the premium is not paid then the contract lapses. Consider two cases, the first when the premium is paid out continuously day by day, and the second, more realistic case, when the premium is paid at discrete intervals. Part of the valuation is to decide whether or not it is worth paying the premium, or whether the contract should be allowed to lapse.

First, consider the case of continuous payment of a premium. If we pay out a constant rate $L\,dt$ in a timestep dt to keep the contract alive then we must solve

$$\frac{\partial V}{\partial t} + \tfrac{1}{2}\sigma^2 S^2 \frac{\partial^2 V}{\partial S^2} + rS\frac{\partial V}{\partial S} - rV - L \leq 0.$$

The term L represents the continual input of cash. But we would only pay the premium if it is, in some sense, 'worth it'. As long as the contract value is positive, we should maintain the payments. If the contract value ever goes negative, we should let the contract lapse. However, we can do better than this. If we impose the constraint

$$V(S, t) \geq 0,$$

with continuity of the delta, and let the contract lapse if ever $V = 0$ then we give our contract the *highest value possible*. This is very much like the American option problem, but now we must optimally cease to pay the premium (instead of optimally exercising).

Now let us consider the more realistic discrete payment case. Suppose that payments of L (not $L\,dt$) are made discretely at time t_i. The value of the contract must increase in value from before the premium is paid to just after it is paid. The reason for this is clear. Once we have paid the premium on date t_i we do not have to worry about handing over any more money until time t_{i+1}. The rise in value exactly balances the premium, L:

$$V(S, t_i^-) = V(S, t_i^+) - L,$$

where the superscripts $+$ and $-$ refer to times just after and just before the premium is paid. But we would only hand over L if the contract would be worth more than L at time t_i^+. Thus we arrive at the jump condition

$$V(S, t_i^-) = \max(V(S, t_i^+) - L, 0).$$

If $V(S, t_i^+) \leq L$ then it is optimal to discontinue payment of the premia.

In practice, the premium L is chosen so that the value of the contract at initiation is exactly equal to L. This means that the start date is just like any other payment date.

9.9.3 Numerical Solution

Although free boundary problems must usually be solved numerically, this is not difficult, as we shall see in later chapters. We solve the relevant equation by either a finite-difference method or the binomial method.

The other numerical method that I describe is the Monte Carlo simulation. If there is any early-exercise feature in a contract this makes solution by Monte Carlo very complicated indeed. I discuss this issue in Chapter 66.

9.10 **SUMMARY**

This chapter raised a lot of issues that will be important for much of the rest of the book. The reader should familiarize himself with these concepts. Most importantly, free boundary problems and optimal strategies occur in many guises. So, even if a contract is not explicitly called 'American', these modeling issues could well be present.

The exercise of American options at times other than 'optimal' is discussed at length in Chapter 35.

FURTHER READING

- See Merton (1992) and Duffie (1992) for further discussion of the 'high contact' condition.

- Solutions for American option problems can be found in Roll (1977), Whaley (1981), Johnson (1983) and Barone-Adesi & Whaley (1987).

- See Rupf, Dewynne, Howison & Wilmott (1993) for the local solution of the American call problem and Barles, Burdeau, Romano & Samsen (1995) for the put. Kruske & Keller (1998) also study the local solution of the put problem and go to a higher order of accuracy.

- The exercise strategy of the holder of an American option and its effect on the profit of both the holder and the writer are discussed in Ahn & Wilmott (1998).

CHAPTER 10
probability density functions and first exit times

In this Chapter...

- the transition probability density function
- how to derive the forward and backward equations for the transition probability density function
- how to use the transition probability density function to solve a variety of problems
- first exit times and their relevance to American options

10.1 INTRODUCTION

Modern finance theory, especially derivatives theory, is based on the random movement of financial quantities. In the main, the building block is the Wiener process and Normal distributions. I have shown how to derive deterministic equations for the values of options in this random world, but I have said little about the way that the future may actually evolve, in which direction a stock is expected to move, or what the probability is of the option expiring in the money. This may seem perverse, but the majority of derivative theory uses ideas of hedging and no arbitrage so as to avoid dealing with the issue of randomness; uncertainty is bad. Nevertheless, it is important to acknowledge the underlying randomness, to study it, to determine properties about possible future outcomes, if one is to have a thorough understanding of financial markets.

10.2 THE TRANSITION PROBABILITY DENSITY FUNCTION

The results of this chapter will be useful for equities, currencies, interest rates or anything that evolves according to a stochastic differential equation. For that reason, I will describe the theories in terms of the general stochastic differential equation

$$dy = A(y, t) \, dt + B(y, t) \, dX \tag{10.1}$$

for the variable y. In our lognormal equity world we would have $A = \mu y$ and $B = \sigma y$, and then we would write S in place of y.

To analyze the probabilistic properties of the random walk, I will introduce the **transition probability density function** $p(y, t; y', t')$ defined by

$$\text{Prob}(a < y < b \text{ at time } t' | y \text{ at time } t) = \int_a^b p(y, t; y', t') \, dy'.$$

In words this is 'the probability that the random variable y lies between a and b at time t' in the future, given that it started out with value y at time t'.

Think of y and t as being current values with y' and t' being future values. The transition probability density function can be used to answer the question, 'What is the probability of the variable y being in a certain range at time t' given that it started out with value y at time t?'

The transition probability density function $p(y, t; y', t')$ satisfies two equations, one involving derivatives with respect to the future state and time (y' and t') and called the forward equation, and the other involving derivatives with respect to the current state and time (y and t) and called the backward equation. These two equations are parabolic partial differential equations not dissimilar to the Black–Scholes equation.[1]

I derive these two equations in the next few sections, using a simple trinomial approximation to the random walk for y.

10.3 A TRINOMIAL MODEL FOR THE RANDOM WALK

By far the easiest and most straightforward way to derive the forward and backward equations is via a trinomial approximation to the continuous-time random walk. This approximation is shown in Figure 10.1.

The variable y can either rise, fall or take the same value after a timestep δt. These movements have certain probabilities associated with them. I am going to choose the size of the rise and the fall to be the same, with probabilities such that the mean and standard deviation of the discrete-time approximation are the same as the mean and standard deviation of the continuous-time model over the same timestep. I have three quantities to play with here, the

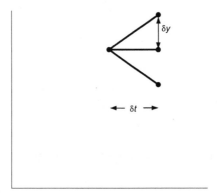

Figure 10.1 The trinomial approximation to the random walk for y.

---

[1] One of them, the backward equation, is *very* similar to the Black–Scholes equation.

jump size δy, the probability of a rise and the probability of a fall, but only two quantities to fix, the mean and standard deviation. The probability of not moving is such that the three probabilities sum to one. I will thus carry around the quantity δy which will drop out from the final equation.

I will use $\phi^+(y, t)$ and $\phi^-(y, t)$ to be the probabilities of a rise and fall respectively. The *mean* of the change in y after the timestep is thus

$$\phi^+ \delta y + (1 - \phi^+ - \phi^-) \cdot 0 + \phi^-(-\delta y) = (\phi^+ - \phi^-)\delta y.$$

Since $\text{Var}[\cdot] = E[(\cdot)^2] - E[\cdot]^2$ the variance is

$$\phi^+ \delta y^2 + (1 - \phi^+ - \phi^-) \cdot 0^2 + \phi^-(-\delta y)^2 - ((\phi^+ - \phi^-)\delta y)^2$$
$$= \delta y^2(\phi^+ + \phi^- - (\phi^+ - \phi^-)^2).$$

In both of these the arguments of ϕ^+ and ϕ^- are y and t.

The mean of the change in the continuous-time version of the random walk is, from Equation (10.1),

$$A(y, t)\,\delta t$$

and the variance is

$$B(y, t)^2\,\delta t.$$

(These are correct only to leading order, the discrete versions are exact.)

To match the mean and standard deviation we choose

$$\phi^+(y, t) = \tfrac{1}{2} \frac{\delta t}{\delta y^2}(B(y, t)^2 + A(y, t)\,\delta y)$$

and

$$\phi^-(y, t) = \tfrac{1}{2} \frac{\delta t}{\delta y^2}(B(y, t)^2 - A(y, t)\,\delta y).$$

Although I said that we have two equations for three unknowns, we must have

$$\delta y = O(\sqrt{\delta t}),$$

otherwise the diffusive properties of the problem are lost.

Now we are set to find the equations for the transition probability density function.

10.4 THE FORWARD EQUATION

In Figure 10.2 is shown a trinomial representation of the random walk. The variable y takes the value y' at time t', but how did it get there?

In our trinomial walk we can only get to the point y' from the three values $y' + \delta y$, y' and $y' - \delta y$. The probability of being at y' at time t' is related to the probabilities of being at the previous three values and *moving in the right direction*:

$$p(y, t; y', t') = \phi^-(y' + \delta y, t' - \delta t)p(y, t; y' + \delta y, t' - \delta t)$$
$$+ (1 - \phi^-(y', t' - \delta t) - \phi^+(y', t' - \delta t))p(y, t; y', t' - \delta t)$$
$$+ \phi^+(y' - \delta y, t' - \delta t)p(y, t; y' - \delta y, t' - \delta t).$$

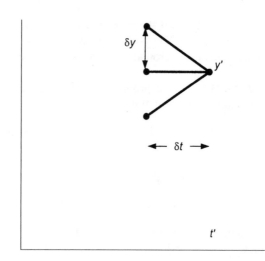

Figure 10.2 The trinomial approximation to the random walk used in finding the forward equation.

We can easily expand each of the terms in Taylor series about the point y', t'. For example,

$$p(y, t; y' + \delta y, t' - \delta t) \approx p(y, t; y', t') + \delta y \frac{\partial p}{\partial y'} + \tfrac{1}{2} \delta y^2 \frac{\partial^2 p}{\partial y'^2} - \delta t \frac{\partial p}{\partial t'} + \cdots.$$

I will omit the rest of the details, but the result is

$$\frac{\partial p}{\partial t'} = \tfrac{1}{2} \frac{\partial^2}{\partial y'^2} (B(y', t')^2 p)$$

$$- \frac{\partial}{\partial y'} (A(y', t')p). \tag{10.2}$$

This is the **Fokker–Planck** or **forward Kolmogorov equation**. It is a forward parabolic partial differential equation, requiring initial conditions at time t and to be solved for $t' > t$.

This equation is to be used if there is some special state now and you want to know what could happen later. For example, you know the current value of y and want to know the distribution of values at some later date.

Example

The most important example to us is that of the distribution of equity prices in the future. If we have the random walk

$$dS = \mu S \, dt + \sigma S \, dX$$

then the forward equation becomes

$$\frac{\partial p}{\partial t'} = \tfrac{1}{2} \frac{\partial^2}{\partial S'^2} (\sigma^2 S'^2 p) - \frac{\partial}{\partial S'} (\mu S' p).$$

A special solution of this is the one having a delta function initial condition

$$p(S, t; S', t) = \delta(S' - S),$$

representing a variable that begins with certainty with value S at time t. The solution of this problem is

$$p(S, t; S', t') = \frac{1}{\sigma S'\sqrt{2\pi(t' - t)}} e^{-\left(\log(S/S') + (\mu - (1/2)\sigma^2)(t' - t)\right)^2 / 2\sigma^2(t' - t)}. \tag{10.3}$$

This is plotted as a function of S' in Figure 10.3 and as a function of both S' and t' in Figure 10.4.

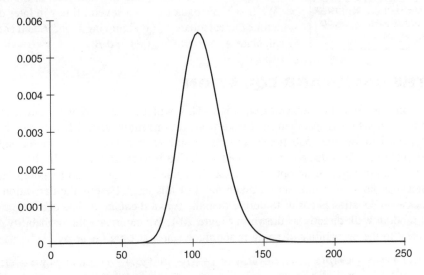

Figure 10.3   The probability density function for the lognormal random walk.

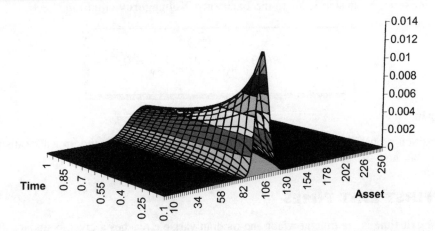

Figure 10.4   The probability density function for the lognormal random walk evolving through time.

10.5 **THE STEADY-STATE DISTRIBUTION**

Some random walks have a steady-state distribution. That is, in the long run as $t' \to \infty$ the distribution $p(y, t; y', t')$ as a function of y' settles down to be independent of the starting state y and time t. Loosely speaking, this requires at least that the random walk is time-homogeneous, i.e. that A and B are independent of t, asymptotically. Some random walks have no such steady state even though they have a time-independent equation; the lognormal random walk either grows without bound or decays to zero.

KNOWING THE STEADY-STATE HELPS US TO TELL WHETHER WE'VE GOT A GOOD MODEL

If there is a steady-state distribution $p_\infty(y')$ then it satisfies

$$\frac{1}{2}\frac{d^2}{dy'^2}(B_\infty^2 p_\infty) - \frac{d}{dy'}(A_\infty p_\infty) = 0.$$

In this equation A_∞ and B_∞ are the functions in the limit $t \to \infty$. We'll see this equation used several times in later chapters, sometimes to calculate $p_\infty(y')$ knowing A and B and sometimes to calculate A knowing $p_\infty(y')$ and B.

10.6 **THE BACKWARD EQUATION**

Now we come to find the backward equation. This will be useful if we want to calculate probabilities of reaching a specified final state from various initial states. It will be a backward parabolic partial differential equation requiring conditions imposed in the future, and solved backwards in time. The equation I am about to derive is very similar to the Black–Scholes equation for the fair value of an option; indeed, the value of an option can be interpreted as an expectation over possible future states, much more of this later. Here is the derivation.

The backward equation is easier to derive than the forward equation. The derivation uses the trinomial random walk directly as drawn in Figure 10.1. We can relate the probability of being at y at time t to the probability of being at the three states at time $t + \delta t$ by

$$p(y, t; y', t') = \phi^+(y, t)p(y + \delta y, t + \delta t; y', t') + (1 - \phi^+(y, t) - \phi^-(y, t))p(y, t + \delta t; y', t')$$

$$+ \phi^-(y, t)p(y - \delta y, t + \delta t; y', t').$$

The Taylor series expansion leads to the **backward Kolmogorov equation**

$$\frac{\partial p}{\partial t} + \frac{1}{2}B(y, t)^2\frac{\partial^2 p}{\partial y^2} + A(y, t)\frac{\partial p}{\partial y} = 0. \tag{10.4}$$

Example

The transition probability density function (10.3) for the lognormal random walk satisfies this equation, but note the different independent variables.

10.7 **FIRST EXIT TIMES**

The **first exit time** is the time at which the random variable reaches a given boundary. Perhaps we want to know how long it will be before a certain level is reached or how long before an

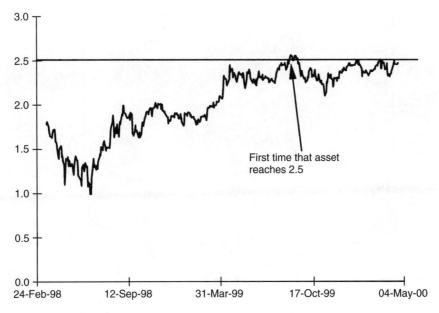

Figure 10.5 An example of a first exit time.

American option should be optimally exercised. An example of a first exit time is given in Figure 10.5.

Questions to ask about first exit times are 'What is the probability of an asset level being reached before a certain time?', or 'How long do we expect it to take for an interest rate to fall to a given level?' I will address these problems now.

10.8 CUMULATIVE DISTRIBUTION FUNCTIONS FOR FIRST EXIT TIMES

What is the probability of your favourite asset doubling or halving in value in the next year? This is a question that can be answered by the solution of a simple diffusion equation. It is an example of the more general question, 'What is the probability of a random variable leaving a given range before a given time?' This question is illustrated in Figure 10.6.

Let me introduce the function $C(y, t; t')$ as the probability of the variable y leaving the region Ω before time t'. This function can be thought of as a cumulative distribution function. This function also satisfies the backward equation

$$\frac{\partial C}{\partial t} + \tfrac{1}{2}B(y, t)^2 \frac{\partial^2 C}{\partial y^2} + A(y, t)\frac{\partial C}{\partial y} = 0.$$

What makes the problem different from that for the transition probability density function are the boundary and final conditions. If the variable y is actually *on* the boundary of the region Ω then clearly the probability of exiting is one:

$$C(y, t, t') = 1 \quad \text{on the edge of } \Omega.$$

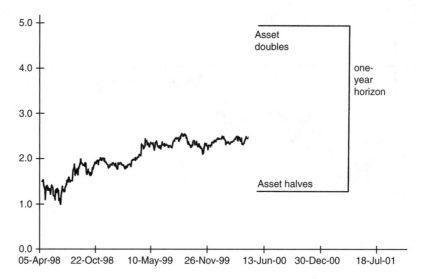

Figure 10.6   What is the probability of the asset leaving the region before the given time?

On the other hand if we are inside the region Ω at time t' then there is no time left for the variable to leave the region and so the probability is zero. Thus we have

$$C(y, t', t') = 0.$$

10.9   EXPECTED FIRST EXIT TIMES

In the previous section I showed how to calculate the probability of leaving a given region. We can use this function to find the *expected* time to exit. Once we have found C then it is simple to find the **expected first exit time**. Let me call the expected first exit time $u(y, t)$. It is a function of where we start out, y and t.

Because C is a cumulative distribution function the expected first exit time can be written as

$$u(y, t) = \int_t^\infty (t' - t) \frac{\partial C}{\partial t'} \, dt'.$$

After an integration by parts we get

$$u(y, t) = \int_t^\infty 1 - C(y, t; t') \, dt'.$$

The function C satisfies the backward equation in y and t so that, after differentiating under the integral sign, we find that u satisfies the equation

$$\frac{\partial u}{\partial t} + \tfrac{1}{2} B(y, t)^2 \frac{\partial^2 u}{\partial y^2} + A(y, t) \frac{\partial u}{\partial y} = -1. \tag{10.5}$$

Since C is one on the boundary of Ω, u must be zero around the boundary of the region. What about the final condition? Typically one solves over a region Ω that is bounded in time, for example as shown in Figure 10.7.

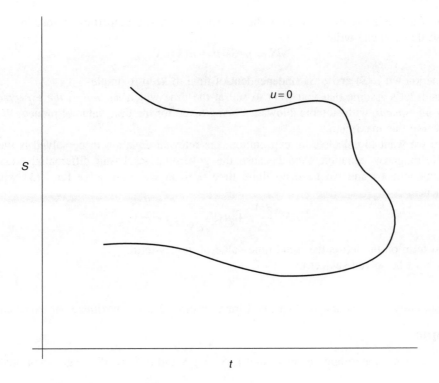

$u = 0$

S

t

Figure 10.7 The first exit time problem.

Example

When the stochastic differential equation is independent of time, that is, both A and B are functions of y only, and the region Ω is also time-homogeneous, then there may be a steady-state solution of (10.5). Returning to the logarithmic asset problem, what is the expected time for the asset to leave the range (S_0, S_1)? The answer to this question is the solution of

$$\tfrac{1}{2}\sigma^2 S^2 \frac{d^2u}{dS^2} + \mu S \frac{du}{dS} = -1,$$

with

$$u(S_0) = u(S_1) = 0,$$

and is

$$u(S) = \frac{1}{\tfrac{1}{2}\sigma^2 - \mu}\left(\log\left(\frac{S}{S_0}\right) - \frac{1 - \left(\frac{S}{S_0}\right)^{1-2\mu/\sigma^2}}{1 - \left(\frac{S_1}{S_0}\right)^{1-2\mu/\sigma^2}} \log\left(\frac{S_1}{S_0}\right) \right).$$

10.10 ANOTHER EXAMPLE OF OPTIMAL STOPPING

You hold some investment, it goes up in value, it goes down in value. Generally speaking, it's going nowhere fast. When should you sell it? Let's make two big assumptions.

First of all, let's say that you know the statistical/stochastic properties of your investment's value so that you can write

$$dS = \mu(S)\,dt + \sigma(S)\,dX$$

for some known $\mu(S)$ and $\sigma(S)$ (independent of time to keep it simple).

Second, let's assume that you want to sell at the time which *maximizes the expected value of your investment*, with suitable allowance being taken for the time value of money. We'll use V to denote this maximum.

Since we want to calculate an expectation, the relevant equation to be solved is the backward Kolmogorov equation. And because the governing stochastic differential equation is time-homogeneous and we have no finite time horizon we must solve for $V(S)$ with no t dependence:

$$\tfrac{1}{2}\sigma(S)^2\frac{d^2V}{dS^2} + \mu(S)\frac{dV}{dS} - rV = 0.$$

The last term on the left is the usual time-value-of-money term.

This must be solved subject to

$$V \geq S$$

with continuity of V and dV/dS. This constraint ensures that we maximize our expected value.

Example

When we have a logarithmic asset so that $\mu(S) = \mu S$ and $\sigma(S) = \sigma S$ we get some interesting results.

The general solution of the second order ordinary differential equation is then

$$AS^{\alpha+} + BS^{\alpha-}$$

where A and B are arbitrary constants and

$$\alpha^{\pm} = \frac{1}{\sigma^2}\left(-\mu + \tfrac{1}{2}\sigma^2 \pm \sqrt{\left(\mu - \tfrac{1}{2}\sigma^2\right)^2 + 2r\sigma^2}\right).$$

If $\mu > r$ then there is no finite solution for V (with a finite time horizon) and one should never sell the asset. If $\mu < r$ then $V = S$ and one should immediately sell the asset. (Remember that these simple results only apply under the second assumption; the risk inherent in the position has not been accounted for.)

How does all this change when there are dividends on the asset?

10.11 EXPECTATIONS AND BLACK–SCHOLES

The transition probability density $p(S, t; S', t')$ for S following the random walk

$$dS = \mu S\,dt + \sigma S\,dX$$

satisfies

$$\frac{\partial p}{\partial t} + \tfrac{1}{2}\sigma^2 S^2\frac{\partial^2 p}{\partial S^2} + \mu S\frac{\partial p}{\partial S} = 0. \tag{10.6}$$

This is the backward Kolmogorov equation.

To calculate the expected value of some function $F(S)$ at time T we must solve (10.6) for the function $p_F(S, t)$ with

$$p_F(S, T) = F(S).$$

If the function $F(S)$ represents an amount of money received at time T then it is natural to ask what is the present value of the expected amount received. In other words, what is the expected amount of an option's payoff? To calculate this present value we simply multiply by the discount factor, which gives

$$e^{-r(T-t)} p_F(S, t) \tag{10.7}$$

when interest rates are constant. In this we have calculated the present value today of the expected payoff received at time T, given that today, time t, the asset value is S. Call the function (10.7) $V(S, t)$, what equation does it satisfy? Substituting

$$p_F(S, t) = e^{r(T-t)} V(S, t)$$

into (10.6) we find that $V(S, t)$ satisfies

$$\frac{\partial V}{\partial t} + \tfrac{1}{2}\sigma^2 S^2 \frac{\partial^2 V}{\partial S^2} + \mu S \frac{\partial V}{\partial S} - rV = 0.$$

This looks very like the Black–Scholes equation, with one small difference. In the Black–Scholes equation there is no μ.

Now forget that there is such a quantity as μ and replace it in this equation with r. The resulting equation is *exactly* the Black–Scholes equation. There must be something special about the random walk in which μ is replaced by r:

$$dS = rS\,dt + \sigma S\,dX. \tag{10.8}$$

This is called the **risk-neutral random walk**. It has the same drift as money in the bank.

Our conclusion is that

> The fair value of an option is the present value of the expected payoff at expiry under a risk-neutral random walk for the underlying

THIS RESULT IS THE BASIS OF PRICING BY SIMULATION

We can write

$$\text{option value} = e^{-r(T-t)} E\,[\text{payoff}(S)]$$

provided that the expectation is with respect to the risk-neutral random walk, not the *real* one.

In this expression we see the short-term interest rate playing two distinct roles. First, it is used for discounting the payoff to the present. This is the term $e^{-r(T-t)}$ outside the expectation. Second, the return on the asset in the risk-neutral world is expected to be $rS\,dt$ in a timestep dt.

10.12 **SUMMARY**

Although probability theory underpins most finance theory, it is possible to go a long way without even knowing what a transition probability density function is. But it is important in many circumstances to remember the foundation of uncertainty and to examine the future in a probabilistic sense. A couple of obvious examples spring to mind. First, if you own an American option when do you expect to exercise it? The value depends theoretically on the parameter σ in the asset price random walk but the expected time to exercise also depends on μ; the payoff may be certain because of hedging but you cannot be certain whether you will still hold the option at expiry. The second example concerns speculation with options. What if you *don't* hedge? In that case your final payoff is at the mercy of the markets, it is uncertain and can only be described probabilistically. Both of these problems can be addressed via transition probability density functions. In Chapter 31 we explore the random behavior of unhedged option positions.

FURTHER READING

- A very good book on probability theory that carefully explains the derivation, meaning and use of transition probability density functions is Cox & Miller (1965).

- The two key references that rigorously relate option values and results from probability theory are by Harrison & Kreps (1979) and Harrison & Pliska (1981).

- Atkinson & Wilmott (1993) discuss transition densities for moving averages of asset prices.

CHAPTER 11
multi-asset options

In this Chapter...

- how to model the behavior of many assets simultaneously
- estimating correlation between asset price movements
- how to value and hedge options on many underlying assets in the Black–Scholes framework
- the pricing formula for European non-path-dependent options on dividend-paying assets
- how to price and hedge quantos, and the role of correlation

11.1 INTRODUCTION

In this chapter I introduce the idea of higher dimensionality by describing the Black–Scholes theory for options on more than one underlying asset. This theory is perfectly straightforward; the only new idea is that of correlated random walks and the corresponding multi-factor version of Itô's lemma.

Although the modeling and mathematics is easy, the final step of the pricing and hedging, the 'solution', can be extremely hard indeed. I explain what makes a problem easy, and what makes it hard, from the numerical analysis point of view.

11.2 MULTI-DIMENSIONAL LOGNORMAL RANDOM WALKS

The basic building block for option pricing with one underlying is the lognormal random walk

$$dS = \mu S \, dt + \sigma S \, dX.$$

This is readily extended to a world containing many assets via models for each underlying

$$dS_i = \mu_i S_i \, dt + \sigma_i S_i \, dX.$$

Here S_i is the price of the ith asset, $i = 1, \ldots, d$, μ_i and σ_i are the drift and volatility of that asset respectively and dX_i is the increment of a Wiener process. We can still continue to think of dX_i as a random number drawn from a Normal distribution with mean zero and standard deviation $dt^{1/2}$ so that

$$E[dX_i] = 0 \quad \text{and} \quad E[dX_i^2] = dt$$

but the random numbers dX_i and dX_j are **correlated**:

$$E[dX_i\, dX_j] = \rho_{ij}\, dt.$$

Here ρ_{ij} is the correlation coefficient between the ith and jth random walks. The symmetric matrix with ρ_{ij} as the entry in the ith row and jth column is called the **correlation matrix**. For example, if we have seven underlyings $d = 7$ and the correlation matrix will look like this:

$$\Sigma = \begin{pmatrix}
1 & \rho_{12} & \rho_{13} & \rho_{14} & \rho_{15} & \rho_{16} & \rho_{17} \\
\rho_{21} & 1 & \rho_{23} & \rho_{24} & \rho_{25} & \rho_{26} & \rho_{27} \\
\rho_{31} & \rho_{32} & 1 & \rho_{34} & \rho_{35} & \rho_{36} & \rho_{37} \\
\rho_{41} & \rho_{42} & \rho_{43} & 1 & \rho_{45} & \rho_{46} & \rho_{47} \\
\rho_{51} & \rho_{52} & \rho_{53} & \rho_{54} & 1 & \rho_{56} & \rho_{57} \\
\rho_{61} & \rho_{62} & \rho_{63} & \rho_{64} & \rho_{65} & 1 & \rho_{67} \\
\rho_{71} & \rho_{72} & \rho_{73} & \rho_{74} & \rho_{75} & \rho_{76} & 1
\end{pmatrix}$$

Note that $\rho_{ii} = 1$ and $\rho_{ij} = \rho_{ji}$. The correlation matrix is positive definite, so that $\mathbf{y}^T \Sigma \mathbf{y} \geq 0$. The **covariance matrix** is simply

$$\mathbf{M \Sigma M},$$

where $\mathbf{M}$ is the matrix with the σ_i along the diagonal and zeros everywhere else.

To be able to manipulate functions of many random variables we need a multidimensional version of Itô's lemma. If we have a function of the variables $S_1, \ldots, S_d$ and t, $V(S_1, \ldots, S_d, t)$, then

$$dV = \left(\frac{\partial V}{\partial t} + \frac{1}{2} \sum_{i=1}^{d} \sum_{j=1}^{d} \sigma_i \sigma_j \rho_{ij} S_i S_j \frac{\partial^2 V}{\partial S_i \partial S_j} \right) dt + \sum_{i=1}^{d} \frac{\partial V}{\partial S_i} dS_i.$$

We can get to this same result by using Taylor series and the rules of thumb:

$$dX_i^2 = dt \quad \text{and} \quad dX_i\, dX_j = \rho_{ij}\, dt.$$

11.3 MEASURING CORRELATIONS

If you have time series data at intervals of δt for all d assets you can calculate the correlation between the returns as follows. First, take the price series for each asset and calculate the return over each period. The return on the ith asset at the kth data point in the time series is simply

$$R_i(t_k) = \frac{S_i(t_k + \delta t) - S_i(t_k)}{S_i(t_k)}.$$

The historical volatility of the ith asset is

$$\sigma_i = \sqrt{\frac{1}{\delta t (M - 1)} \sum_{k=1}^{M} (R_i(t_k) - \overline{R}_i)^2}$$

where M is the number of data points in the return series and $\overline{R}_i$ is the mean of all the returns in the series.

The covariance between the returns on assets i and j is given by

$$\frac{1}{\delta t(M-1)} \sum_{k=1}^{M} (R_i(t_k) - \overline{R}_i)(R_j(t_k) - \overline{R}_j).$$

The correlation is then

$$\frac{1}{\delta t(M-1)\sigma_i\sigma_j} \sum_{k=1}^{M} (R_i(t_k) - \overline{R}_i)(R_j(t_k) - \overline{R}_j).$$

In Excel correlation between two time series can be found using the CORREL worksheet function, or Tools | Data Analysis | Correlation.

Figure 11.1 shows the correlation matrix for Marks & Spencer, Tesco, Sainsbury and IBM.

Correlations measured from financial time series data are notoriously unstable. If you split your data into two equal groups, up to one date and beyond that date, and calculate the correlations for each group you may find that they differ quite markedly. You could calculate a 60-day correlation, say, from several years' data and the result would look something like Figure 11.2. You might want to use a historical 60-day correlation if you have

PERSONALLY, I DON'T BELIEVE IN CORRELATION.... AND I'M NOT THAT SURE OF VOL, EITHER

<HELP> for explanation.   P140 Comdty CORR
Start date shifted.   <PAGE> for t-Statistics -or- <MENU> to Update Matrix.
CORRELATION MATRIX

RANGE 3/ 8/99 TO 9/ 7/99   PERIOD D   PAGE 1 OF 3
Name:STORES   -OR- CHOOSE ONE   0 - Use PDF Default
                                           0   1 - Year to Date
                                                        2 - 3 Month
Observations = 132   3 - 6 Month
                                                        4 - 12 Month
                                                        5 - 5 Year
                        * - observations limited due to this securiy
                    Correlation Coefficients

		GBp MKS	GBp TSCO	GBp SBRY	USD * IBM						
GBp	MKS	1.000	0.362	0.239	-0.463	n.a.	n.a.	n.a.	n.a.	n.a.	n.a.
GBp	TSCO	0.362	1.000	0.596	0.091	n.a.	n.a.	n.a.	n.a.	n.a.	n.a.
GBp	SBRY	0.239	0.596	1.000	0.457	n.a.	n.a.	n.a.	n.a.	n.a.	n.a.
USD	IBM	-0.463	0.091	0.457	1.000	n.a.	n.a.	n.a.	n.a.	n.a.	n.a.
		n.a.	n.a.	n.a.	n.a.	n.a.	n.a.	n.a.	n.a.	n.a.	n.a.
		n.a.	n.a.	n.a.	n.a.	n.a.	n.a.	n.a.	n.a.	n.a.	n.a.
		n.a.	n.a.	n.a.	n.a.	n.a.	n.a.	n.a.	n.a.	n.a.	n.a.
		n.a.	n.a.	n.a.	n.a.	n.a.	n.a.	n.a.	n.a.	n.a.	n.a.
		n.a.	n.a.	n.a.	n.a.	n.a.	n.a.	n.a.	n.a.	n.a.	n.a.
		n.a.	n.a.	n.a.	n.a.	n.a.	n.a.	n.a.	n.a.	n.a.	n.a.

Figure 11.1   Some correlations. Source: Bloomberg L.P.

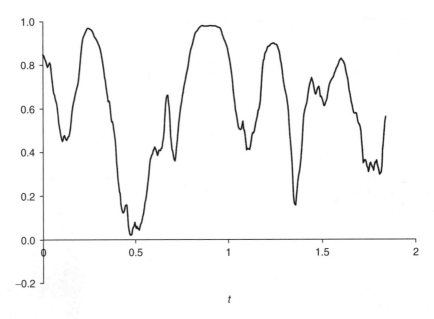

Figure 11.2 A correlation time series.

a contract of that maturity. But, as can be seen from the figure, such a historical correlation should be used with care; correlations are even more unstable than volatilities.

The other possibility is to back out an **implied correlation** from the quoted price of an instrument. The idea behind that approach is the same as with implied volatility, it gives an estimate of the market's perception of correlation.

11.4 OPTIONS ON MANY UNDERLYINGS

Options with many underlyings are called **basket options, options on baskets** or **rainbow options**. The theoretical side of pricing and hedging is straightforward, following the Black–Scholes arguments but now in higher dimensions.

Set up a portfolio consisting of one basket option and short a number Δ_i of each of the assets S_i:

$$\Pi = V(S_1, \ldots, S_d, t) - \sum_{i=1}^{d} \Delta_i S_i.$$

The change in this portfolio is given by

$$d\Pi = \left(\frac{\partial V}{\partial t} + \frac{1}{2} \sum_{i=1}^{d} \sum_{j=1}^{d} \sigma_i \sigma_j \rho_{ij} S_i S_j \frac{\partial^2 V}{\partial S_i \partial S_j} \right) dt + \sum_{i=1}^{d} \left(\frac{\partial V}{\partial S_i} - \Delta_i \right) dS_i.$$

If we choose

$$\Delta_i = \frac{\partial V}{\partial S_i}$$

for each i, then the portfolio is hedged, and risk-free. Setting the return equal to the risk-free rate we arrive at

$$\frac{\partial V}{\partial t} + \frac{1}{2} \sum_{i=1}^{d} \sum_{j=1}^{d} \sigma_i \sigma_j \rho_{ij} S_i S_j \frac{\partial^2 V}{\partial S_i \partial S_j} + r \sum_{i=1}^{d} S_i \frac{\partial V}{\partial S_i} - rV = 0. \tag{11.1}$$

This is the multidimensional version of the Black–Scholes equation. The modifications that need to be made for dividends are obvious. When there is a dividend yield of D_i on the ith asset we have

$$\frac{\partial V}{\partial t} + \frac{1}{2} \sum_{i=1}^{d} \sum_{j=1}^{d} \sigma_i \sigma_j \rho_{ij} S_i S_j \frac{\partial^2 V}{\partial S_i \partial S_j} + \sum_{i=1}^{d} (r - D_i) S_i \frac{\partial V}{\partial S_i} - rV = 0$$

11.5 THE PRICING FORMULA FOR EUROPEAN NON-PATH-DEPENDENT OPTIONS ON DIVIDEND-PAYING ASSETS

Because there is a Green's function for this problem (see Chapter 6) we can write down the value of a European non-path-dependent option with payoff of Payoff$(S_1, \ldots, S_d)$ at time T:

$$V = e^{-r(T-t)} (2\pi(T-t))^{-d/2} (\text{Det } \boldsymbol{\Sigma})^{-1/2} (\sigma_1 \cdots \sigma_d)^{-1}$$
$$\int_0^\infty \cdots \int_0^\infty \frac{\text{Payoff}(S_1' \cdots S_d')}{S_1' \cdots S_d'} \exp\left(-\tfrac{1}{2} \boldsymbol{\alpha}^T \boldsymbol{\Sigma}^{-1} \boldsymbol{\alpha}\right) dS_1' \cdots dS_d'.$$
$$\alpha_i = \frac{1}{\sigma_i (T-t)^{1/2}} \left(\log\left(\frac{S_i}{S_i'}\right) + \left(r - D_i - \frac{\sigma_i^2}{2}\right)(T-t)\right) \tag{11.2}$$

This has included a constant continuous dividend yield of D_i on each asset.

11.6 EXCHANGING ONE ASSET FOR ANOTHER: A SIMILARITY SOLUTION

An **exchange option** gives the holder the right to exchange one asset for another, in some ratio. The payoff for this contract at expiry is

$$\max(q_1 S_1 - q_2 S_2, 0)$$

where q_1 and q_2 are constants.

The partial differential equation satisfied by this option in a Black–Scholes world is

$$\frac{\partial V}{\partial t} + \frac{1}{2} \sum_{i=1}^{2} \sum_{j=1}^{2} \sigma_i \sigma_j \rho_{ij} S_i S_j \frac{\partial^2 V}{\partial S_i \partial S_j} + \sum_{i=1}^{2} (r - D_i) S_i \frac{\partial V}{\partial S_i} - rV = 0.$$

A dividend yield has been included for both assets. Since there are only two underlyings the summations in these only go up to two.

This contract is special in that there is a similarity reduction. Let's postulate that the solution takes the form

$$V(S_1, S_2, t) = q_1 S_2 H(\xi, t),$$

where the new variable is

$$\xi = \frac{S_1}{S_2}.$$

If this is the case, then instead of finding a function V of three variables, we only need find a function H of two variables, a much easier task.

Changing variables from S_1, S_2 to ξ we must use the following for the derivatives.

$$\frac{\partial}{\partial S_1} = \frac{1}{S_2} \frac{\partial}{\partial \xi}, \quad \frac{\partial}{\partial S_2} = -\frac{\xi}{S_2} \frac{\partial}{\partial \xi},$$

$$\frac{\partial^2}{\partial S_1^2} = \frac{1}{S_2^2} \frac{\partial^2}{\partial \xi^2}, \quad \frac{\partial^2}{\partial S_2^2} = \frac{\xi^2}{S_2^2} \frac{\partial^2}{\partial \xi^2} + \frac{2\xi}{S_2^2} \frac{\partial}{\partial \xi}, \quad \frac{\partial^2}{\partial S_1 \partial S_2} = -\frac{\xi}{S_2^2} \frac{\partial^2}{\partial \xi^2} - \frac{1}{S_2^2} \frac{\partial}{\partial \xi}.$$

The time derivative is unchanged. The partial differential equation now becomes

$$\frac{\partial H}{\partial t} + \frac{1}{2} \sigma'^2 \xi^2 \frac{\partial^2 H}{\partial \xi^2} + (D_2 - D_1) \xi \frac{\partial H}{\partial \xi} - D_2 H = 0$$

where

$$\sigma' = \sqrt{\sigma_1^2 - 2\rho_{12}\sigma_1\sigma_2 + \sigma_2^2}.$$

You will recognize this equation as being the Black–Scholes equation for a single stock with D_2 in place of r, D_1 in place of the dividend yield on the single stock and with a volatility of σ'.

From this it follows, retracing our steps and writing the result in the original variables, that

$$V(S_1, S_2, t) = q_1 S_1 e^{-D_1(T-t)} N(d_1') - q_2 S_2 e^{-D_2(T-t)} N(d_2')$$

where

$$d_1' = \frac{\log\left(\dfrac{q_1 S_1}{q_2 S_2}\right) + \left(D_2 - D_1 + \frac{1}{2}\sigma'^2\right)(T-t)}{\sigma'\sqrt{T-t}} \quad \text{and} \quad d_2' = d_1' - \sigma'\sqrt{T-t}.$$

11.7 QUANTOS

There is one special, and very important, type of multi-asset option. This is the cross-currency contract called a **quanto**. The quanto has a payoff defined with respect to an asset or an index (or an interest rate) in one country, but then the payoff is converted to another currency for payment. An example of such a contract would be a call on the Nikkei Dow index but paid in US dollars. This contract is exposed to the dollar-yen exchange rate and the Nikkei Dow index. We could write down the differential equation directly assuming that the underlyings satisfy lognormal random walks with correlation ρ. But we will build up the problem from first principles to demonstrate what hedging must take place.

Define $S_\$$ to be the yen-dollar exchange rate (number of dollars per yen[1]) and S_N is the level of the Nikkei Dow index. We assume that they satisfy

$$dS_\$ = \mu_\$ S_\$ \, dt + \sigma_\$ S_\$ \, dX_\$ \quad \text{and} \quad dS_N = \mu_N S_N \, dt + \sigma_N S_N \, dX_N,$$

with a correlation coefficient ρ between them.

Construct a portfolio consisting of the quanto in question, hedged with yen and the Nikkei Dow index:

$$\Pi = V(S_\$, S_N, t) - \Delta_\$ S_\$ - \Delta_N S_N S_\$.$$

Note that every term in this equation is measured in dollars. $\Delta_\$$ is the number of yen we hold short, so $-\Delta_\$ S_\$$ is the dollar value of that yen. Similarly, with the term $-\Delta_N S_N S_\$$ we have converted the yen-denominated index S_N into dollars, and Δ_N is the amount of the index held short.

The change in the value of the portfolio is due to the change in the value of its components and the interest received on the yen:

$$d\Pi = \left(\frac{\partial V}{\partial t} + \tfrac{1}{2} \sigma_\$^2 S_\$^2 \frac{\partial^2 V}{\partial S_\$^2} + \rho \sigma_\$ \sigma_N S_\$ S_N \frac{\partial^2 V}{\partial S_\$ \partial S_N} + \tfrac{1}{2} \sigma_N^2 S_N^2 \frac{\partial^2 V}{\partial S_N^2} - \rho \sigma_\$ \sigma_N \Delta_N S_\$ S_N - r_f \Delta_\$ S_\$ \right) dt$$

$$+ \left(\frac{\partial V}{\partial S_\$} - \Delta_\$ - \Delta_N S_N \right) dS_\$ + \left(\frac{\partial V}{\partial S_N} - \Delta_N S_\$ \right) dS_N.$$

There is a term in the above that we have not seen before, the $-\rho \sigma_\$ \sigma_N S_\$ S_N$. This is due to the increment of the product $-\Delta_N S_N S_\$$. There is also the interest received by the yen holding; we *have* seen such a term before. We now choose

$$\Delta_\$ = \frac{\partial V}{\partial S_\$} - \frac{S_N}{S_\$} \frac{\partial V}{\partial S_N} \quad \text{and} \quad \Delta_N = \frac{1}{S_\$} \frac{\partial V}{\partial S_N}$$

to eliminate the risk in the portfolio. Setting the return on this riskless portfolio equal to the *US risk-free rate of interest* $r_\$$, since Π is measured entirely in dollars, yields

$$\frac{\partial V}{\partial t} + \tfrac{1}{2} \sigma_\$^2 S_\$^2 \frac{\partial^2 V}{\partial S_\$^2} + \rho \sigma_\$ \sigma_N S_\$ S_N \frac{\partial^2 V}{\partial S_\$ \partial S_N} + \tfrac{1}{2} \sigma_N^2 S_N^2 \frac{\partial^2 V}{\partial S_N^2}$$

$$+ S_\$ \frac{\partial V}{\partial S_\$} (r_\$ - r_f) + S_N \frac{\partial V}{\partial S_N} (r_f - \rho \sigma_\$ \sigma_N) - r_\$ V = 0.$$

This completes the formulation of the pricing equation. The equation is valid for any contract with underlying measured in one currency but paid in another. To fully specify our particular quanto we must give the final conditions on $t = T$:

$$V(S_\$, S_N, T) = \max(S_N - E, 0).$$

Note that as far as the payoff is concerned we don't much care what $S_\$$ is (only that we can hedge with it somehow). We do *not* multiply this by the exchange rate. Because of the simple form of the payoff we can look for a solution that is independent of the exchange rate. Trying

$$V(S_\$, S_N, t) = W(S_N, t)$$

---

[1] This is the opposite of market convention; currencies are usually quoted against the dollar with pounds sterling being an exception. I use this way around for this problem just to simplify the algebra.

we find that

$$\frac{\partial W}{\partial t} + \tfrac{1}{2}\sigma_N^2 S_N^2 \frac{\partial^2 W}{\partial S_N^2} + S_N \frac{\partial W}{\partial S_N}(r_f - \rho\sigma_\$\sigma_N) - r_\$ V = 0.$$

This is the simple one-factor Black–Scholes equation. If we compare this equation with the Black–Scholes equation with a constant dividend yield we see that pricing the quanto is equivalent to using a dividend yield of

$$r_\$ - r_f + \rho\sigma_\$\sigma_N.$$

The only noticeable effect of the cross-currency feature on the option value is an adjustment to a dividend yield. This yield depends on the volatility of the exchange rate and the correlation between the underlying and the exchange rate. This is a common result for the simpler quantos.

11.8 **TWO EXAMPLES**

In Figure 11.3 is shown the termsheet for 'La Tricolore' Capital-guaranteed Note. This contract pays off the *second* best performing of three currencies against the French franc, but only if the second-best performing has appreciated against the franc, otherwise it pays off at par. This contract does not have any unusual features, and has a value that can be written as a three-dimensional integral, of the form (11.2). But what would be the payoff function? You wouldn't

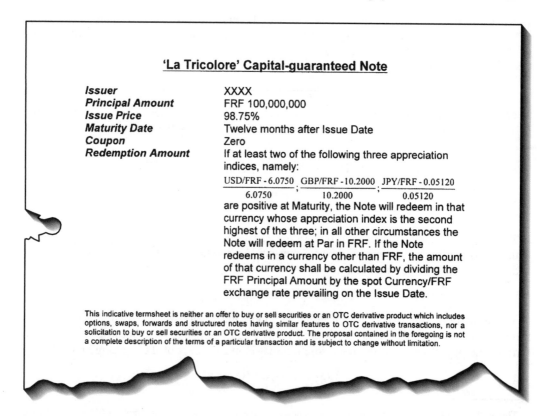

'La Tricolore' Capital-guaranteed Note

Issuer	XXXX
Principal Amount	FRF 100,000,000
Issue Price	98.75%
Maturity Date	Twelve months after Issue Date
Coupon	Zero
Redemption Amount	If at least two of the following three appreciation indices, namely:

$$\frac{\text{USD/FRF} - 6.0750}{6.0750} \; ; \; \frac{\text{GBP/FRF} - 10.2000}{10.2000} \; ; \; \frac{\text{JPY/FRF} - 0.05120}{0.05120}$$

are positive at Maturity, the Note will redeem in that currency whose appreciation index is the second highest of the three; in all other circumstances the Note will redeem at Par in FRF. If the Note redeems in a currency other than FRF, the amount of that currency shall be calculated by dividing the FRF Principal Amount by the spot Currency/FRF exchange rate prevailing on the Issue Date.

Figure 11.3   Termsheet for 'La Tricolore' capital-guaranteed note.

use a partial differential equation to price this contract. Instead you would estimate the multiple integral directly by the methods of Chapter 66.

The next example, whose termsheet is shown in Figure 11.4, is of basket equity swap. This rather complex, high-dimensional contract is for a swap of interest payment based on three-month LIBOR and the level of an index. The index is made up of the weighted average

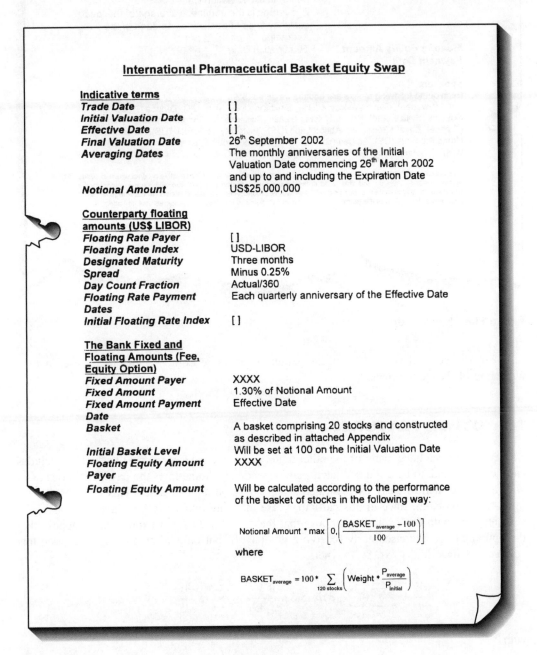

Figure 11.4 Termsheet for a basket equity swap.

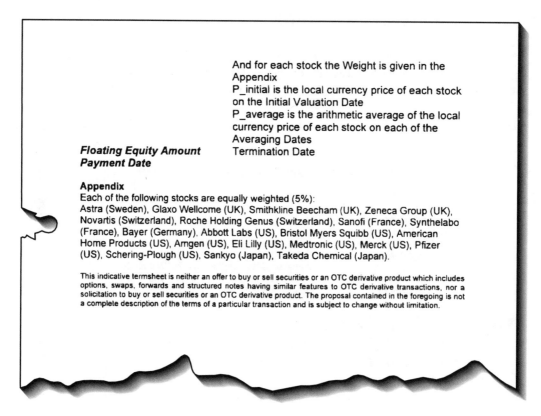

And for each stock the Weight is given in the Appendix
P_initial is the local currency price of each stock on the Initial Valuation Date
P_average is the arithmetic average of the local currency price of each stock on each of the Averaging Dates

Floating Equity Amount   Termination Date
Payment Date

Appendix
Each of the following stocks are equally weighted (5%):
Astra (Sweden), Glaxo Wellcome (UK), Smithkline Beecham (UK), Zeneca Group (UK), Novartis (Switzerland), Roche Holding Genus (Switzerland), Sanofi (France), Synthelabo (France), Bayer (Germany), Abbott Labs (US), Bristol Myers Squibb (US), American Home Products (US), Amgen (US), Eli Lilly (US), Medtronic (US), Merck (US), Pfizer (US), Schering-Plough (US), Sankyo (Japan), Takeda Chemical (Japan).

This indicative termsheet is neither an offer to buy or sell securities or an OTC derivative product which includes options, swaps, forwards and structured notes having similar features to OTC derivative transactions, nor a solicitation to buy or sell securities or an OTC derivative product. The proposal contained in the foregoing is not a complete description of the terms of a particular transaction and is subject to change without limitation.

Figure 11.4 *(Continued).*

of 20 pharmaceutical stocks. To make matters even more complex, the index uses a time averaging of the stock prices.

11.9 OTHER FEATURES

Basket options can have many of the other features that we have seen or will see. This includes early exercise and, to be discussed later, path dependency. Sometimes the payoff is in one asset with a feature such as early exercise being dependent on another asset. This would still be a multi-factor problem, since in this particular case there are two sources of randomness.

Continuing with this example, suppose that the payoff is $P(S_1)$ at expiry. Also suppose that the option can be exercised early receiving this payoff, but only when $S_2 > E_2$. To price this contract we must find $V(S_1, S_2, t)$ where

$$\frac{\partial V}{\partial t} + \frac{1}{2}\sum_{i=1}^{2}\sum_{j=1}^{2}\sigma_i\sigma_j\rho_{ij}S_iS_j\frac{\partial^2 V}{\partial S_i\partial S_j} + \sum_{i=1}^{2}rS_i\frac{\partial V}{\partial S_i} - rV = 0,$$

with

$$V(S_1, S_2, T) = P(S_1).$$

subject to

$$V(S_1, S_2, t) \geq P(S_1) \quad \text{for} \quad S_2 > E_2,$$

and continuity of V and its first derivatives.

If we can price and hedge an option on a single asset with a set of characteristics, then we can price and hedge a multi-asset version also, theoretically. The practice of pricing and hedging may be much harder as I mention below.

11.10 **REALITIES OF PRICING BASKET OPTIONS**

The factors that determine the ease or difficulty of pricing and hedging multi-asset options are

- existence of a closed-form solution
- number of underlying assets, the dimensionality
- path dependency
- early exercise

We have seen all of these except path dependency, which is the subject of Part Two.

The solution technique that we use will generally be one of

- finite-difference solution of a partial differential equation
- numerical integration
- Monte Carlo simulation

These methods are the subjects of Part Six.

11.10.1 Easy Problems

If we have a closed-form solution then our work is done; we can easily find values and hedge ratios. This is provided that the solution is in terms of sufficiently simple functions for which there are spreadsheet functions or other libraries. If the contract is European with no path-dependency then the solution may be of the form (11.2). If this is the case, then we often have to do the integration numerically. This is not difficult. Several methods are described in Chapter 66, including Monte Carlo integration and the use of low-discrepancy sequences.

11.10.2 Medium Problems

If we have low dimensionality, less than three or four, say, the finite-difference methods are the obvious choice. They cope well with early exercise and many path-dependent features can be incorporated, though usually at the cost of an extra dimension.

For higher dimensions, Monte Carlo simulations are good. They cope with all path-dependent features. Unfortunately, they are not very efficient for American-style early exercise.

11.10.3 Hard Problems

The hardest problems to solve are those with both high dimensionality, for which we would like to use Monte Carlo simulation, and early exercise, for which we would like to use finite-difference methods. There is currently no numerical method that copes well with such a problem.

11.11 REALITIES OF HEDGING BASKET OPTIONS

Even if we can find option values and the greeks, they are often very sensitive to the level of the correlation. But as I have said, the correlation is a very difficult quantity to measure. So the hedge ratios are very likely to be inaccurate. If we are delta hedging then we need accurate estimates of the deltas. This makes basket options very difficult to delta hedge successfully.

When we have a contract that is difficult to delta hedge we can try to reduce sensitivity to parameters, and the model, by hedging with other derivatives. This was the basis of vega hedging, mentioned in Chapter 7. We could try to use the same idea to reduce sensitivity to the correlation. Unfortunately, that is also difficult because there just aren't enough contracts traded that depend on the right correlations.

11.12 CORRELATION VERSUS COINTEGRATION

The correlations between financial quantities are notoriously unstable. One could easily argue that a theory should not be built up using parameters that are so unpredictable. I would tend to agree with this point of view. One could propose a stochastic correlation model, but that approach has its own problems.

An alternative statistical measure to correlation is **cointegration**. Very loosely speaking, two time series are cointegrated if a linear combination has constant mean and standard deviation; in other words, if the two series never stray too far from one another. This is probably a more robust measure of the linkage between two financial quantities but as yet there is little derivative theory based on the concept.

11.13 SUMMARY

The new ideas in this chapter were the multi-factor, correlated random walks for assets, and Itô's lemma in higher dimensions. These are both simple concepts, and we will use them often, especially in interest-rate-related topics.

FURTHER READING

- See Hamilton (1994) for further details of the measurement of correlation and cointegration.
- The first solution of the exchange option problem was by Margrabe (1978).
- For analytical results, formulae or numerical algorithm for the pricing of some other multi-factor options see Stulz (1982), Johnson (1987), Boyle, Evnine & Gibbs (1989), Boyle & Tse (1990), Rubinstein (1991) and Rich & Chance (1993).
- For details of cointegration, what it means and how it works, see the papers by Alexander & Johnson (1992, 1994).

CHAPTER 12
the binomial model

In this Chapter...

- a simple model for an asset price random walk
- delta hedging
- no arbitrage
- the basics of the binomial method for valuing options
- risk neutrality

12.1 INTRODUCTION

We have seen in Chapter 3 a model for equities and other assets that is based on the mathematical theory of stochastic calculus. There is another, equally popular, approach that leads to the same partial differential equation, the Black–Scholes equation, in a way that some people find more 'accessible', which can be made equally 'rigorous'. This approach, via the **binomial model** for equities, is the subject of this chapter.

Undoubtedly, one of the reasons for the popularity of this model is that it can be implemented without any higher mathematics (such as differential calculus) and there is actually no need to derive a partial differential equation before this implementation. This is a positive point, however the downside is that it is harder to attain greater levels of sophistication or numerical analysis in this setting.

Before I describe this model I want to stress that the binomial model may be thought of as being either a genuine *model* for the behavior of equities or, alternatively, as a numerical method for the solution of the Black–Scholes equation.[1] Most importantly, we see the ideas of delta hedging, risk elimination and risk-neutral valuation occurring in another setting.

The binomial model is very important because it shows how to get away from a reliance on closed-form solutions. Indeed, it is extremely important to have a way of valuing options that only relies on a simple model and fast, accurate numerical methods. Often in real life a contract may contain features that make analytic solution very hard or impossible. Some of these features may be just a minor modification to some other, easily-priced, contract but even minor changes to a contract can have important effects on the value and especially on the method of solution. The classic example is of the American put. Early exercise may seem to be a small change to a contract but the difference between the values of a European and an American put can

---

[1] In this case, it is very similar to an explicit finite-difference method, of which more later.

be large and certainly there is no simple closed-form solution for the American option and its value must be found numerically.

12.2 EQUITIES CAN GO DOWN AS WELL AS UP

In the binomial model we assume that the asset, which initially has the value S, can, during a timestep δt, either rise to a value uS or fall to a value vS, with $0 < v < 1 < u$. The probability of a rise is p and so the probability of a fall is $1 - p$. This behavior is shown in Figure 12.1.

The three constants u, v and p are chosen to give the binomial walk the same drift and standard deviation as that given by the stochastic differential Equation (3.6). Having only these two equations for the three parameters gives us one degree of freedom in this choice. This degree of freedom is often used to give the random walk the further property that after an up and a down movement (or a down followed by an up) the asset returns to its starting value, S.[2] This gives us the requirement that

$$v(uS) = u(vS) = S$$

i.e.

$$uv = 1. \tag{12.1}$$

Our starting point, the lognormal random walk,

$$dS = \mu S \, dt + \sigma S \, dX$$

has the solution, found in Section 4.13.2,

$$S(t) = S(0)e^{(\mu - (1/2)\sigma^2)t + \sigma\phi\sqrt{t}}$$

where ϕ is a standardized Normal random variable.

For the binomial random walk to have the correct drift over a time period of δt we need

$$puS + (1 - p)vS = SE\left[e^{(\mu - (1/2)\sigma^2)\delta t + \sigma\phi\sqrt{\delta t}}\right] = Se^{\mu\,\delta t},$$

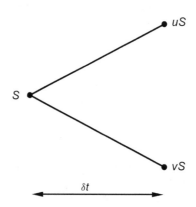

Figure 12.1 A schematic diagram of one timestep in the life of a binomial walk.

---

[2] Other choices are possible. For example, sometimes the probability of an up move is set equal to the probability of a down move i.e. $p = 1/2$.

i.e.

$$pu + (1-p)v = e^{\mu\,\delta t}.$$

Rearranging this equation we get

$$p = \frac{e^{\mu\,\delta t} - v}{u - v}. \tag{12.2}$$

Then for the binomial random walk to have the correct variance we need (details omitted)

$$pu^2 + (1-p)v^2 = e^{(2\mu+\sigma^2)\,\delta t}. \tag{12.3}$$

Equations (12.1), (12.2) and (12.3) can be solved to give

$$u = \tfrac{1}{2}(e^{-\mu\,\delta t} + e^{(\mu+\sigma^2)\,\delta t}) + \tfrac{1}{2}\sqrt{(e^{-\mu\,\delta t} + e^{(\mu+\sigma^2)\,\delta t})^2 - 4}. \tag{12.4}$$

Approximations that are good enough for most purposes are

$$u \approx 1 + \sigma\,\delta t^{1/2} + \tfrac{1}{2}\sigma^2\,\delta t,$$

$$v \approx 1 - \sigma\,\delta t^{1/2} + \tfrac{1}{2}\sigma^2\,\delta t$$

and

$$p \approx \tfrac{1}{2} + \frac{\left(\mu - \tfrac{1}{2}\sigma^2\right)\,\delta t^{1/2}}{2\sigma}.$$

12.3 **THE BINOMIAL TREE**

The binomial model, just introduced, allows the stock to move up or down a prescribed amount over the next timestep. If the stock starts out with value S then it will take either the value uS or vS after the next timestep. We can extend the random walk to the next timestep. After two timesteps the asset will be at either u^2S, if there were two up moves, uvS, if an up was followed by a down or vice versa, or v^2S, if there were two consecutive down moves. After three timesteps the asset can be at u^3S, u^2vS, etc. One can imagine extending this random walk out all the way until expiry. The resulting structure looks like Figure 12.2 where the nodes represent the values taken by the asset. This structure is called the **binomial tree**. Observe how

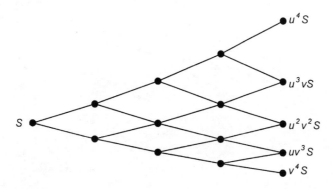

Figure 12.2 The binomial tree.

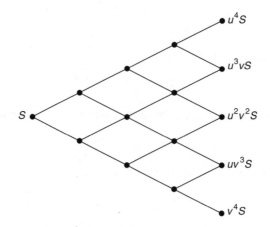

Figure 12.3   The binomial tree: a schematic version.

the tree bends due to the geometric nature of the asset growth. Often this tree is drawn as in Figure 12.3 because it is easier to draw, but this doesn't quite capture the correct structure.

The top and bottom branches of the tree at expiry can only be reached by one path each, either all up or all down moves, whereas there will be several paths possible for each of the intermediate values at expiry. Therefore the intermediate values are more likely to be reached than the end values if one were doing a simulation. The binomial tree therefore contains within it an approximation to the probability density function for the lognormal random walk.

12.4   **THE ASSET PRICE DISTRIBUTION**

The probability of reaching a particular node in the binomial tree depends on the number of distinct paths to that node and the probabilities of the up and down moves. Since up and down moves are approximately equally likely and since there are more paths to the interior prices than to the two extremes we will find that the probability distribution of future prices is roughly bell shaped. In Figure 12.4 is shown the number of paths to each node after four timesteps

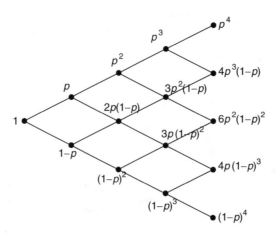

Figure 12.4   Counting paths.

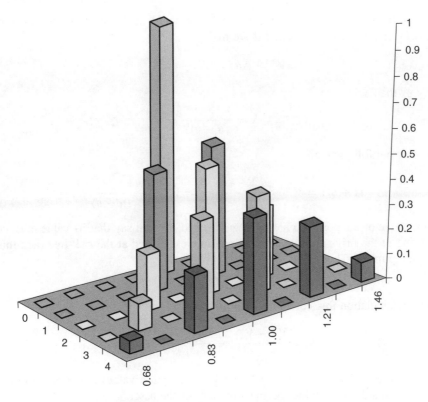

Figure 12.5 The probability distribution of future asset prices.

and the probability of getting to each. In Figure 12.5 this is interpreted as probability density functions at a sequence of times.

12.5 AN EQUATION FOR THE VALUE OF AN OPTION

Suppose, for the moment, that we know the value of the option at the time $t + \delta t$. For example, this time may be the expiry of the option, say. Now construct a portfolio at time t consisting of one option and a short position in a quantity Δ of the underlying. At time t this portfolio has value

$$\Pi = V - \Delta S,$$

where the value V is to be determined.

At time $t + \delta t$ the portfolio takes one of two values, depending on whether the asset rises or falls. These two values are

$$V^+ - \Delta uS \quad \text{and} \quad V^- - \Delta vS.$$

Since we assume that we know V^+, V^-, u, v, S and Δ, the values of both of these expressions are known and, in particular, depend on Δ.

Having the freedom to choose Δ, we can make the value of this portfolio the same whether the asset rises or falls. This is ensured if we make

$$V^+ - \Delta u S = V^- - \Delta v S.$$

This gives us the choice

$$\Delta = \frac{V^+ - V^-}{(u - v)S}, \tag{12.5}$$

when the new portfolio value is

$$\Pi + \delta \Pi = V^+ - \frac{u(V^+ - V^-)}{(u - v)} = V^- - \frac{v(V^+ - V^-)}{(u - v)}.$$

Since the value of the portfolio has been guaranteed, we can say that its value must coincide with the value of the original portfolio plus any interest earned at the risk-free rate; this is the no-arbitrage argument. Thus

$$\delta \Pi = r \Pi \, \delta t.$$

After some manipulation this equation becomes

$$V = \frac{V^+ - V^-}{u - v} + \frac{u V^- - v V^+}{(1 + r \, \delta t)(u - v)}. \tag{12.6}$$

This, then, is an equation for V given V^+ and V^-, the option values at the next timestep, and the parameters u and v describing the random walk of the asset.

To $O(\delta t)$ we can write (12.6) as

$$e^{r \, \delta t} V = p' V^+ + (1 - p') V^-, \tag{12.7}$$

where

$$p' = \frac{e^{r \, \delta t} - v}{u - v}. \tag{12.8}$$

Expressions (12.2) and (12.8) differ in that where one has the interest rate r the other has the drift μ, but are otherwise the same. Interpreting p' as a probability, this is just risk neutrality again. And (12.7) is the statement that the option value at time t is the present value of the risk-neutral expected value at any later time.

Supposing that we know V^+ and V^- we can use (12.7) to find V. But do we know V^+ and V^-?

12.6   **VALUING BACK DOWN THE TREE**

We certainly know V^+ and V^- at expiry, time T, because we know the option value as a function of the asset then; this is the payoff function. If we know the value of the option at expiry we can use Equation (12.7) to find the option value at the time $T - \delta t$ for all values of S on the tree. But knowing these values means that we can find the option values one step further back in time. Thus we work our way back down the tree until we get to the root. This root is the current time and asset value, and thus we find the option value today.

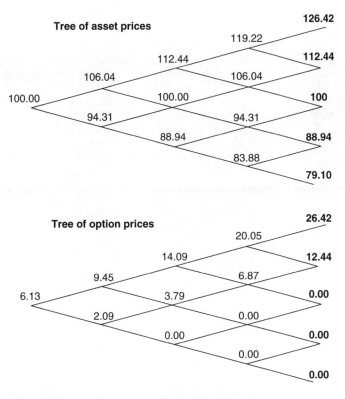

Figure 12.6 The binomial tree and corresponding option prices.

This method is illustrated in Figure 12.6. Here we are valuing a European call option with strike price 100 and maturity in four months' time. Today's asset price is 100, and the volatility is 20%. To make things as simple as possible the interest rate is zero.

I use a timestep of one month so that there are four steps until expiry. Using these numbers we have $\delta t = 1/12 = 0.08333$, $u = 1.0604$, $v = 0.9431$ and $p' = 0.5567$. As an example, after one timestep the asset takes either the value $100 \times 1.0604 = 106.04$ or $100 \times 0.9431 = 94.31$. Working back from expiry, the option value at the timestep before expiry when $S = 119.22$ is given by

$$e^{-0.1 \times 0.833}(0.5567 \times 26.42 + (1 - 0.5567) \times 12.44) = 20.05.$$

Working right back down the tree to the present time, the option value when the asset is 100 is 6.14. Compare this with the theoretical, continuous-time solution (given by the Black–Scholes call value) of 6.35. The difference is entirely due to the size and number of the timesteps. The larger the number of timesteps, the greater the accuracy. I come back to the issue of accuracy in later chapters.

In practice, the binomial method is programmed rather than done on a spreadsheet. Here is a function that takes inputs for the underlying and the option, using an externally-defined payoff function. Key points to note about this program concern the building up of the arrays for the asset S() and the option V(). First of all, the asset array is built up only in order to find the final values of the asset at each node at the final timestep, expiry. The asset values on other nodes are never used. Second, the argument j refers to how far up the asset is from the lowest node *at that timestep*.

CODE IMPLEMENTING THE BINOMIAL METHOD FOR A EUROPEAN OPTION

```
Function Price(Asset As Double, Volatility As Double, _
               IntRate As Double, Strike As _
               Double, Expiry As Double, _
               NoSteps As Integer)
ReDim S(0 To NoSteps)
ReDim V(0 To NoSteps)
timestep = Expiry / NoSteps
DiscountFactor = Exp(-IntRate * timestep)
temp1 = Exp((IntRate + Volatility * Volatility) _
                                         * timestep)
temp2 = 0.5 * (DiscountFactor + temp1)
u = temp2 + Sqr(temp2 * temp2 - 1)
d = 1 / u
p = (Exp(IntRate * timestep) - d) / (u - d)

S(0) = Asset
For n = 1 To NoSteps
    For j = n To 1 Step -1
        S(j) = u * S(j - 1)
    Next j
        S(0) = d * S(0)
Next n

For j = 0 To NoSteps
    V(j) = Payoff(S(j), Strike)
Next j

For n = NoSteps To 1 Step -1
    For j = 0 To n - 1
        V(j) = (p * V(j + 1) + (1 - p) * V(j)) _
                * DiscountFactor
    Next j
Next n
Price = V(0)
End Function
```

Here is the externally-defined payoff function `Payoff(S, Strike)` for a call.

```
Function Payoff(S, K)
Payoff = 0
If S > K Then Payoff = S - K
End Function
```

Because I never use the asset nodes other than at expiry I could have used only the one array in the above, with the same array being used for both *S* and *V*. I have kept them separate to make the program more transparent. Also, I could have saved the values of *V* at all of the nodes, whereas in the above I have only used the node at the present time. Saving all the values will be important if you want to see how the option value changes with the asset price and time, if you want to calculate greeks for example.

In Figure 12.7 I show a plot of the calculated option price against the number of timesteps using this algorithm. The inset figure is a close-up. Observe the oscillation. In this example, an

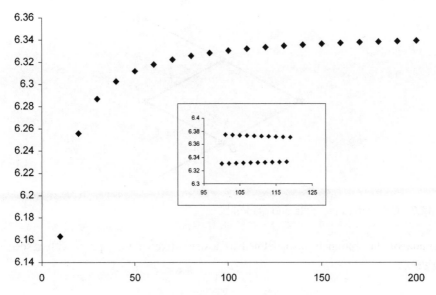

Figure 12.7 Option price as a function of number of timesteps.

odd number of timesteps gives an answer that is too high and an even number an answer that is too low.

12.7 **THE GREEKS**

The greeks are defined as derivatives of the option value with respect to various variables and parameters. It is important to distinguish whether the differentiation is with respect to a variable or a parameter (it could, of course, be with respect to both). If the differentiation is only with respect to the asset price and/or time then there is sufficient information in our binomial tree to estimate the derivative. It may not be an accurate estimate, but it will be *an* estimate. The option's delta, gamma and theta can all be estimated from the tree.

On the other hand, if you want to examine the sensitivity of the option with respect to one of the parameters, then you must perform another binomial calculation. This applies to the option's vega and rho for example.

Let me take these two cases in turn.

From the binomial model the option's delta is defined by

$$\frac{V^+ - V^-}{(u - v)S}.$$

We can calculate this quantity directly from the tree. Referring to Figure 12.8, the delta uses the option value at the two points marked 'D', together with today's asset price and the parameters u and v. This is a simple calculation.

In the limit as the timestep approaches zero, the delta becomes

$$\frac{\partial V}{\partial S}.$$

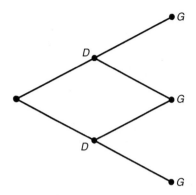

Figure 12.8   Calculating the delta and gamma.

The gamma of the option is also defined as a derivative of the option with respect to the underlying:

$$\frac{\partial^2 V}{\partial S^2}.$$

To estimate this quantity using our tree is not so clear. It will be much easier when we use a finite-difference grid. However, gamma is a measure of how much we must rehedge at the next timestep. But we can calculate the delta at points marked with a D in Figure 12.8 from the option value one timestep further in the future. The gamma is then just the change in the delta from one of these to the other divided by the distance between them. This calculation uses the points marked 'G' in Figure 12.8.

The theta of the option is the sensitivity of the option price to time, assuming that the asset price does not change. Again, this is easier to calculate from a finite-difference grid. An obvious choice for the discrete-time definition of theta is to interpolate between V^+ and V^- to find a theoretical option value *had the asset not changed* and use this to estimate

$$\frac{\partial V}{\partial t}.$$

This results in

$$\frac{\frac{1}{2}(V^+ + V^-) - V}{\delta t}.$$

As the timestep gets smaller and smaller these greeks approach the Black–Scholes continuous-time values.

Estimating the other type of greeks, the ones involving differentiation with respect to parameters, is slightly harder. They are harder to calculate in the sense that you must perform a second binomial calculation. I will illustrate this with the calculation of the option's vega.

The vega is the sensitivity of the option value to the volatility

$$\frac{\partial V}{\partial \sigma}.$$

Suppose we want to find the option value and vega when the volatility is 20%. The most efficient way to do this is to calculate the option price twice, using a binomial tree, with two different values of σ. Calculate the option value using a volatility of $\sigma \pm \varepsilon$, for a small number ε, and call the values you find $V_\pm$. The option value is approximated by the average value

$$V = \tfrac{1}{2}(V_+ + V_-)$$

and the vega is approximated by

$$\frac{V_+ - V_-}{2\varepsilon}.$$

The idea can be applied to other greeks.

12.8 EARLY EXERCISE

American-style exercise is easy to implement in a binomial setting. The algorithm is identical to that for European exercise with one exception. We use the same binomial tree, with the same u, v and p, but there is a slight difference in the formula for V. We must ensure that there are no arbitrage opportunities at any of the nodes.

For reasons which will become apparent, I'm going to change my notation now, making it more complex but more informative. Introduce the notation S_j^n to mean the asset price at the nth timestep, at the node j from the bottom, $0 \le j \le n$. This notation is consistent with the code above. In our lognormal world we have

$$S_j^n = S u^j v^{n-j},$$

where S is the current asset price. Also introduce V_j^n as the option value at the same node. Our ultimate goal is to find V_0^0 knowing the payoff, i.e. knowing V_j^N for all $0 \le j \le M$ where M is the number of timesteps.

Returning to the American option problem, arbitrage occurs if the option value goes below the payoff at any time. If our theoretical value falls below the payoff then it is time to exercise. If we do then exercise the option, its value and the payoff must be the same. If we find that

$$\frac{V_{j+1}^{n+1} - V_j^{n+1}}{u - v} + e^{-r\,\delta t}\frac{uV_j^{n+1} - vV_{j+1}^{n+1}}{u - v} \ge \text{Payoff}(S_j^n)$$

then we use this as our new value. But if

$$\frac{V_{j+1}^{n+1} - V_j^{n+1}}{u - v} + e^{-r\,\delta t}\frac{uV_j^{n+1} - vV_{j+1}^{n+1}}{u - v} < \text{Payoff}(S_j^n)$$

we should exercise, giving us a better value of

$$V_j^n = \text{Payoff}(S_j^n).$$

We can put these two together to get

$$V_j^n = \max\left(\frac{V_{j+1}^{n+1} - V_j^{n+1}}{u - v} + e^{-r\,\delta t}\frac{uV_j^{n+1} - vV_{j+1}^{n+1}}{u - v}, \; \text{Payoff}(S_j^n)\right)$$

instead of (12.6). This ensures that there are no arbitrage opportunities. This modification is easy to code, but note that the payoff is a function of the asset price at the node in question.

This is new, not seen in the European problem for which we did not have to keep track of the asset values on each of the nodes.

Below is a function for calculating the value of an American-style option. Note the differences between this program and the one for European-style exercise. The code is the same except that we keep track of more information and the line that updates the option value incorporates the no-arbitrage condition.

CODE FOR A US OPTION, IT'S NOT THAT MUCH DIFFERENT FROM THE EURO PROGRAM

```
Function USPrice(Asset As Double, Volatility As _
                Double, IntRate As Double, _
                Strike As Double, Expiry As _
                Double, NoSteps As Integer)
ReDim S(0 To NoSteps, 0 To NoSteps)
ReDim V(0 To NoSteps, 0 To NoSteps)
timestep = Expiry / NoSteps
DiscountFactor = Exp(-IntRate * timestep)
temp1 = Exp((IntRate + Volatility * Volatility) _
                                    * timestep)
temp2 = 0.5 * (DiscountFactor + temp1)
u = temp2 + Sqr(temp2 * temp2 - 1)
d = 1 / u
p = (Exp(IntRate * timestep) - d) / (u - d)

S(0, 0) = Asset
For n = 1 To NoSteps
    For j = n To 1 Step -1
        S(j, n) = u * S(j - 1, n - 1)
    Next j
        S(0, n) = d * S(0, n - 1)
Next n

For j = 0 To NoSteps
    V(j, NoSteps) = Payoff(S(j, NoSteps), Strike)
Next j

For n = NoSteps To 1 Step -1
    For j = 0 To NoSteps - 1
        V(j, n - 1) = max((p * V(j + 1, n) _
        + (1 - p) * V(j, n)) _
        * DiscountFactor, Payoff(S(j, n - 1), Strike))
    Next j
Next n
USPrice = V(0, 0)
End Function
```

12.9 **THE CONTINUOUS-TIME LIMIT**

Equation (12.6) and the Black–Scholes Equation (5.6) are more closely related than they may at first seem. Recalling that the Black–Scholes equation is in continuous time, we examine (12.6) as $\delta t \to 0$.

First of all, from (12.4) we have that

$$u \sim 1 + \sigma\sqrt{\delta t} + \tfrac{1}{2}\sigma^2\,\delta t + \cdots$$

and

$$v \sim 1 - \sigma\sqrt{\delta t} + \tfrac{1}{2}\sigma^2\,\delta t + \cdots.$$

Next we write

$$V = V(S, t), \quad V^+ = V(uS, t + \delta t) \quad \text{and} \quad V^- = V(vS, t + \delta t).$$

Expanding these expressions in Taylor series for small δt and substituting into (12.5) we find that

$$\Delta \sim \frac{\partial V}{\partial S} \quad \text{as} \quad \delta t \to 0.$$

Thus the binomial delta becomes, in the limit, the Black–Scholes delta.

Similarly, we can substitute the expressions for V, V^+ and V^- into (12.6) to find

$$\frac{\partial V}{\partial t} + \tfrac{1}{2}\sigma^2 S^2 \frac{\partial^2 V}{\partial S^2} + r\frac{\partial V}{\partial S} - rV = 0.$$

This is the Black–Scholes equation. Again, the drift rate μ has disappeared from the equation.

12.10 NO ARBITRAGE IN THE BINOMIAL, BLACK–SCHOLES AND 'OTHER' WORLDS

With the binomial discrete-time model, as with the Black–Scholes continuous-time model, we have been able to eliminate uncertainty in the value of a portfolio by a judicious choice of a hedge. In both cases we find that it does not matter how the underlying asset moves, the resulting value of the portfolio is the same. This is especially clear in the above binomial model. This hedging is only possible in these two simple, popular models. For consider a trivial generalization: the trinomial random walk.

In Figure 12.9 we see a representation of a trinomial random walk. After a timestep δt the asset could have risen to uS, fallen to vS or not moved from S.

What happens if we try to hedge an option under this scenario? As before, we can 'hedge' with $-\Delta$ of the underlying but this time we would like to choose Δ so that the value of the portfolio (of one option and $-\Delta$ of the asset) is the same at time $t + \delta t$ no matter to which value the asset moves. In other words, we want the portfolio to have the same value for all *three* possible outcomes. Unfortunately, we cannot choose a value for Δ that ensures this will be the case: this amounts to solving two equations (first portfolio value = second portfolio value = third portfolio value) with just one unknown (the delta). Hedging is not possible in the

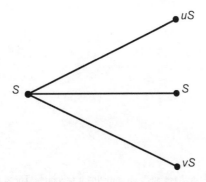

Figure 12.9 The trinomial tree. Perfect risk-free hedging is not possible under this scenario.

trinomial world. Indeed, perfect hedging, and thus the application of the 'no-arbitrage principle' is only possible in the two special cases: the Black–Scholes continuous time/continuous asset world, and the binomial world. And in the far more complex 'real' world, delta hedging is *not* possible.[3]

12.11 **SUMMARY**

In this chapter I described the basics of the binomial model, deriving pricing equations and algorithms for both European- and American-style exercise. The method can be extended in many ways, to incorporate dividends, to allow Bermudan exercise, to value path-dependent contracts and to price contracts depending on other stochastic variables such as interest rates. I have not gone into the method in any detail for the simple reason that the binomial method is just a simple version of an explicit finite-difference scheme. As such it will be discussed in depth in Part Seven. Finite-difference methods have an obvious advantage over the binomial method, because they are far more flexible.

FURTHER READING

- The original binomial concept is due to Cox, Ross & Rubinstein (1979).
- Almost every book on options describes the binomial method in more depth than I do. One of the best is Hull (1997) who also describes its use in the fixed-income world.

---

[3] Is it good for the popular models to have such an unrealistic property? These models are at least a good *starting* point.

CHAPTER 13
predicting the markets?

In this Chapter...

- some of the commonly used technical methods for predicting market direction
- some modern approaches to modeling markets and their microstructure

13.1 INTRODUCTION

People have been making predictions about the future since the dawn of time. And predicting the future of the financial markets has been especially popular. Despite the claims of many 'legendary' investors it is not clear whether there is any validity in any of the methods they use, or whether the claims are examples of survivor bias. The big losers tend to keep quiet.

WE'VE ALL DONE IT, THO' IT'S NOTHING TO BE PROUD OF

In this chapter we look at some of the traditional methods for determining trends, technical analysis, and also some of the more recent methods, often emanating from physics. I won't be describing some of the more dubious ideas, such as astrology, but then we Scorpios tend to be sceptical.

13.2 TECHNICAL ANALYSIS

Technical analysis is a way of predicting future price movements based only on observing the past history of prices. This price history may also include other quantities such as volume of trade. These methods contrast with **fundamental analysis** in which prediction is made based on an examination of the factors underlying the stock or other instrument. This may include general economic or political analysis, or analysis of factors specific to the stock, such as the effect of global warming on snowfall in the Alps, if one is concerned with a travel company. In practice, most traders will use a combination of both technical and fundamental analysis.

Technical analysis is also called **charting** because the graphical representation of prices etc. plays an important part. Technical analysis is thought to be particularly good for timing market

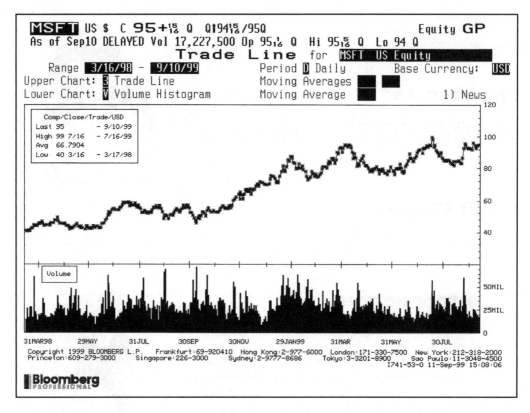

Figure 13.1 Price and volume. Source: Bloomberg L.P.

moves; fundamental analysis may get the direction right, but not necessarily when the move will happen.

13.2.1 Plotting

The simplest chart types just join together the prices from one day to the next, with time along the horizontal axis. These are the sort of plots we have seen throughout this book. Sometimes a logarithmic scale is used for the vertical price axis to represent return rather than absolute level. Later on we'll see some more complicated types of plotting. Sometimes you will see trading volume on the same graph; this is also used for prediction but I won't go into any details here, see Figure 13.1.

13.2.2 Support and Resistance

Resistance is a price level above which an asset seems to have difficulty rising. This may be a previously realized highest value, or it may be a psychologically important (round) number. **Support** is a level below which an asset price seems to be reluctant to fall. There may be sufficient demand at this low price to stop it falling any further. Examples of support and resistance are shown in Figure 13.2.

When a support or resistance level finally breaks it is said to do so quite dramatically.

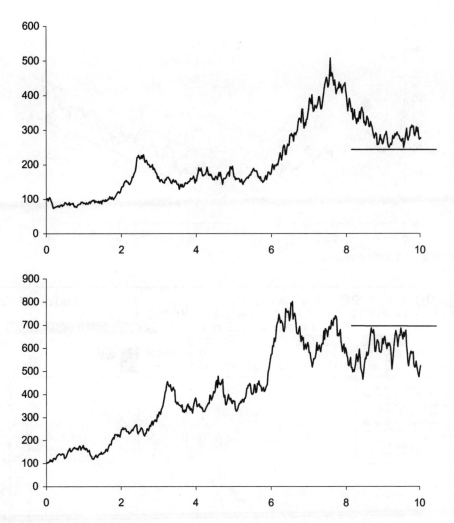

Figure 13.2 Support and resistance.

13.2.3 Trendlines

Similar to support and resistance are **trendlines**. These are formed by joining together successive peaks and/or troughs in the price history to form a rising or falling support or resistance level. An example is shown in Figure 13.3.

13.2.4 Moving Averages

Moving averages are calculated in many ways. Different time windows can be used, or even exponentially-weighted averages can be calculated. Moving averages are supposed to distil out the basic trend in a price by smoothing the random noise.

Sometimes two moving averages are calculated, say a ten-day and a 250-day average. The crossing of these two would signify a change in the underlying trend and a time to buy or sell.

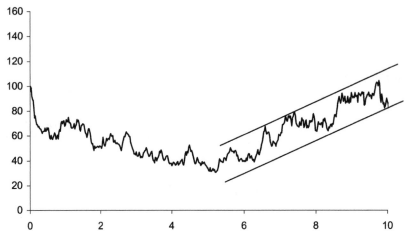

Figure 13.3 A trending stock.

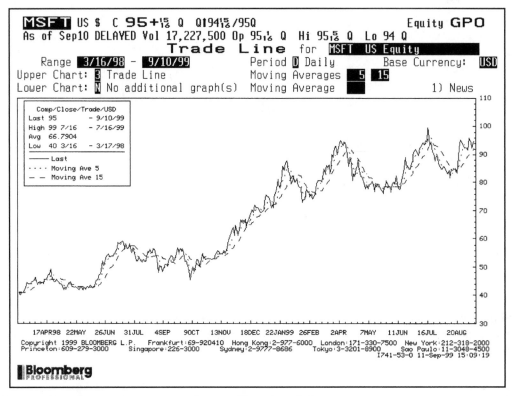

Figure 13.4 Two moving averages. Source: Bloomberg L.P.

Although I'm not the greatest fan of technical analysis, there is some evidence that there may be predictive power in moving averages.

Figure 13.4 shows a Bloomberg screen with Microsoft share price, five- and 15-day moving averages.

13.2.5 Relative Strength

The **relative strength index** is the percentage of up moves in the last N days. A number higher than 70% is said to be overbought and therefore likely to fall and below 30% is said to be oversold and should rise.

13.2.6 Oscillators

An **oscillator** is another indicator of over/underbought conditions. One way of calculating it is as follows.
   Define k by

$$100 \times \frac{\text{Current close} - \text{lowest over } n \text{ periods}}{\text{Highest over } n \text{ periods} - \text{lowest over } n \text{ periods}}.$$

Now take a moving average of the last three days, say. This average is plotted against time and any move outside the range 30–70% could be an indication of a move in the asset. See Figure 13.5.

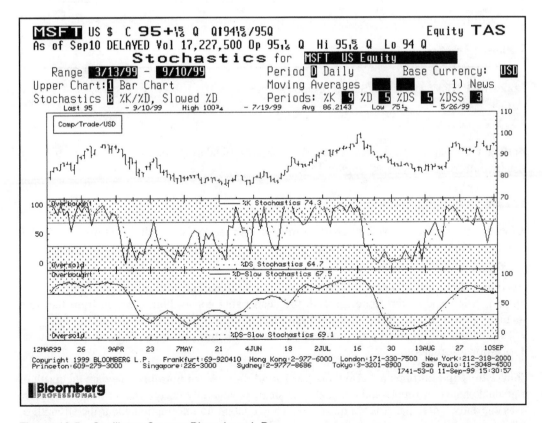

Figure 13.5 Oscillator. Source: Bloomberg L.P.

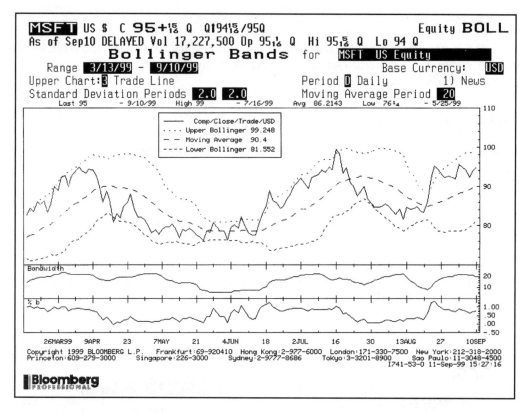

Figure 13.6 Bollinger Bands. Source: Bloomberg L.P.

13.2.7 Bollinger Bands

Bollinger Bands are plots of a specified number of standard deviations above and below a specified moving average; see Figure 13.6.

13.2.8 Miscellaneous Patterns

As well the 'quantitative' side of charting there is also the 'artistic' side. Practitioners say that certain patterns anticipate certain future moves. It's rather like your grandmother reading tea leaves.

Head and shoulders is a common pattern and is best described with reference to Figure 13.7. There is a left and a right shoulder with the head rising above. Following on from the right shoulder should be a dramatic decline in the asset price.

This pattern is supposed to be one of the most reliable predictors. It is also seen in an upside-down formation.

Saucer tops and bottoms are also known as **rounding tops** and **bottoms**. They are the result of a gradual change in supply and demand. The shape is generally fairly symmetrical as the price rises and falls. These patterns are quite rare. They contain no information about the strength of the new trend.

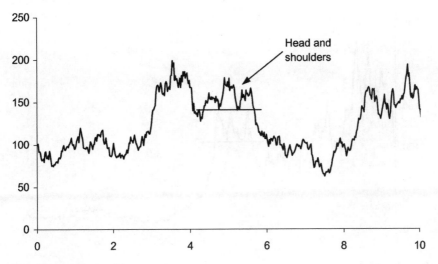

Figure 13.7 Head and shoulders.

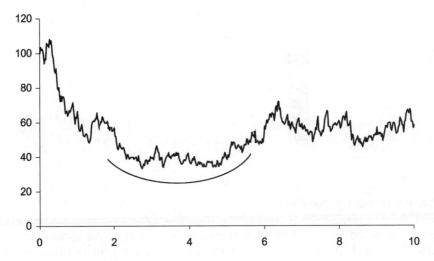

Figure 13.8 Saucer bottom.

Double and triple tops and bottoms are quite rare patterns, the triple being even rarer than the double. The double top looks like an 'M' and a double bottom like a 'W'. The triple top is similar but with three peaks, as shown in Figure 13.9. The key point about the peaks and troughs is that they should all be at approximately the same level.

13.2.9 Japanese Candlesticks

Japanese candlesticks contain more information than the simple plots described so far. They record the opening and closing prices as well as the day's high and low. A rectangle is drawn extending from the close to the open, and is colored white if close is above open and black if close is below open. The high-low range is marked by a continuous line.

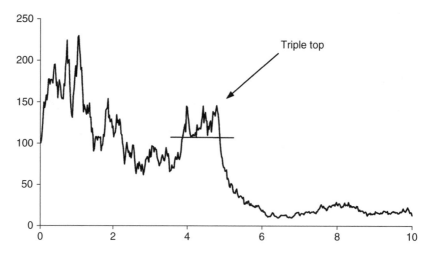

Figure 13.9 A triple top.

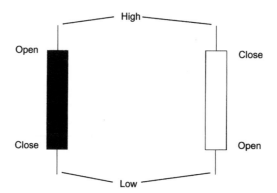

Figure 13.10 Japanese candlesticks.

Certain combinations of candlesticks appearing consecutively have special meanings and names like 'Hanging Man' and 'Upside Gap Two Crows'. See Figure 13.10 for the two types of candlestick and see Figure 13.11 for candlesticks in action. On this chart are shown 'HR' = Bearish Harami, 'D' = Doji (representing indecision), 'BH' = Bullish Harami, 'EL' = Bearish Engulfing Line, and 'H' = Hanging Man (representing reversal after a trend).

Figure 13.12 shows some of the possible candlestick shapes and their interpretation.

13.2.10 Point and Figure Charts

Point and figure charts are different from the charts described above in that they do not have any explicit timescale on the horizontal axis. Figure 13.13 is an example of a point and figure chart. Each box on the chart represents a prespecified asset price move. The boxes are a way of discretizing asset price moves, instead of discretizing in time. For each consecutive asset price rise of the box size draw an 'X' in the box, in a rising column, one above the other. When this uptrend finishes, and the asset falls, start putting 'O' in a descending column, to the right of the previous rising Xs.

Figure 13.11 A candlestick chart. Source: Bloomberg L.P.

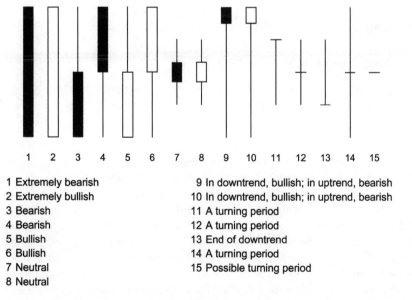

1 Extremely bearish	9 In downtrend, bullish; in uptrend, bearish
2 Extremely bullish	10 In downtrend, bullish; in uptrend, bearish
3 Bearish	11 A turning period
4 Bearish	12 A turning period
5 Bullish	13 End of downtrend
6 Bullish	14 A turning period
7 Neutral	15 Possible turning period
8 Neutral	

Figure 13.12 The meanings of the various candlesticks.

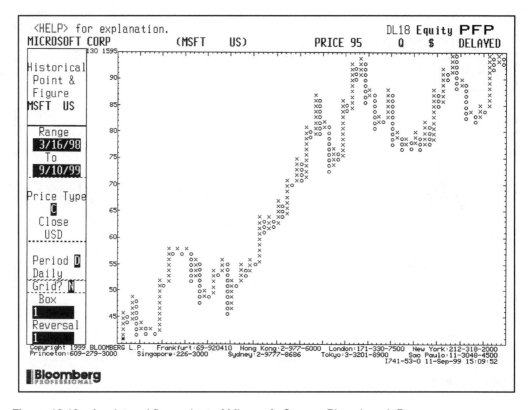

Figure 13.13   A point and figure chart of Microsoft. Source: Bloomberg L.P.

- A long column of Xs denotes demand exceeding supply.
- A long column of Os denotes supply exceeding demand.
- Short up and down columns denote a balance of supply and demand.

13.3   **WAVE THEORY**

As well as plotting and spotting trends in price movements there have been some theories for price prediction based on market cycles or waves. Below, I briefly mention a couple.

13.3.1   Elliott Waves and Fibonacci Numbers

Ralph N. Elliott observed repetitive patterns, waves or cycles in prices. Roughly speaking, there are supposed to be five points in a bullish wave and then three in a bearish one (see Figure 13.14). Within this **Elliott wave theory** there is also supposed to be some predictive ability in terms of the sizes of the peaks in each wave. For some reason, the ratio of peaks in a trend is supposed to be fairly constant; the ratio of second peak to first should be approximately 1.618 and of the third to the second 2.618. Unfortunately, the number 1.618 is approximately the **Golden ratio** of the ancient Greeks; $\frac{1}{2}\sqrt{5}$. It is also the ratio of successive numbers in the **Fibonacci series** given by $a_n = a_{n-1} + a_{n-2}$ for large n. I say 'unfortunately', because people extrapolate wildly from this. And if it's a coincidence then... Figure 13.15 shows the key levels coming from the Fibonacci series.

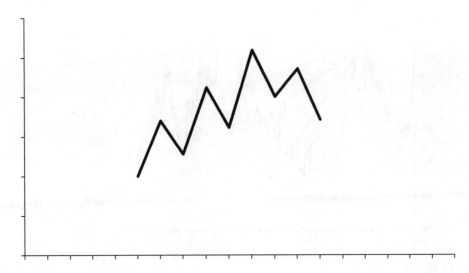

Figure 13.14   Elliot waves.

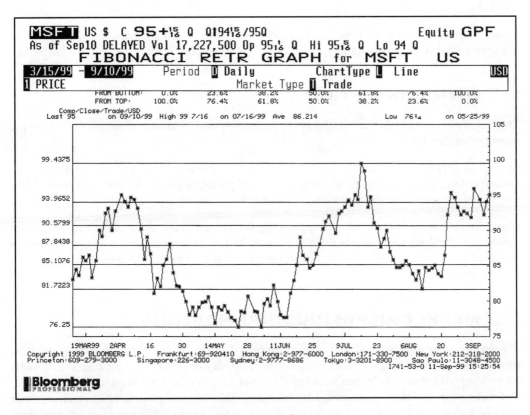

Figure 13.15   Fibonacci lines. Source: Bloomberg L.P.

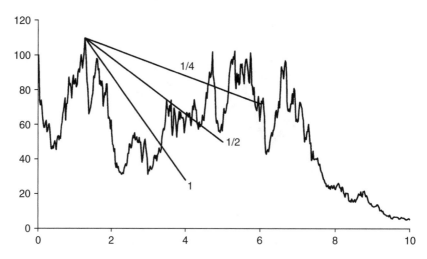

Figure 13.16 Gann charts.

13.3.2 Gann Charts

Figure 13.16 shows a Gann chart. The lines all have slopes which are fractions of the slope of the lowest line. Need I say more?

13.4 **OTHER ANALYTICS**

There's an almost endless number of ways that chartists analyze data. I'll mention just a couple more before moving on.

Volume is simply the number of contracts traded in a given period. A rising price and high volume mean a strong, upwardly trending market. But a rising price with low volume could be a sign that the market is about to turn.

Open interest is the number of still outstanding futures contracts, those which have not been closed out. Because there are equal numbers of buyers and sellers, open interest does not necessarily give any directional info, but an increase in open interest can mean that an existing trend is strong.

13.5 **MARKET MICROSTRUCTURE MODELING**

The financial markets are made up of many types of players. There are the 'producers' who manufacture or produce or sell various goods and who may be involved in the financial markets for hedging. There are the 'speculators' who try and spot trends in the market, to exploit them and make money. These speculators may be using technical analysis methods, such as those described above, or fundamental analysis, whereby they examine the records and future plans of firms to determine whether stocks are under- or overpriced. Almost all traders use technical analysis at some time. Then there are the market makers who buy and sell financial instruments, holding them for a very short time, often minutes, and profit on bid-offer spreads.

There have been many attempts to model the interaction of these agents, sometimes in a game theoretic way, to try and model the asset price movements that in this book we have taken for granted. For example, can the dynamics induced by the actions of a combination of these three types of agent result in Brownian motion and lognormal random walks?

Below are just a very few examples of work in this area.

13.5.1 Effect of Demand on Price

Buying and selling assets moves their prices. Market makers respond to demand by increasing price, and reduce prices when the market is selling. If one can model the relationship between demand and price then it should be possible to analyze the effect that various types of technical trading rule have on the evolution of prices, and eventually to model the dynamics of prices.

A common starting point is to assume that there are two types of trader and one market maker. One trader follows a technical trading rule such as watching a moving average and the other is a **noise trader** who randomly buys or sells.

Interesting results follow from such models. For example,

- trend followers can induce patterns in asset price time series,
- these artificially induced patterns can only be exploited for gain by someone following a suitably different trend,
- the more people following the same trend as you, the more money you will lose.

There are good reasons for there being genuine trends in the market: there is a slow diffusion of information from the knowledgeable to the less knowledgeable; the piece-by-piece secret acquisition of a company will gradually move a stock price upwards.

On the other hand, if there is no genuine reason for a trend, if it is simply a case of trend followers begetting a trend, then it may be beneficial to be a contrarian.

13.5.2 Combining Market Microstructure and Option Theory

Arbitrage does exist, and many people make money from its existence. Yet the action of arbitragers will, via a demand/price relationship, remove the arbitrage. But there will be a timescale associated with this removal. What is the optimal way to exploit the arbitrage opportunity while knowing that your actions will to some extent be self-defeating?

13.5.3 Imitation

Another approach to market microstructure modeling is based on the true observation that people copy each other. In these models there are a number of traders who act partly in response to private information about a stock, partly randomly as noise traders, and partly to imitate their nearest neighbors. These models can result in market bubbles or market crashes.

13.6 CRISIS PREDICTION

There has been some work on analyzing data over various timescales to determine the likelihood of a market crash. Some ideas from earthquake modeling have been used to derive a 'Richter'-like measure of market moves. Of course, an effective predictor of market crashes could either:

- increase the chance or size of a crash as everyone panics, or
- reduce the chance or size of the crash since everyone gets advance warning and can calmly and logically act accordingly.

13.7 **SUMMARY**

I started out in finance many years ago plotting all of the technical indicators. I was not very successful at it. I could only get directions right for those assets with obvious seasonality effects, such as some commodities.

There is only one technical indicator that I believe in. There is definitely a strong correlation between hemlines and the state of the economy. The shorter the skirts, the better the economy.

FURTHER READING

- The book on technical analysis written by the news agency Reuters (1999) is excellent, as is Meyers (1994).
- Farmer & Joshi (1999) discusses and models trend following, and the creation of trends. He also demonstrates properties of the relationship between demand and price that prevent arbitrage.
- Dewynne & Wilmott (1999) show how to optimally exploit an arbitrage opportunity while moving the market as little as possible.
- Bhamra (1999) has worked on imitation in financial markets.
- Olsen & Associates www.olsen.ch are currently working in the area of crisis modeling and prediction.
- Johnson (1999) models self-organized segregation of traders, and concludes that cautious traders perform poorly.
- The above is only a brief description of a very few examples from an expanding field. See O'Hara (1995) for a wide-ranging discussion of market microstructure models.
- Bernstein (1998) has a whole chapter on the Golden Ratio.
- Elton & Gruber (1995) describe the efficient market hypothesis and criticize technical analysis.
- Prast (2000a,b) discusses 'herding' in the financial markets.

CHAPTER 14
a trading game

In this Chapter...

- A simple game simulating trading in options

14.1 INTRODUCTION

A lot of people reading this book will never have traded options or even stocks. In this chapter I describe a very simple trading game so that a group of people can try out their skill without losing their shirts. The game is based on that invented by one of my ex-students, David Epstein.

14.2 AIMS

The aims of this game are to familiarize students with the basic market-traded derivative contracts and to promote an understanding of the concepts involved in trading, such as bid, offer, arbitrage and liquidity.

14.3 OBJECT OF THE GAME

To make more money than your opponents. After the final round of trading, each player sums up their profits and losses. The player who has made the most profit is the winner.

14.4 RULES OF THE GAME

1. One person (possibly a lecturer) is the game organizer and in charge of choosing the types of contracts available for trading, the number and length of the trading rounds, judging any disputes and jollying the game along during slack periods.
2. The trading game takes place over a number of rounds. At the end of each round, a six-sided die is thrown. After the last round, the 'share price' is deemed to be the sum of all of the die rolls.
3. Traded contracts may include some or all of forwards, calls and puts at the discretion of the organizer. The organizer must also decide what exercise prices are available for call or put options.
4. All contracts expire at the end of the final round. The settlement value of each traded contract can then be determined by substituting the share price into the appropriate formula. A player's profits and losses on each trade can then be calculated and the resultant profit/loss is their final score.

5. During a round, players can offer to buy or sell any of the traded contracts. If another player chooses to take them up on their offer, then the deal is agreed and both parties must record the transaction on their trading sheet.

6. A deal on a contract must include the following information:

   • Forward: Forward price and quantity
   • Call or Put: Type of option (call or put), exercise price, cost and quantity

   The organizer chooses the types of contracts available and the strike prices.
   The forward price or option cost and the quantity in a deal are chosen by the players.

For beginners, play three games in succession, with the following structures:

1. Play with just the forward contract.
2. Play with the forward contract and the call option with exercise price 15.
3. Play with the forward, the call and put options with exercise price 15.

All three games take place over five rounds, each five minutes in length.

14.5 **NOTES**

1. Depending on the level of prior knowledge of the players, the organizer may need to explain the characteristics of the various traded contracts. It will be instructive to emphasize that the forward contract has no cost initially.

2. There will probably be times when the organizer has to act as a 'market maker' and promote trading; for instance, asking the group if anyone wants to buy shares or at what price someone is willing to do so.

3. For more advanced students, consider introducing some of the following ideas:

   • Increase the number of rounds.
   • Decrease the length of each round.
   • Include extra calls and puts with different exercise prices or which either come into existence or expire at different times. You must fix the details of these extra contracts in advance of the game.
   • Include other contracts e.g. Asian options or barriers.
   • Include a second die for a second underlying share price.

4. Including futures with 'daily' marking to market can be tried, but slows down the game. Nevertheless, it does illustrate the importance of margin, especially if the students have a limit on how far 'in debt' they are allowed to go.

14.6 **HOW TO FILL IN YOUR TRADING SHEET**

14.6.1 During a Trading Round

In the 'Contract' column, fill in the specifications of the instrument that you have bought/sold. Specify the forward price or exercise price if applicable (e.g. if there is more than one contract of this type in the game).

In the 'Buy/Sell' column, fill in whether you have bought or sold the contract, and the quantity.
In the 'Cost per contract' column, fill in the cost of a single contract.

14.6.2 At the End of the Game

In the 'Settlement Value' column, fill in the value of a single contract with the final share price.
In the 'Profit/Loss per contract' column, fill in the profit/loss for a single contract.
In the 'Total Profit/Loss' column, fill in the total profit/loss for the trade (= profit/loss × quantity).

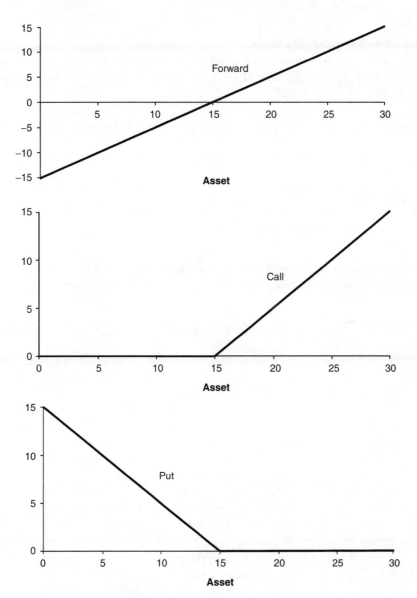

Figure 14.1 Available contracts.

Example

During a round, your transactions are:

Buy 10 call options, with exercise price 20, at a cost of $2 each. Sell 1 put option, with exercise price 15, at a cost of $1. Buy 5 forwards, with forward price 19.

Your trading sheet should be filled in as below...

Contract	Buy/Sell	Cost per contract	Total cost	Settlement value	Profit/loss
Call 20	Buy 10	2			
Put 15	Sell 1	1			
Forward 19	Buy 5	—			

At the end of the game, the final share price is 21. Consequently, the trading sheet is completed as follows.

Contract	Buy/Sell	Cost per contract	Total cost	Settlement value	Profit/loss
Call 20	Buy 10	2	$21 - 20 = 1$	$1 - 2 = -1$	$-1 \times 10 = -10$
Put 15	Sell 1	1	0	$1 - 0 = +1$	$+1 \times 1 = +1$
Forward 19	Buy 5	—	$21 - 19 = 2$	$+2$	$+2 \times 5 = +10$

The total profit and loss for the trader is therefore $-10 + 1 + 10 = +1$.

Remember that

- If you buy a contract, your profit/loss = settlement value − cost per contract
- If you sell a contract, your profit/loss = cost per contract − settlement value

Trading sheet

Contract	Buy/Sell	Cost per contract	Total cost	Settlement value	Profit/loss

The Trading Game–designed by David Epstein, 1999.

PART TWO
path dependency

The second part of the book builds upon both the mathematics and the finance of Part One. The mathematical tools are extended to examine and model path dependency. The financial contracts that we look at now include path-dependent contracts such as barriers, Asians and lookbacks.

Chapter 15: An Introduction to Exotic and Path-dependent Options This is just an overview of exotic options in which I unsuccessfully try to classify various kinds of exotics. I describe several important features to look out for.

Chapter 16: Barrier Options The commonest type of path-dependent option is the barrier option. It is only weakly path-dependent (a concept I will explain) and slots very easily into the Black–Scholes framework.

Chapter 17: Strongly Path-dependent Options Some strongly path-dependent options are harder to price for technical reasons. However, with a little bit of ingenuity, we can price these contracts quite easily.

Chapter 18: Asian Options Asian options depend on the realized average of an asset price path; they are considered in some depth.

Chapter 19: Lookback Options Lookback options depend on the realized maximum or minimum of the asset price path. They are explained in depth.

Chapter 20: Derivatives and Stochastic Control Some recent exotic contracts have an important element of optimal decision-making about them. In other words, the option holder has to make some decisions during the life of the contract. This may involve when to do something, such as exercise, or what quantity of the underlying to trade, or which cashflows to take. To analyze many of these contracts requires some knowledge of the mathematics of stochastic control.

Chapter 21: Miscellaneous Exotics My failure to classify all exotic option means that I am left with a chapter of miscellaneous contracts. I describe some more tricks of the trade for their valuation.

CHAPTER 15
an introduction to exotic and path-dependent options

In this Chapter...

- how to classify options according to important features
- how to think about derivatives in a way that makes it easy to compare and contrast different contracts
- the names and contract details for many basic types of exotic options

15.1 INTRODUCTION

The contracts we have seen so far are the most basic, and most important, derivative contracts but they only hint at the features that can be found in the more interesting products. In this chapter I prepare the way for the complex products I will be discussing in the next few chapters.

It is an impossible task to classify all options. The best that we can reasonably achieve is a rough characterization of the most popular of the features to be found in derivative products. I list some of these features in this chapter and give several examples. In the following few chapters I go into more detail in the description of the options and their pricing and hedging. The features that I describe now are discrete cashflows, early exercise, weak path dependence and strong path dependence, time dependence and dimensionality. Finally, I comment on the 'order' of an option.

Exotic options are interesting for several reasons. They are harder to price, sometimes being very model-dependent. The risks inherent in the contracts are usually more obscure and can lead to unexpected losses. Careful hedging becomes important, whether delta hedging or some form of static hedging, to minimize cashflows. Actually, how to hedge exotics is all that really matters. A trader may have a good idea of a reasonable price for an instrument, either from experience or by looking at the prices of similar instruments. But he may not be so sure about the risks in the contract or how to hedge them away successfully.

We are going to continue with the Black–Scholes theme for the moment and show how to price and hedge exotics in their framework. Their assumptions will be relaxed in later chapters.

15.2 **DISCRETE CASHFLOWS**

Imagine a contract that pays the holder an amount q at time t_q. The contract could be a bond and the payment a coupon. If we use $V(t)$ to denote the contract value (ignoring any dependence on any underlying asset) and t_q^- and t_q^+ to denote just before and just after the cashflow date then simple arbitrage considerations lead to

$$V(t_q^-) = V(t_q^+) + q.$$

NO ARBITRAGE LEADS TO THIS SIMPLE JUMP CONDITION WHEN MONEY CHANGES HANDS

This is a jump condition. The value of the contract jumps by the amount of the cashflow. If this were not the case then there would be an arbitrage opportunity. The behavior of the contract value across the payment date is shown in Figure 15.1.

If the contract is contingent on an underlying variable so that we have $V(S, t)$ then we can accommodate cashflows that depend on the level of the asset S, i.e. we could have $q(S)$. Furthermore, this also allows us to lump all our options on the same underlying into one large portfolio. Then, across the expiry of each option, there will be a jump in the value of our whole portfolio of the amount of the payoff for that option.

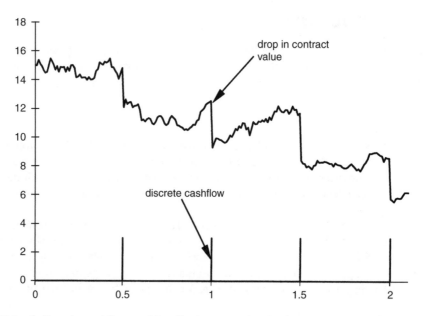

Figure 15.1 A discrete cashflow and its effect on a contract value.

There is one small technical requirement here, that the cashflow must be a deterministic function of time and the underlying asset. For example, the contract holder could receive a payment of S^2, for some asset with price S. The above argument would not be valid if, for example, the cashflow depended on the toss of a coin; one dollar is received if heads is thrown and nothing otherwise. The jump condition does not necessarily apply, because the cashflow is not deterministic.

If the cashflow is not deterministic the modeling is not so straightforward. There is no 'no arbitrage' argument to appeal to, and the result could easily depend on an individual's risk preferences. Nevertheless, we could say, for example, that the jump condition would be that the change in value of the contract would be the *expected* value of the cashflow:

$$V(t_q^-) = V(t_q^+) + E[q].$$

Such a condition would not, however, allow for the risk inherent in the uncertain cashflow.

15.3 **EARLY EXERCISE**

We have seen early exercise in the American option problem. Early exercise is a common feature of other contracts, perhaps going by other names. For example, the conversion of convertible bonds, discussed in Chapter 43, is mathematically identical to the early exercise of an American option. The key point about early exercise is that the holder of this valuable right should ideally act *optimally*, i.e. they must decide *when* to exercise or convert. In the partial differential equation framework that has been set up, this optimality is achieved by solving a free boundary problem, with a constraint on the option value, together with a smoothness condition. It is this smoothness condition, that the derivative of the option value with respect to the underlying is continuous, that ensures optimality i.e. maximization of the option value with respect to the exercise or conversion strategy. It is perfectly possible for there to be more than one early-exercise region.

One rarely-mentioned aspect of American options and, generally speaking, contracts with early exercise-type characteristics, is that they are path-dependent. Whether the owner of the option still holds the option at expiry depends on whether or not he has exercised the option, and thus on the path taken by the underlying. For American-type options this path dependence is weak, in the sense that the partial differential equation to be solved has no more independent variables than a similar, European, contract.

15.4 **WEAK PATH DEPENDENCE**

The next most common reason for weak path dependence in a contract is a **barrier**. Barrier (or knock-in, or knock-out) options are triggered by the action of the underlying hitting a prescribed value at some time before expiry. For example, as long as the asset remains below 150, the contract will have a call payoff at expiry. However, should the asset reach this level before expiry

WEAK PATH DEPENDENCE MEANS THAT THE OPTION DEPENDS ONLY ON ASSET AND TIME

then the option becomes worthless; the option has 'knocked out'. This contract is clearly path-dependent, for consider the two paths in Figure 15.2; one has a payoff at expiry because the

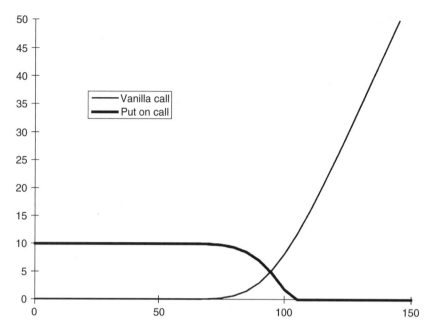

Figure 15.2 Two paths, having the same value at expiry, but with completely different payoffs.

barrier was not triggered, the other is worthless, yet both have the same value of the underlying at expiry.

We shall see in Chapter 16 that such a contract is only weakly path-dependent: we still solve a partial differential equation in the two variables, the underlying and time.

15.5 STRONG PATH DEPENDENCE

Of particular interest, mathematical and practical, are the strongly path-dependent contracts. These have payoffs that depend on some property of the asset price path in addition to the value of the underlying at the present moment in time; in the equity option language, we cannot write the value as $V(S, t)$. The contract value is a function of at least one more independent variable. This is best illustrated by an example.

The Asian option has a payoff that depends on the average value of the underlying asset from inception to expiry. We must keep track of more information about the asset price path than simply its present position. The extra information that we need is contained in the 'running average'. This is the average of the asset price from inception until the present, when we are valuing the option. No other information is needed. This running average is then used as a new independent variable, the option value is a function of this as well as the usual underlying and time, and a derivative of the option value with respect to the running average appears in the governing equation.

There are many such contracts in existence, and I show how to put many of them into the same general framework.

15.6 **TIME DEPENDENCE**

We have seen time dependence in parameters, and have shown how to apply the Black–Scholes formulae when interest rates, dividends and volatility vary in time (in a known, deterministic, way). Here we are concerned with time dependence in the option contract. We can add such time dependence to any of the features described above. For example, early exercise might only be permitted on certain dates or during certain periods.

HOW DOES THE CONTRACTS SPEC DEPEND ON TIME?

This intermittent early exercise is a characteristic of **Bermudan options**. Similarly, the position of the barrier in a knock-out option may change with time. Every month it may be reset at a higher level than the month before. Or we can readily imagine a knock-out option in which the barrier is only active (i.e. can be triggered) during the last week of every month. These contracts are referred to as time-inhomogeneous.

15.7 **DIMENSIONALITY**

Dimensionality refers to the number of underlying independent variables. The vanilla option has two independent variables, S and t, and is thus two-dimensional. The weakly path-dependent contracts have the same number of dimensions as their non-path-dependent cousins, i.e. a barrier call option has the same two dimensions as a vanilla call. For these contracts the roles of the asset dimension and the time dimension are

NUMBER OF DIMENSIONS IS IMPORTANT.... THE MORE THERE ARE, THE HARDER TO SOLVE

quite different from each other, as discussed in Chapter 6 on the diffusion equation. This is because the governing equation, the Black–Scholes equation, contains a second asset-price derivative but only a first time derivative.

We can have two types of three-dimensional problem. The first occurs when we have a second source of randomness, such as a second underlying asset. We might, for example, have an option on the maximum of two equities. Both of these underlyings are stochastic, each with a volatility, and there will be a correlation between them. In the governing equation we will see a second derivative of the option value with respect to each asset. We say that there is diffusion in both S_1 and S_2.

The other type of problem that is also three-dimensional is the strongly path-dependent contract. We will see examples of these in later chapters. Typically, the new independent variable is a measure of the path-dependent quantity on which the option is contingent. The new variable may be the average of the asset price to date, say. In this case, derivatives of the option value with respect to this new variable are only of the first order. Thus the new variable acts more like another time-like variable.

15.8 **THE ORDER OF AN OPTION**

The final classification that we make is the **order** of an option. Not only is this a classification but the idea also introduces fundamental modeling issues.

ORDER WILL AFFECT THE SENSITIVITY OF THE THEORETICAL PRICE TO THE MODEL

The basic, vanilla options are of first order. Their payoffs depend only on the underlying asset, the quantity that we are *directly* modeling. Other, path-dependent, contracts can still be of first order if the payoff depends only on properties of the asset price path. 'Higher order' refers to options whose payoff, and hence value, is contingent on the value of *another* option. The obvious second-order options are compound options, for example, a call option giving the holder the right to buy a put option. The compound option expires at some date T_1 and the option on which it is contingent expires at a later time T_2. Technically speaking, such an option is weakly path-dependent. The *theoretical* pricing of such a contract is straightforward, as we shall see.

From a practical point of view, the compound option raises some important modeling issues: the payoff for the compound option depends on the *market* value of the underlying option, and not on the theoretical price. If you hold a compound option, and want to exercise the first option then you must take possession of the underlying option. If that option is worth less than you think it should be worth (because your model says so) then there is not much you can do about it. High order option values are very sensitive to the basic pricing model and should be handled with care. This issue of not only modeling the underlying, but also modeling what the market does (regardless of whether it is 'correct' or not) will be seen in other parts of this book.

15.9 **DECISIONS, DECISIONS**

Holding an American option you are faced with the decision whether and when to exercise your rights. The American option is the most common contract that contains within it a decision feature. Other contracts require more subtle and interesting decisions to be made. We'll be seeing several examples of these later and I'll mention just the one now.

The passport option, discussed in depth in Chapter 21, is an option on a trading account. You buy and sell some asset. If you are in profit on the expiry of the option you keep the money, if you have made a loss it is written off. The decisions to be made here are when to buy, sell or hold, and how much to buy, sell or hold.

15.10 **CLASSIFICATION TABLES**

Watch out for tables like the following for the classification of special contracts.

Classification	Option Name
Time dependence	Do details vary with time? E.g. discrete sampling.
Cashflow	Does money change hands during life of contract?
Decisions	Does holder and/or writer have to make decisions?
Path dependence	Weak or Strong?
Dimension	2, 3, 4, ... ?
Order	First, second, ...?

15.11 **COMPOUNDS AND CHOOSERS**

Compound and **chooser options** are simply options on options. The compound option gives the holder the right to buy (call) or sell (put) another option. Thus we can imagine owning a call on a put, for example. This gives us the right to buy a put option for a specified amount on a specified date. If we exercise the option then we will own a put option which gives us the right to sell the underlying. This compound option is second order because the compound option gives us rights over another derivative. Although the Black–Scholes model can theoretically cope with second-order contracts it is not so clear that the model is completely satisfactory in practice; when we exercise the contract we get an option at the market price, not at our theoretical price.

In the Black–Scholes framework the compound option is priced as follows. There are two steps: first price the underlying option and then price the compound option. Suppose that the underlying option has a payoff of $F(S)$ at time T, and that the compound option can be exercised at time $T_{Co} < T$ to get $G(V(S, T_{Co}))$ where $V(S, t)$ is the value of the underlying option. Step one is to price the underlying option i.e. to find $V(S, t)$. This satisfies

$$\frac{\partial V}{\partial t} + \tfrac{1}{2}\sigma^2 S^2 \frac{\partial^2 V}{\partial S^2} + rS\frac{\partial V}{\partial S} - rV = 0 \quad \text{with} \quad V(S, T) = F(S).$$

Solve this problem so that you have found $V(S, T_{Co})$. This is the (theoretical) value of the underlying option at time T_{Co}, which is the time at which you can exercise your compound option. Now comes the second step, to value the compound option. The value of this is $Co(S, t)$ which satisfies

$$\frac{\partial Co}{\partial t} + \tfrac{1}{2}\sigma^2 S^2 \frac{\partial^2 Co}{\partial S^2} + rS\frac{\partial Co}{\partial S} - rCo = 0 \quad \text{with} \quad Co(S, T_{Co}) = G(V(S, T_{Co})).$$

As an example, if we have a call on a call with exercise prices E for the underlying and E_{Co} for the compound option, then we have

$$F(S) = \max(S - E, 0) \quad \text{and} \quad G(V) = \max(V - E_{Co}, 0).$$

In Figure 15.3 is shown the value of a vanilla call option at the time of expiry of a put option on this call. This is obviously some time before the expiry of the underlying call. In the same figure is the payoff for the put on this option. This is the final condition for the Black–Scholes partial differential equation.

It is possible to find analytical formulae for the price of basic compound options in the Black–Scholes framework when volatility is constant. These formulae involve the cumulative distribution function for a bivariate Normal variable. However, because of the second-order nature of compound options and thus their sensitivity to the precise nature of the asset price random walk, these formulae are dangerous to use in practice. Practitioners use either

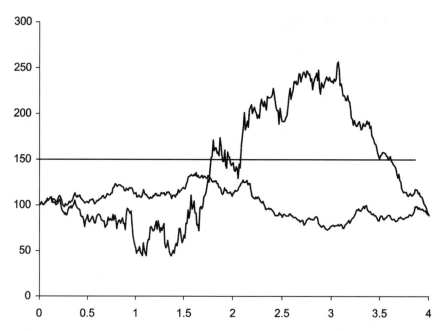

Figure 15.3 The value of a vanilla call option some time before expiry and the payoff for a put on this option.

a stochastic volatility model or an implied volatility surface, two subjects I cover in later chapters.

Chooser options are similar to compounds in that they give the holder the right to buy a further option. With the chooser option the holder can choose whether to receive a call or a put, for example. Generally, we can write the value of the chooser option as $Ch(S, t)$ and the value of the underlying options as $V_1(S, t)$ and $V_2(S, t)$ (or more). Now

$$\frac{\partial Ch}{\partial t} + \tfrac{1}{2}\sigma^2 S^2 \frac{\partial^2 Ch}{\partial S^2} + rS\frac{\partial Ch}{\partial S} - rCh = 0,$$

$$\frac{\partial V_1}{\partial t} + \tfrac{1}{2}\sigma^2 S^2 \frac{\partial^2 V_1}{\partial S^2} + rS\frac{\partial V_1}{\partial S} - rV_1 = 0$$

and
$$\frac{\partial V_2}{\partial t} + \tfrac{1}{2}\sigma^2 S^2 \frac{\partial^2 V_2}{\partial S^2} + rS\frac{\partial V_2}{\partial S} - rV_2 = 0.$$

Final conditions are the usual payoffs for the underlying options at their expiry dates and

$$Ch(S, T_{Ch}) = \max(V_1(S, T_{Ch}) - E_1, V_2(S, T_{Ch}) - E_2, 0),$$

with the obvious notation.

The practical problems with pricing choosers are the same as for compounds.

Classification	Compound/Chooser
Time dependence	No
Cashflow	No
Decisions	No (or trivial)
Path dependence	Weak
Dimension	2
Order	Second

Option classification table for Compounds and Choosers

In Figure 15.4 is shown the values of a vanilla call and a vanilla put some time before expiry. In the same figure is the payoff for a call on the best of these two options (less an exercise price). This is the final condition for the Black–Scholes partial differential equation.

Extendible options are very, very similar to compounds and choosers. At some specified time the holder can choose to accept the payoff for the original option or to extend the option's

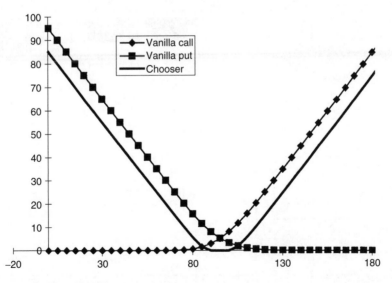

Figure 15.4 Value of a vanilla call option and a vanilla put option some time before expiry and the payoff for the best of these two.

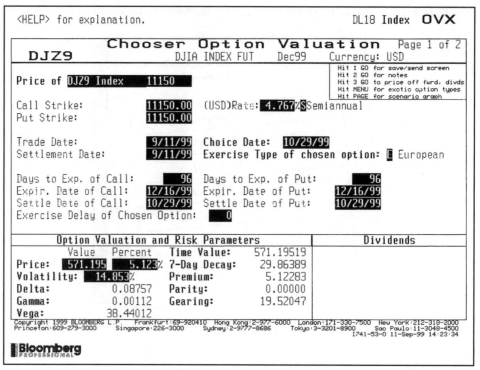

Figure 15.5 Bloomberg chooser option valuation screen. Source: Bloomberg L.P.

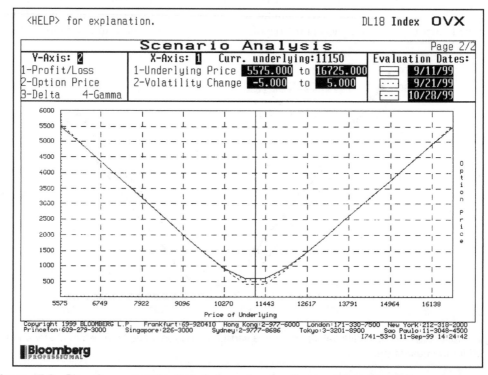

Figure 15.6 Bloomberg scenario analysis for a chooser. Source: Bloomberg L.P.

life and even change the strike. Sometimes it is the writer who has these powers of extension. The reader has sufficient knowledge to be able to model these contracts in the Black–Scholes framework.

Figures 15.5 and 15.6 show the Bloomberg screens for valuing chooser options.

15.12   RANGE NOTES

Range notes are very popular contracts, existing on the 'lognormal' assets such as equities and currencies, and as fixed-income products. In its basic, equity derivative form, the range note pays at a rate of L all the time that the underlying lies within a given range, $S_l \leq S \leq S_u$. That is, for every dt that the asset is in the range you receive $L\,dt$. Introducing $\mathcal{I}(S)$ as the function taking the value 1 when $S_l \leq S \leq S_u$ and zero otherwise, the range note satisfies

$$\frac{\partial V}{\partial t} + \tfrac{1}{2}\sigma^2 S^2 \frac{\partial^2 V}{\partial S^2} + rS\frac{\partial V}{\partial S} - rV + L\mathcal{I}(S) = 0.$$

In Figure 15.7 is shown the termsheet for a range note on the Mexican peso-US dollar exchange rate. This contract pays out the positive part of the difference between the number of days the exchange rate is inside the range less the number of days outside the range. This payment is received at expiry. (This contract is subtly different, and more complicated than the basic range note described above. Why? When you have finished Part Two you should be able to price this contract.)

Classification	Range Note
Time dependence	No
Cashflow	Yes (continuous)
Decisions	No
Path dependence	Weak
Dimension	2
Order	first

Option classification table for a Range Note

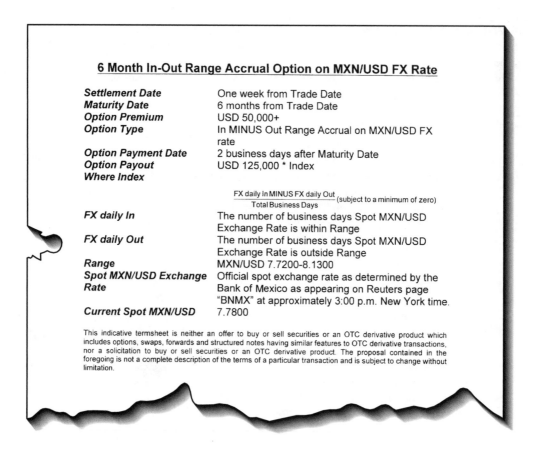

6 Month In-Out Range Accrual Option on MXN/USD FX Rate

Settlement Date	One week from Trade Date
Maturity Date	6 months from Trade Date
Option Premium	USD 50,000+
Option Type	In MINUS Out Range Accrual on MXN/USD FX rate
Option Payment Date	2 business days after Maturity Date
Option Payout	USD 125,000 * Index
Where Index	$\frac{\text{FX daily In MINUS FX daily Out}}{\text{Total Business Days}}$ (subject to a minimum of zero)
FX daily In	The number of business days Spot MXN/USD Exchange Rate is within Range
FX daily Out	The number of business days Spot MXN/USD Exchange Rate is outside Range
Range	MXN/USD 7.7200-8.1300
Spot MXN/USD Exchange Rate	Official spot exchange rate as determined by the Bank of Mexico as appearing on Reuters page "BNMX" at approximately 3:00 p.m. New York time.
Current Spot MXN/USD	7.7800

This indicative termsheet is neither an offer to buy or sell securities or an OTC derivative product which includes options, swaps, forwards and structured notes having similar features to OTC derivative transactions, nor a solicitation to buy or sell securities or an OTC derivative product. The proposal contained in the foregoing is not a complete description of the terms of a particular transaction and is subject to change without limitation.

Figure 15.7 Termsheet for an in-out range accrual note on MXN/USD.

15.13 BARRIER OPTIONS

Barrier options have a payoff that is contingent on the underlying asset reaching some specified level before expiry. The critical level is called the barrier; there may be more than one. Barrier options are weakly path-dependent. Barrier options are discussed in depth in Chapter 16.

Barrier options come in two main varieties, the 'in' barrier option (or **knock-in**) and the 'out' barrier option (or **knock-out**). The former only have a payoff if the barrier level is reached before expiry and the latter only have a payoff if the barrier is *not* reached before expiry. These contracts are weakly path-dependent, meaning that the price depends only on the current level of the asset and the time to expiry. They satisfy the Black–Scholes equation, with special boundary conditions as we shall see.

15.14 ASIAN OPTIONS

Asian options have a payoff that depends on the average value of the underlying asset over some period before expiry. They are the first strongly path-dependent contract we examine. They are strongly path dependent because their value prior to expiry depends on the path taken and not just on where they have reached. Their value depends on the *average to date* of the asset. This average to date will be very important to us, we introduce something like it as a

Classification	Knock-out
Time dependence	No
Cashflow	No
Decisions	No
Path dependence	Weak
Dimension	2
Order	first

Classification option table for Knock-out

Classification	Knock-in
Time dependence	No
Cashflow	No
Decisions	No
Path dependence	Weak
Dimension	2
Order	Second?

Classification option table for Knock-in

new state variable. We shall see how to derive a partial differential equation for the value of this Asian contract, but now the differential equation will have *three* independent variables.

The average used in the calculation of the option's payoff can be defined in many different ways. It can be an arithmetic average or a geometric average, for example. The data could be continuously sampled, so that every realized asset price over the given period is used. More commonly, for practical and legal reasons, the data is usually sampled discretely; the calculated average may only use every Friday's closing price, for example. We shall see in Chapter 18 how to price contracts with a wide range of definitions for the average and with either continuous or discrete sampling.

Classification	Asian
Time dependence	Yes – if discrete / No – if continuous
Cashflow	No
Decisions	No
Path dependence	Strong
Dimension	3 (possibn. reduction)
Order	first

Classification option table for Asian Options

15.15 LOOKBACK OPTIONS

Lookback options have a payoff that depends on the realized maximum or minimum of the underlying asset over some period prior to expiry. An extreme example that captures the flavor of these contracts is the option that pays off the difference between the maximum realized value of the asset and the minimum value over the next year. Thus it enables the holder to buy at the lowest price and sell at the highest, every trader's dream. Of course, this payoff comes at a price. And for such a contract that price would be very high.

Again the maximum or minimum can be calculated continuously or discretely, using every realized asset price or just a subset. In practice the maximum or minimum is measured discretely.

Classification	Lookback
Time dependence	Yes – if discrete / No – if continuous
Cashflow	No
Decisions	No
Path dependence	Strong
Dimension	3 (possibn. reduction)
Order	first

Classification option table for Lookback

15.16 **SUMMARY**

This chapter suggests ways to think about derivative contracts that make their analysis simpler. To be able to make comparisons between different contracts is a big step forward in understanding them. After digesting this and the next few chapters, you will be able to tell very quickly whether a particular contract is easy or difficult to price and hedge. And you will know whether the Black–Scholes framework is suitable, or whether it may be dangerous to apply it directly.

In this chapter we also began to look at some rather more complicated contracts than we have seen so far. We examine some of these contracts in depth in the next few chapters, considering them from both a theoretical and a practical viewpoint.

FURTHER READING

- Geske (1979) discusses the valuation of compound options.

- See Taleb (1997) for more details of classifications of the type I have described. This book is an excellent, and entertaining, read.

- The book by Zhang (1997) is a discussion of many types of exotic options, with many formulae.

CHAPTER 16
barrier options

In this Chapter...

- the different types of barrier options
- how to price many barrier contracts in the partial differential equation frame-work
- some of the practical problems with the pricing and hedging of barriers

16.1 INTRODUCTION

I mentioned barrier options briefly in the previous chapter. In this chapter we study them in detail, from both a theoretical and a practical perspective. **Barrier options** are path-dependent options. They have a payoff that is dependent on the realized asset path via its level; certain aspects of the contract are triggered if the asset price becomes too high or too low. For example, an up-and-out call option pays off the usual $\max(S - E, 0)$ at expiry unless at any time previously the underlying asset has traded at a value S_u or higher. In this example, if the asset reaches this level (from below, obviously) then it is said to 'knock out', becoming worthless. Apart from 'out' options like this, there are also 'in' options which only receive a payoff if a level is reached, otherwise they expire worthless.

Barrier options are popular for a number of reasons. Perhaps the purchaser uses them to hedge very specific cashflows with similar properties. Usually, the purchaser has very precise views about the direction of the market. If he wants the payoff from a call option but does not want to pay for all the upside potential, believing that the upward movement of the underlying will be limited prior to expiry, then he may choose to buy an up-and-out call. It will be cheaper than a similar vanilla call, since the upside is severely limited. If he is right and the barrier is not triggered he gets the payoff he wanted. The closer that the barrier is to the current asset price then the greater the likelihood of the option being knocked out, and thus the cheaper the contract.

Conversely, an 'in' option will be bought by someone who believes that the barrier level will be realized. Again, the option is cheaper than the equivalent vanilla option.

16.2 **DIFFERENT TYPES OF BARRIER OPTION**

There are two main types of barrier option:

- The **out option**, that only pays off if a level is *not* reached. If the barrier is reached then the option is said to have **knocked out**.

- The **in option**, that pays off as long as a level is reached before expiry. If the barrier is reached then the option is said to have **knocked in**.

Then we further characterize the barrier option by the position of the barrier relative to the initial value of the underlying:

- If the barrier is above the initial asset value, we have an **up** option.
- If the barrier is below the initial asset value, we have a **down** option.

Finally, we describe the payoff received at expiry:

- The payoffs are all the usual suspects, call, put, binary, etc.

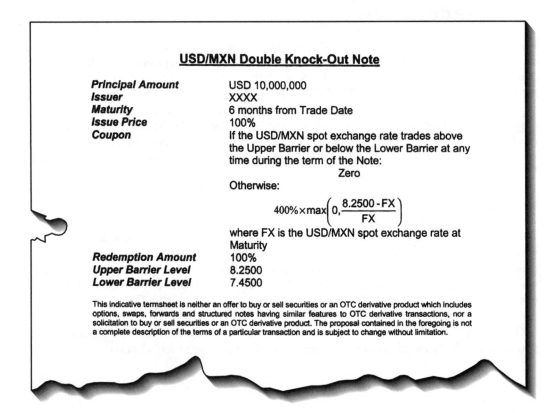

Figure 16.1   Termsheet for a USD/MXN double knock-out note.

The above classifies the commonest barrier options. In all of these contracts the position of the barrier could be time-dependent. The level may begin at one level and then rise, say. Usually the level is a piecewise-constant function of time.

Another style of barrier option is the **double barrier**. Here there is both an upper and a lower barrier, the first above and the second below the current asset price. In a double 'out' option the contract becomes worthless if *either* of the barriers is reached. In a double 'in' option one of the barriers must be reached before expiry, otherwise the option expires worthless. Other possibilities can be imagined; one barrier is an 'in' and the other an 'out', at expiry the contract could have either an 'in' or an 'out' payoff.

Sometimes a **rebate** is paid if the barrier level is reached. This is often the case for 'out' barriers in which case the rebate can be thought of as cushioning the blow of losing the rest of the payoff. The rebate may be paid as soon as the barrier is triggered or not until expiry.

In Figure 16.1 is shown the termsheet for a double knock-out option on the Mexican peso, US dollar exchange rate. The upper barrier is set at 8.25 and the lower barrier at 7.45. If the exchange rate trades inside this range until expiry then there is a payment. This is a very vanilla example of a barrier contract.

16.3 PRICING BARRIERS IN THE PARTIAL DIFFERENTIAL EQUATION FRAMEWORK

Barrier options are path-dependent. Their payoff, and therefore value, depends on the path taken by the asset up to expiry. Yet that dependence is weak. We only have to know whether or not the barrier has been triggered, we do not need any other information about the path. This is in contrast to some of the contracts we will be seeing shortly, such as the Asian option, that are strongly path-dependent. I use $V(S, t)$ to denote the value of the barrier contract *before the barrier has been triggered*. This value still satisfies the Black–Scholes equation

$$\frac{\partial V}{\partial t} + \tfrac{1}{2}\sigma^2 S^2 \frac{\partial^2 V}{\partial S^2} + rS\frac{\partial V}{\partial S} - rV = 0.$$

The details of the barrier feature come in through the specification of the boundary conditions.

16.3.1 'Out' Barriers

If the underlying asset reaches the barrier in an 'out' barrier option then the contract becomes worthless. This leads to the boundary condition

$$V(S_u, t) = 0 \quad \text{for } t < T,$$

for an up-barrier option with the barrier level at $S = S_u$. We must solve the Black–Scholes equation for $0 \leq S \leq S_u$ with this condition on $S = S_u$ and a final condition corresponding to the payoff received if the barrier is not triggered. For a call option we would have

$$V(S, T) = \max(S - E, 0).$$

If we have a down-and-out option with a barrier at S_d then we solve for $S_d \leq S < \infty$ with

$$V(S_d, t) = 0,$$

and the relevant final condition at expiry.

The boundary conditions are easily changed to accommodate rebates. If a rebate of R is paid when the barrier is hit then

$$V(S_d, t) = R.$$

16.3.2 'In' Barriers

An 'in' option only has a payoff if the barrier is triggered. If the barrier is not triggered then the option expires worthless

$$V(S, T) = 0.$$

The value in the option is in the potential to hit the barrier. If the option is an up-and-in contract then on the upper barrier the contract must have the same value as a vanilla contract:

$$V(S_u, t) = \text{value of vanilla contract, a function of } t.$$

Using the notation $V_v(S, t)$ for value of the equivalent vanilla contract (a vanilla call, if we have an up-and-in call option) then we must have

$$V(S_u, t) = V_v(S_u, t) \quad \text{for } t < T.$$

A similar boundary condition holds for a down-and-in option.

The contract we receive when the barrier is triggered is a derivative itself, and therefore the 'in' option is a second-order contract.

In solving for the value of an 'in' option completely numerically we must solve for the value of the vanilla option first, before solving for the value of the barrier option. The solution therefore takes roughly twice as long as the solution of the 'out' option.[1]

16.3.3 Some Formulae When Volatility is Constant

When volatility is constant we can solve for the theoretical price of many types of barrier contract. Some examples are given here and lots more can be found at the end of the chapter. (However, such formulae are rarely used in practice for reasons to be discussed below.)

I continue to use $V_v(S, t)$ for the value of the equivalent vanilla contract.

Down-and-out Call Option As the first example, consider the down-and-out call option with barrier level S_d below the strike price E. The function $V_v(S, t)$ is the Black–Scholes value of a vanilla option with the same maturity and payoff as our barrier option. The value of the down-and-out option is then given by

$$V(S, t) = V_v(S, t) - \left(\frac{S}{S_d}\right)^{1-(2r/\sigma^2)} V_v \frac{S_d^2}{S, t}.$$

Let us confirm that this is indeed the solution. First, does it satisfy the Black–Scholes equation? Clearly, the first term on the right-hand side does. The second term does also. Actually, if we have any solution, V_{BS}, of the Black–Scholes equation it is easy to show that

$$S^{1-(2r/\sigma^2)} V_{BS} \frac{X}{S, t}$$

is also a solution for any X.

---

[1] And, of course, the vanilla option must be solved for $0 \leq S < \infty$.

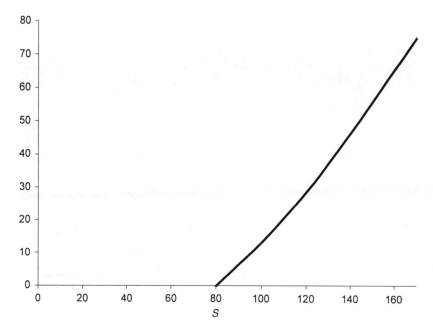

Figure 16.2 Value of a down-and-out call option.

What about the condition that the option value must be zero on $S = S_d$? Substitute $S = S_d$ in the above to confirm that this is the case. And the final condition? Since $S_d^2/S < E$ for $S > S_d$ the value of $V_v(S_d^2/S, T)$ is zero. Thus the final condition is satisfied.

The value of this option is shown as a function of S in Figure 16.2.

Down-and-in Call Option In the absence of any rebates the relationship between an 'in' barrier option and an 'out' barrier option (with same payoff and same barrier level) is very simple:

$$\text{in} + \text{out} = \text{vanilla}.$$

If the 'in' barrier is triggered then so is the 'out' barrier, so whether or not the barrier is triggered we still get the vanilla payoff at expiry.

Thus, the value of a down-and-in call option is

$$V(S, t) = \left(\frac{S}{S_d} \right)^{1-(2r/\sigma^2)} V_v \frac{S_d^2}{S, t}.$$

The value of this option is shown as a function of S in Figure 16.3. Also shown is the value of the vanilla call. Note that the two values coincide at the barrier.

Up-and-out Call Option The barrier S_u for an up-and-out call option must be above the strike price E (otherwise the option would be valueless). This makes the solution for the price more complicated, and I just quote it here. The value of an up-and-out call option is

$$S(N(d_1) - N(d_3) - b(N(d_6) - N(d_8))) - Ee^{-r(T-t)}(N(d_2) - N(d_4) - a(N(d_5) - N(d_7))),$$

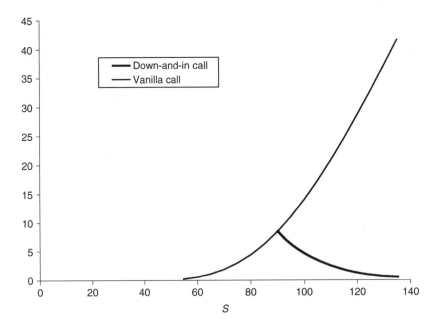

Figure 16.3 Value of a down-and-in call option.

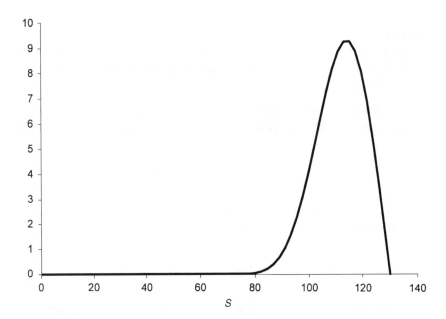

Figure 16.4 Value of an up-and-out call option.

where $N(\cdot)$ is the cumulative distribution function for a standardized Normal variable and a, b and the ds are given at the end of the chapter.

The value of this option is shown as a function of S in Figure 16.4. In Figure 16.5 is shown the delta.

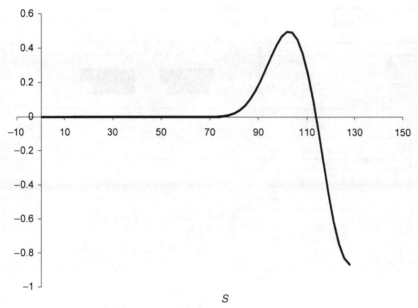

Figure 16.5 Delta of an up-and-out call option.

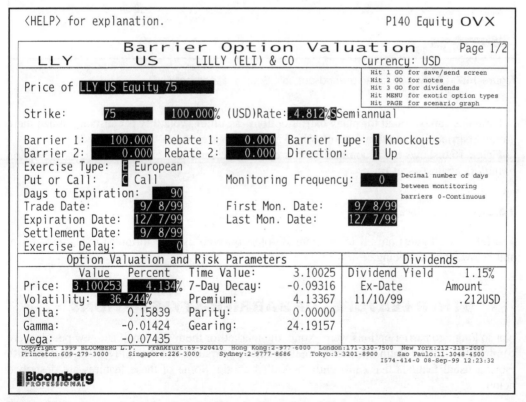

Figure 16.6 An up-and-out call again. Source: Bloomberg L.P.

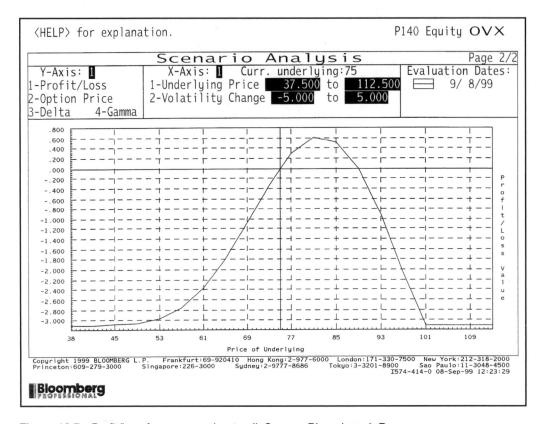

Figure 16.7 Profit/loss for an up-and-out call. Source: Bloomberg L.P.

Formulae can be found for many barrier options (assuming volatility is constant). When there are two barriers the solution can often be found by Fourier series, see Chapter 6.

Figure 16.6 shows the Bloomberg barrier option calculator and Figure 16.7 shows the option *profit/loss* against asset price.

16.3.4 Some More Examples

The following figures are all taken from Bloomberg, who use the formulae explained above, and below, for the pricing.

16.4 OTHER FEATURES OF BARRIER-STYLE OPTIONS

Not so long ago barrier options were exotic, the market for them was small and few people were comfortable pricing them. Nowadays they are heavily traded and it is only the contracts with more unusual features that can rightly be called exotic. Some of these features are described below.

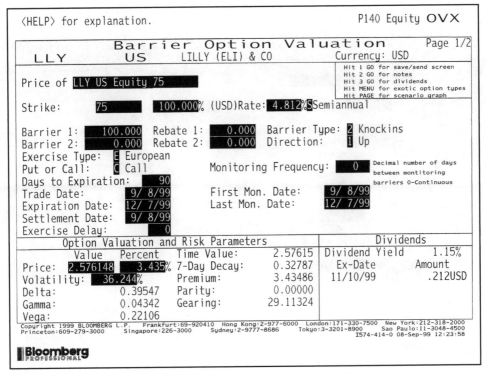

Figure 16.8 Calculator for an up-and-in call. Source: Bloomberg L.P.

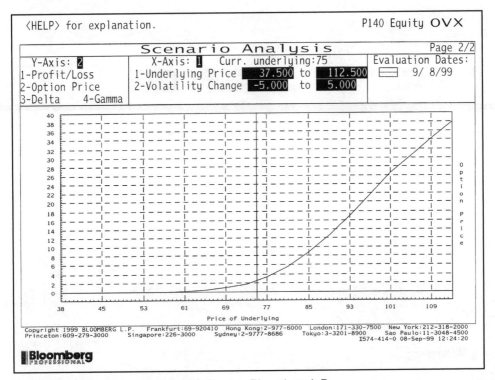

Figure 16.9 Value of an up-and-in call. Source: Bloomberg L.P.

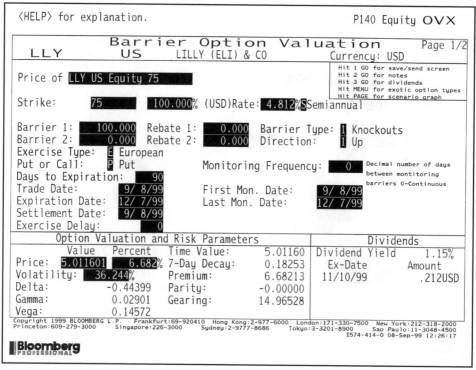

Figure 16.10 Calculator for an up-and-out put. Source: Bloomberg L.P.

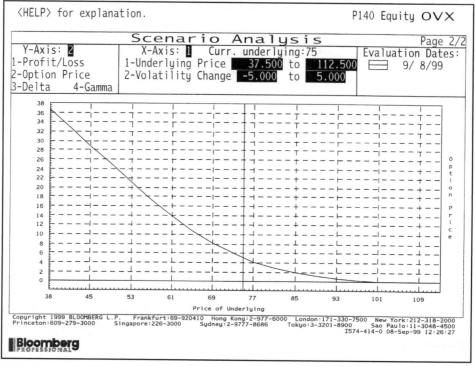

Figure 16.11 Value of an up-and-out put. Source: Bloomberg L.P.

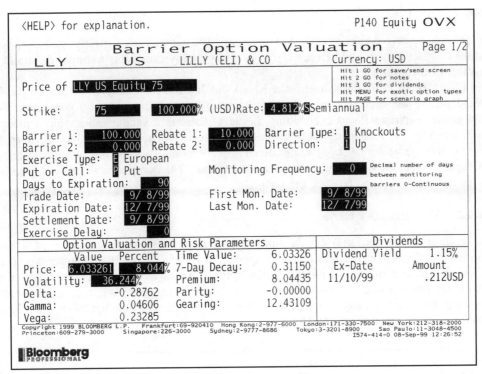

Figure 16.12 Calculator for an up-and-out put with a rebate on the upper barrier. Source: Bloomberg L.P.

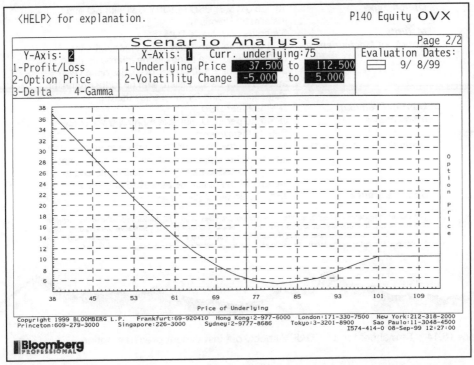

Figure 16.13 Value of an up-and-out put with a rebate on the upper barrier. Source: Bloomberg L.P.

16.4.1 Early Exercise

It is possible to have American-style early exercise. The contract must specify what the payoff is if the contract is exercised before expiry. As always, early exercise is a simple constraint on the value of the option.

In Figure 16.14 is the termsheet for a Knock-out Installment Premium Option on the US dollar-Japanese yen exchange rate. This knocks out if the exchange rate ever goes above 140. If the option expires without ever hitting this level there is a vanilla call payoff. I mention this contract in the section on early exercise because it has a similar feature. To keep the contract alive the holder must pay in installments every month. We saw this installment feature in Chapter 9 where it was likened to American exercise. The question is when to stop paying the installments? This can be done optimally.

16.4.2 The Intermittent Barrier

The position of the barrier(s) can be time-dependent. A more extreme version of a time-dependent barrier is to have a barrier that disappears altogether for specified time periods. These options are called **protected** or **partial** barrier options. An example is shown in Figure 16.15.

There are two types of such contract. In one the barrier is triggered as long as the asset price is beyond the barrier on days on which the barrier is active. The solution of this problem is shown schematically in Figure 16.16.

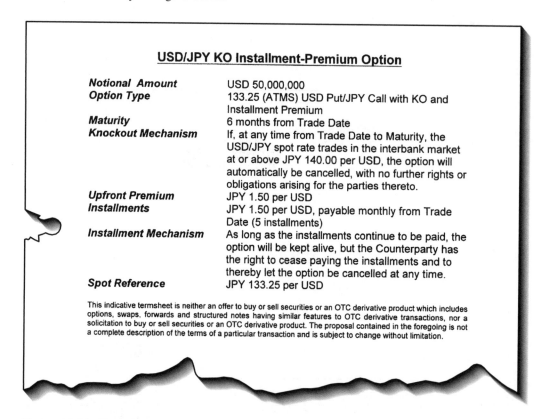

USD/JPY KO Installment-Premium Option

Notional Amount	USD 50,000,000
Option Type	133.25 (ATMS) USD Put/JPY Call with KO and Installment Premium
Maturity	6 months from Trade Date
Knockout Mechanism	If, at any time from Trade Date to Maturity, the USD/JPY spot rate trades in the interbank market at or above JPY 140.00 per USD, the option will automatically be cancelled, with no further rights or obligations arising for the parties thereto.
Upfront Premium	JPY 1.50 per USD
Installments	JPY 1.50 per USD, payable monthly from Trade Date (5 installments)
Installment Mechanism	As long as the installments continue to be paid, the option will be kept alive, but the Counterparty has the right to cease paying the installments and to thereby let the option be cancelled at any time.
Spot Reference	JPY 133.25 per USD

This indicative termsheet is neither an offer to buy or sell securities or an OTC derivative product which includes options, swaps, forwards and structured notes having similar features to OTC derivative transactions, nor a solicitation to buy or sell securities or an OTC derivative product. The proposal contained in the foregoing is not a complete description of the terms of a particular transaction and is subject to change without limitation.

Figure 16.14 Termsheet for a USD/JPY knock-out installment premium option.

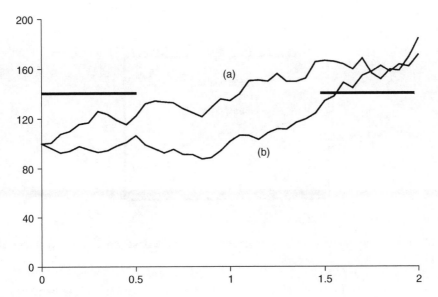

Figure 16.15 The intermittent barrier. Two varieties: barrier triggered if asset outside barrier on active days; barrier only triggered by asset price crossing barrier.

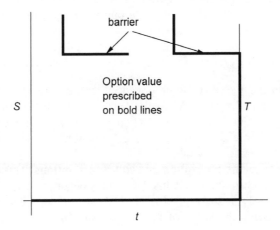

Figure 16.16 The intermittent barrier: barrier triggered if asset outside barrier on active days. Solution procedure.

The second type of intermittent barrier is only triggered if the asset path crosses the barrier on an active day. The barrier will not be triggered if the asset weaves its way through the barriers. The solution of this problem is shown schematically in Figure 16.17.

16.4.3 Repeated Hitting of the Barrier

The double barrier that we have seen above can be made more complicated. Instead of only requiring one hit of either barrier we could insist that *both* barriers are hit before the barrier is triggered.

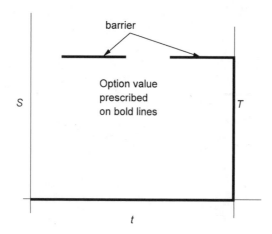

Figure 16.17 The intermittent barrier: barrier only triggered by asset price crossing barrier. Solution procedure.

This contract is easy to value. Observe that the first time that one of the barriers is hit the contract becomes a vanilla barrier option. Thus on the two barriers we solve the Black–Scholes equation with boundary conditions that our double barrier value is equal to an up-barrier option on the lower barrier and a down-barrier option on the upper barrier.

In Chapter 21 we will see the related Parisian option, the payoff of which depends on the length of time that the asset has been beyond the barrier.

16.4.4 Resetting of Barrier

Another type of barrier contract that can be priced by the same two- (or more) step procedure as 'in' barriers is the **reset barrier**. When the barrier is hit the contract turns into another barrier option with a different barrier level. The contract may be time-dependent in the sense that if the barrier is hit before a certain time we get a new barrier option, and if it is hit after a certain time we get the vanilla payoff.

Related to these contracts are the **roll-up** and **roll-down options**. These begin life as vanilla options, but if the asset reaches some predefined level they become a barrier option. For example, with a roll-up put if the roll-up strike level is reached the contract becomes an up-and-out put with the roll-up strike being the strike of the barrier put. The barrier level will then be at a prespecified level.

16.4.5 Outside Barrier Options

Outside or **rainbow barrier options** have payoffs or a trigger feature that depends on a second underlying. Thus the barrier might be triggered by one asset, with the payoff depending on the other. These products are clearly multi-factor contracts.

16.4.6 Soft Barriers

The **soft barrier option** allows the contract to be gradually knocked in or out. The contract specifies two levels, an upper and a lower. In the knock-out option a proportion of the contract is knocked out depending on the distance that the asset has reached between the two barriers.

For example, suppose that the option is an up and out with a soft barrier range of 100 to 120. If the maximum asset value reached before expiry is 105 then 5/20 or 25% of the payoff is lost. These are strongly path-dependent contracts.

16.4.7 Parisian Options

Parisian options have barriers that are triggered only if the underlying has been beyond the barrier level for more than a specified time. This additional feature reduces the possibility of manipulation of the trigger event and makes the dynamic hedging easier. However, this new feature also increases the dimensionality of the problem and so we must wait for a few chapters before we can analyze this contract in detail.

16.5 **FIRST EXIT TIME**

The path dependency in a barrier option arises because the option payoff depends on whether or not the barrier has been triggered. The *value* can be interpreted as the present value of the risk-neutral expected payoff but the likelihood of the barrier being triggered before expiry only has meaning if we calculate the probability using the *real* random walk for the asset. For an up-and-in barrier option, the probability of the barrier being triggered before expiry, $Q(S, t)$, is given by the solution of

$$\frac{\partial Q}{\partial t} + \frac{1}{2}\sigma^2 S^2 \frac{\partial^2 Q}{\partial S^2} + \mu S \frac{\partial Q}{\partial S} = 0$$

with

$$Q(S, T) = 0 \quad \text{and} \quad Q(S_u, t) = 1,$$

as discussed in Chapter 10. Because we are using the real process for S this problem contains the real drift rate μ. The expected time $u(S)$ before the level S_u is hit from below is the solution of

$$\frac{1}{2}\sigma^2 S^2 \frac{d^2 u}{dS^2} + \mu S \frac{du}{dS} = -1$$

i.e.

$$\frac{1}{\frac{1}{2}\sigma^2 - \mu} \log\left(\frac{S}{S_u}\right),$$

but only for $2\mu > \sigma^2$. If $2\mu < \sigma^2$ then the expected first exit time is infinite.

Calculations like these can be used by the speculator who has a view on the direction of the underlying, believing that the barrier will or will not be triggered. His view can be quantified as a probability, for example. Or he can determine whether the first exit time is greater or less than the remaining time to maturity.

The hedger will also find such calculations useful. As we discuss below, delta hedging barrier options is notoriously difficult and usually they are statically hedged as well to some extent. The choice of the static hedge may be influenced by the *real* time at which the barrier is expected to be triggered.

DETERMINING VOL IS TRICKY, AND MANY BARRIER OPTIONS ARE SENSITIVE TO VOL BEHAVIOR

16.6 **MARKET PRACTICE: WHAT VOLATILITY SHOULD I USE?**

Practitioners do not price contracts using a single, constant volatility. Let us see some of the pitfalls with this, and then see what practitioners do.

In Figure 16.18 we see a plot of the value of an up-and-out call option using three different volatilities, 15%, 20% and 25%. I have chosen three very different values to make a point. If we are unsure about the value of the volatility (as we surely are) then which value do we use to price the contract? Observe that at approximately $S = 100$ the option value seems to be insensitive to the volatility; the vega is zero. If S is greater than this value perhaps we should only sell the contract for a volatility of 15% to be on the safe side. If S is less than this, perhaps we should sell the contract for 25%, again to play it safe. Now ask the question, Do I believe that volatility will be one of 15%, 20% or 25%, and will be fixed at that level? Or do I believe that volatility could move around between 15% and 25%? Clearly the latter is closer to the truth. But the measurement of vega, and the plots in Figure 16.18 assume that volatility is fixed until expiry. If we are

I LIKE MODELS THAT DON'T MAKE TOO MANY ASSUMPTIONS ABOUT VOLATILITY

concerned with playing it safe we should assume that the behavior of volatility will be that which gives us the lowest value if we are buying the contract. The worst outcome for volatility is for it to be low around the strike price, and high around the barrier. Financially, this means that if we are near the strike we get a small payoff, but if we are near the barrier we are likely to hit it. Mathematically, the 'worst' choice of volatility path depends on the sign of the gamma at each point. If gamma is positive then low volatility is bad, if gamma is negative then

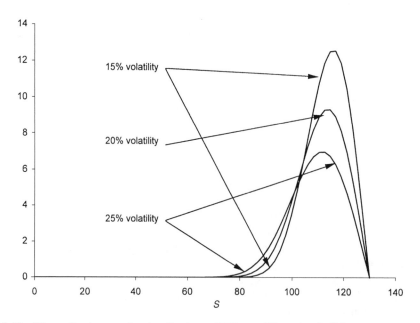

Figure 16.18 Theoretical up-and-out call price with three different volatilities.

high volatility is bad. A better way to price options when the volatility is uncertain is described in Chapter 27. When the gamma is not single-signed, the measurement of vega can be meaningless. Barrier options with non-single-signed gamma include the up-and-out call, down-and-out put and many double-barrier options.

Figures 16.19 through 16.22 show the details of a double knock-out put contract, its price versus the underlying, its gamma versus the underlying and its price versus volatility. This is a contract with a gamma that changes sign as can be seen from Figure 16.21. You must be very careful when pricing such a contract as to what volatility to use. Suppose you wanted to know the implied volatility for this contract when the price was 3.2, what value would you get? Refer to Figure 16.22.

To accommodate problems like this, practitioners have invented a number of 'patches'. One is to use two different volatilities in the option price. For example, one can calculate implied volatilities from vanilla options with the same strike, expiry and payoff as the barrier option and also from American-style one-touch options with the strike at the barrier level. The implied volatility from the vanilla option contains the market's estimate of the value of the payoff, but including all the upside potential that the call has but which is irrelevant for the up-and-out option. The one-touch volatility, however, contains the market's view of the likelihood of the barrier level being reached. These two volatilities can be used to price an up-and-out call by observing that an 'out' option is the same as a vanilla minus an 'in' option. Use the vanilla volatility to price the vanilla call and the one-touch volatility to price the 'in' call.

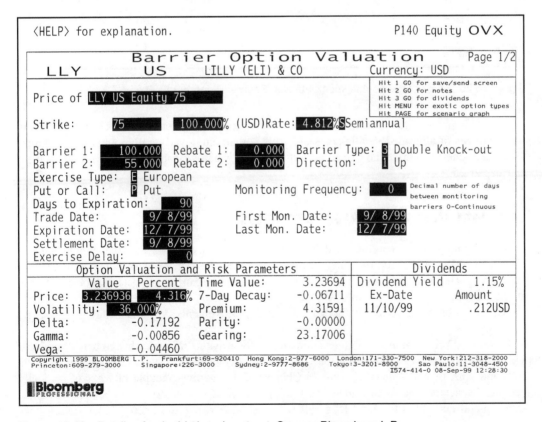

Figure 16.19 Details of a double knock-out put. Source: Bloomberg L.P.

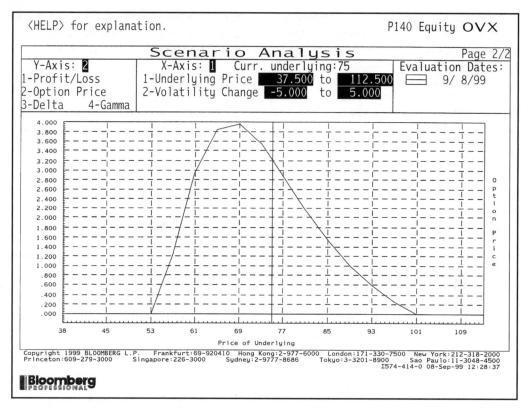

Figure 16.20 Price of the double knock-out put. Source: Bloomberg L.P.

The other practitioner approach to the pricing is to use a volatility surface, implied from market prices of all traded vanilla contracts. This is then employed in a binomial tree or finite-difference scheme to price the barrier option *consistently* across instruments. This is the subject of Chapter 25.

16.7 HEDGING BARRIER OPTIONS

Barrier options have discontinuous delta at the barrier. For a knock-out, the option value is continuous, decreasing approximately linearly towards the barrier then being zero beyond the barrier. This discontinuity in the delta means that the gamma is instantaneously infinite at the barrier. Delta hedging through the barrier is virtually impossible, and certainly very costly. This raises the issue of whether there are improvements on delta hedging for barrier options.

There have been a number of suggestions made for ways to *statically* hedge barrier options. These methods try to mimic as closely as possible the value of a barrier option with vanilla calls and puts, or with binary options. In Chapter 32 I describe a couple of ways of statically hedging barrier options with traded vanilla options. A very common practice for hedging a short up-and-out call is to buy a long call with the same strike and expiry. If the option does knock out then you are fortunate in being left with a long call position.

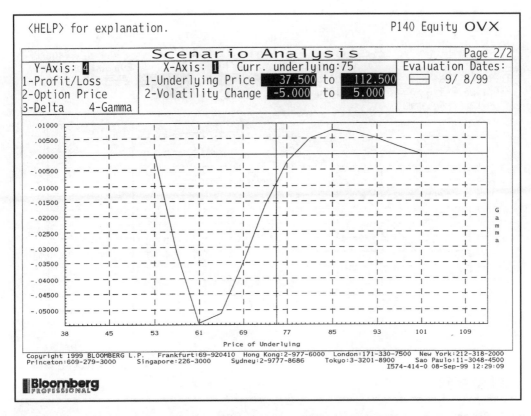

Figure 16.21 Gamma of the double knock-out put. Source: Bloomberg L.P.

I now describe another simple but useful technique, based on the **reflection principle** and **put-call symmetry**. This technique only really works if the barrier and strike lie in the correct order, as we shall see. The method gives an approximate hedge only.

The simplest example of put-call symmetry is actually put-call parity. At all asset levels we have

$$V_C - V_P = S - Ee^{-r(T-t)},$$

where E is the strike of the two options, and C and P refer to call and put. Suppose we have a down-and-in call, how can we use this result? To make things simple for the moment, let's have the barrier and the strike at the same level. Now hedge our down-and-in call with a short position in a vanilla put with the same strike. If the barrier is reached we have a position worth

$$V_C - V_P.$$

The first term is from the down-and-in call and the second from the vanilla put. This is exactly the same as

$$S - Ee^{-r(T-t)} = E(1 - e^{-r(T-t)}),$$

because of put-call parity and since the barrier and the strike are the same. If the barrier is not touched then both options expire worthless. If the interest rate were zero then we would

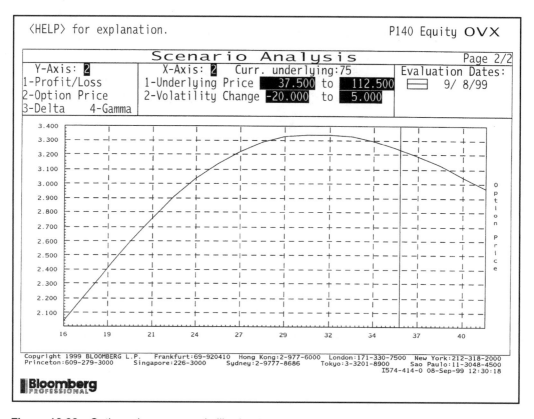

Figure 16.22 Option price versus volatility for the double knock-out put. Source: Bloomberg L.P.

have a perfect hedge. If rates are non-zero what we are left with is a one-touch option with a small and time-dependent value on the barrier. Although this leftover cashflow is non-zero, it is small, bounded and more manageable than the original cashflows.

Now suppose that the strike and the barrier are distinct. Let us continue with the down-and-in call, now with barrier below the strike. The static hedge is not much more complicated than the previous example. All we need to know is the relationship between the value of a call option with strike E when $S = S_d$ and a put option with strike S_d^2/E. It is easy to show from the formulae for calls and puts that if interest rates are zero, the value of this call at $S = S_d$ is equal to a number E/S_d of the puts, valued at S_d. We would therefore hedge our down-and-in call with E/S_d puts struck at S_d^2/E. Note that the geometric average of the strike of the call and the strike of the put is the same as the barrier level; this is where the idea of 'reflection' comes in. The strike of the hedging put is at the reflection in the barrier of the call's strike. When rates are non-zero there is some error in this hedge, but again it is small and manageable, decreasing as we get closer to expiry. If the barrier is not touched then both options expire worthless (the strike of the put is below the barrier, remember).

If the barrier level is above the strike, matters are more complicated since if the barrier is touched we get an in-the-money call. The reflection principle does not work because the put would also be in the money at expiry if the barrier is not touched.

16.7.1 Slippage Costs

The delta of a barrier option is discontinuous at the barrier, whether it is an in- or an out-option. This presents a particular problem to do with **slippage** or **gapping**. Should the underlying move significantly as the barrier is triggered it is likely that it will not be possible to continuously hedge through the barrier. For example, if the contract is knocked out then one finds oneself with a $-\Delta$ holding of the underlying that should have been offloaded sooner. This can have a significant effect on the hedging costs.

It is not too difficult to allow for the *expected* slippage costs, and all that is required is a slight modification to the apparent barrier level.

At the barrier we hold $-\Delta$ of the underlying. The value of this position is $-\Delta X$, since $S = X$ is the barrier level. Suppose that the asset moves by a small fraction k before we can close out our asset position or, equivalently, that there is a transaction charge involved in closing.[2] We thus lose

$$-k\Delta X$$

on the trigger event.

Now refer to Figure 16.23 where we'll look at the specific example of a down-and-out option. Because we lose $-k\Delta X$ we should use the boundary condition

$$V(X, t) = -k\Delta X.$$

After a little bit of Taylor series we find that this is approximately the same as

$$V((1 + k)X, t) = 0.$$

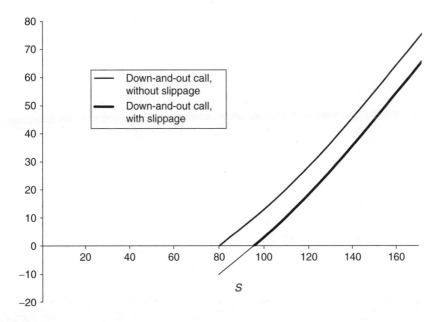

Figure 16.23 Incorporating slippage.

---

[2] Much more about this in Chapter 24.

In other words, we should apply the boundary condition at a slightly higher value of S and so slightly reduce the option's value.

16.8 SUMMARY

In this chapter we have seen a description of many types of barrier option. We have seen how to put these contracts into the partial differential equation framework. Many of these contracts have simple pricing formulae. Unfortunately, the extreme nature of these contracts makes them very difficult to hedge in practice and, in particular, they can be very sensitive to the volatility of the underlying. Worse still, if the gamma of the contract changes sign we cannot play safe by adding a spread to the volatility. Practitioners seem to be most comfortable statically hedging as much of the barrier contract as possible using traded vanilla options and pricing the residual using a full implied volatility surface. The combination of these two principles is crucial. If one were to use a volatility surface without statically hedging then one could make matters worse; the volatility surface implied from vanillas may turn out to give the barrier option an inaccurate value. Less dangerous, but still not ideal, is the static hedging of the barrier option with vanillas and then using a single volatility to price the barrier. If both of these concepts are used together there is an element of consistency across the pricing.

MORE FORMULAE

In the following I use $N(\cdot)$ to denote the cumulative distribution function for a standardized Normal variable. The dividend yield on stocks or the foreign interest rate for FX are denoted by q. Also

$$a = \left(\frac{S_b}{S}\right)^{-1+(2(r-q)/\sigma^2)},$$

$$b = \left(\frac{S_b}{S}\right)^{1+(2(r-q)/\sigma^2)},$$

where S_b is the barrier position (whether S_u or S_d should be obvious from the example),

$$d_1 = \frac{\log\left(\frac{S}{E}\right) + \left(r - q + \frac{1}{2}\sigma^2\right)(T-t)}{\sigma\sqrt{T-t}},$$

$$d_2 = \frac{\log\left(\frac{S}{E}\right) + \left(r - q - \frac{1}{2}\sigma^2\right)(T-t)}{\sigma\sqrt{T-t}},$$

$$d_3 = \frac{\log\left(\frac{S}{S_b}\right) + \left(r - q + \frac{1}{2}\sigma^2\right)(T-t)}{\sigma\sqrt{T-t}},$$

$$d_4 = \frac{\log\left(\frac{S}{S_b}\right) + \left(r - q - \frac{1}{2}\sigma^2\right)(T-t)}{\sigma\sqrt{T-t}},$$

$$d_5 = \frac{\log\left(\dfrac{S}{S_b}\right) - \left(r - q - \frac{1}{2}\sigma^2\right)(T-t)}{\sigma\sqrt{T-t}},$$

$$d_6 = \frac{\log\left(\dfrac{S}{S_b}\right) - \left(r - q + \frac{1}{2}\sigma^2\right)(T-t)}{\sigma\sqrt{T-t}},$$

$$d_7 = \frac{\log\left(\dfrac{SE}{S_b^2}\right) - \left(r - q - \frac{1}{2}\sigma^2\right)(T-t)}{\sigma\sqrt{T-t}},$$

$$d_8 = \frac{\log\left(\dfrac{SE}{S_b^2}\right) - \left(r - q + \frac{1}{2}\sigma^2\right)(T-t)}{\sigma\sqrt{T-t}}.$$

Up-and-out Call

$$Se^{-q(T-t)}(N(d_1) - N(d_3) - b(N(d_6) - N(d_8)))$$
$$- Ee^{-r(T-t)}(N(d_2) - N(d_4) - a(N(d_5) - N(d_7))).$$

Up-and-in Call

$$Se^{-q(T-t)}(N(d_3) + b(N(d_6) - N(d_8))) - Ee^{-r(T-t)}(N(d_4) + a(N(d_5) - N(d_7))).$$

Down-and-out Call

1. $E > S_b$:

$$Se^{-q(T-t)}(N(d_1) - b(1 - N(d_8))) - Ee^{-r(T-t)}(N(d_2) - a(1 - N(d_7))).$$

2. $E < S_b$:

$$Se^{-q(T-t)}(N(d_3) - b(1 - N(d_6))) - Ee^{-r(T-t)}(N(d_4) - a(1 - N(d_5))).$$

Down-and-in Call

1. $E > S_b$:

$$Se^{-q(T-t)}b(1 - N(d_8)) - Ee^{-r(T-t)}a(1 - N(d_7)).$$

2. $E < S_b$:

$$Se^{-q(T-t)}(N(d_1) - N(d_3) + b(1 - N(d_6))) - Ee^{-r(T-t)}(N(d_2) - N(d_4) + a(1 - N(d_5))).$$

Down-and-out Put

$$Ee^{-r(T-t)}(N(d_4) - N(d_2) - a(N(d_7) - N(d_5)))$$
$$- Se^{-q(T-t)}(N(d_3) - N(d_1) - b(N(d_8) - N(d_6))).$$

Down-and-in Put

$$Ee^{-r(T-t)}(1 - N(d_4) + a(N(d_7) - N(d_5))) - Se^{-q(T-t)}(1 - N(d_3) + b(N(d_8) - N(d_6))).$$

Up-and-out Put

1. $E > S_b$:

$$Ee^{-r(T-t)}(1 - N(d_2) - a(N(d_7) - N(d_5))) - Se^{-q(T-t)}(1 - N(d_1) - bN(d_8)).$$

2. $E < S_b$:

$$Ee^{-r(T-t)}(1 - N(d_4) - aN(d_7)) - Se^{-q(T-t)}(1 - N(d_3) - bN(d_6)).$$

Up-and-in Put

1. $E > S_b$:

$$Ee^{-r(T-t)}(N(d_4) - N(d_2) + aN(d_5)) - Se^{-q(T-t)}(N(d_3) - N(d_1) + bN(d_6)).$$

2. $E < S_b$:

$$Ee^{-r(T-t)}(1 - N(d_4) - aN(d_5)) - Se^{-q(T-t)}(1 - N(d_3) - bN(d_6)).$$

FURTHER READING

- Many of the original barrier formulae are due to Reiner & Rubinstein (1991).
- The formulae above are explained in Taleb (1997) and Haug (1998). Taleb discusses barrier options in great detail, including the reality of hedging that I have only touched upon.
- The article by Carr (1995) contains an extensive literature review as well as a detailed discussion of protected barrier options and rainbow barrier options.
- See Derman, Ergener & Kani (1997) for a full description of the static replication of barrier options with vanilla options.
- See Carr (1994) for more details of put-call symmetry.

CHAPTER 17
strongly path-dependent options

In this Chapter...

* strong path dependence
* pricing many strongly path-dependent contracts in the Black–Scholes partial differential equation framework
* how to handle both continuously-sampled and discretely-sampled paths
* jump conditions for differential equations

17.1 INTRODUCTION

To be able to turn the valuation of a derivative contract into the solution of a partial differential equation is a big step forward. The partial differential equation approach is one of the best ways to price a contract because of its flexibility and because of the large body of knowledge that has grown up around the fast and accurate numerical solution of these problems. This body of knowledge was, in the main, based around the solution of differential equations arising in physical applied mathematics but is now starting to be used in the financial world.

In this chapter I show how to generalize the Black–Scholes analysis, delta hedging and no arbitrage, to the pricing of many more derivative contracts, specifically contracts that are strongly path-dependent. I will describe the theory in the abstract, giving brief examples occasionally, but saving the detailed application to specific contracts until later chapters.

17.2 PATH-DEPENDENT QUANTITIES REPRESENTED BY AN INTEGRAL

We start by assuming that the underlying asset follows the lognormal random walk

$$dS = \mu S\,dt + \sigma S\,dX.$$

Imagine a contract that pays off at expiry, T, an amount that is a function of the path taken by the asset between time zero and expiry. Let us suppose that this path-dependent quantity can be represented by an integral of some function of the asset over the period zero to T:

$$I(T) = \int_0^T f(S, \tau)\,d\tau.$$

This is not such a strong assumption, since most of the path-dependent quantities in exotic derivative contracts, such as averages, can be written in this form with a suitable choice of $f(S, t)$.

We are thus assuming that the payoff is given by

$$P(S, I)$$

at time $t = T$.

Prior to expiry we have information about the possible final value of S (at time T) in the present value of S (at time t). For example, the higher S is today, the higher it will probably end up at expiry. Similarly, we have information about the possible final value of I in the value of the integral to date:

$$I(t) = \int_0^t f(S, \tau)\,d\tau. \tag{17.1}$$

As we get closer to expiry, so we become more confident about the final value of I.

One can imagine that the value of the option is therefore not only a function of S and t, but also a function of I; I will be our new independent variable, called a **state variable**. We see in the next section how this observation leads to a pricing equation. In anticipation of an argument that will use Itô's lemma, we need to know the stochastic differential equation satisfied by I. This could not be simpler. Incrementing t by dt in (17.1) we find that

$$dI = f(S, t)\,dt. \tag{17.2}$$

Observe that I is a smooth function (except at discontinuities of f) and from (17.2) we can see that its stochastic differential equation contains no stochastic terms.

17.2.1 Examples

An Asian option has a payoff that depends on the average of the asset price over some period. If that period is from time zero to expiry and the average is arithmetic then we write

$$I = \int_0^t S\,d\tau.$$

The payoff may then be, for example,

$$\max\left(\frac{I}{T} - S, 0\right).$$

This would be an average strike put, of which more later.

If the average is geometric then we write

$$I = \int_0^t \log(S)\, d\tau.$$

As another example, imagine a contract that pays off a function of the square of the underlying asset, but only counts those times for which the asset is below S_u. We write

$$I = \int_0^t S^2 \mathcal{H}(S_u - S)\, d\tau,$$

where $\mathcal{H}(\cdot)$ is the Heaviside function.

We are now ready to price some options.

17.3   CONTINUOUS SAMPLING: THE PRICING EQUATION

I will derive the pricing partial differential equation for a contract that pays some function of our new variable I. The value of the contract is now a function of the three variables, $V(S, I, t)$. Set up a portfolio containing one of the path-dependent options and short a number Δ of the underlying asset:

$$\Pi = V(S, I, t) - \Delta S.$$

The change in the value of this portfolio is given by

$$d\Pi = \left(\frac{\partial V}{\partial t} + \tfrac{1}{2}\sigma^2 S^2 \frac{\partial^2 V}{\partial S^2}\right) dt + \frac{\partial V}{\partial I}\, dI + \left(\frac{\partial V}{\partial S} - \Delta\right) dS.$$

Choosing

$$\Delta = \frac{\partial V}{\partial S}$$

to hedge the risk, and using (17.2), we find that

$$d\Pi = \left(\frac{\partial V}{\partial t} + \tfrac{1}{2}\sigma^2 S^2 \frac{\partial^2 V}{\partial S^2} + f(S, t)\frac{\partial V}{\partial I}\right) dt.$$

This change is risk-free, thus earns the risk-free rate of interest r, leading to the pricing equation

$$\boxed{\frac{\partial V}{\partial t} + \tfrac{1}{2}\sigma^2 S^2 \frac{\partial^2 V}{\partial S^2} + f(S, t)\frac{\partial V}{\partial I} + rS\frac{\partial V}{\partial S} - rV = 0} \qquad (17.3)$$

THE PDE CHANGES SLIGHTLY, BUT IN AN IMPORTANT WAY

This is to be solved subject to

$$V(S, I, T) = P(S, I).$$

This completes the formulation of the valuation problem. The obvious changes can be made to accommodate dividends on the underlying.

17.3.1 Example

Continuing with the arithmetic Asian example, we have

$$I = \int_0^t S \, d\tau,$$

so that the equation to be solved is

$$\frac{\partial V}{\partial t} + \frac{1}{2}\sigma^2 S^2 \frac{\partial^2 V}{\partial S^2} + S \frac{\partial V}{\partial I} + rS \frac{\partial V}{\partial S} - rV = 0.$$

USE THIS FORMULATION FOR CONTRACTS DEPENDING ON DISCRETELY-SAMPLED QUANTITIES

17.4 PATH-DEPENDENT QUANTITIES REPRESENTED BY AN UPDATING RULE

For practical and legal reasons path-dependent quantities are never measured continuously. There is minimum timestep between sampling of the path-dependent quantity. This timestep may be small, one day, say, or much longer. From a practical viewpoint it is difficult to incorporate every single traded price into an average, for example. Data can be unreliable and the exact time of a trade may not be known accurately. From a legal point of view, to avoid disagreements over the value of the path-dependent quantity, it is usual to only use key prices, such as closing prices, that are, in a sense, guaranteed to be a genuine traded price. If the time between samples is small we can confidently use a continuous-sampling model; the error will be small. If the time between samples is long, or the time to expiry itself is short, we must build this into our model. This is the goal of this section.

I introduce the idea of an **updating rule**, an algorithm for defining the path-dependent quantity in terms of the current 'state of the world'. The path-dependent quantity is measured on the **sampling dates** t_i, and takes the value I_i for $t_i \leq t < t_{i+1}$. At the sampling date t_i the quantity I_{i-1} is updated according to a rule such as

$$I_i = F(S(t_i), I_{i-1}, i).$$

Note how, in this simplest example (which can be generalized), the new value of I is determined by only the old value of I and the value of the underlying on the sampling date, and the sampling date.

When we come to concrete examples, as next, I shall change notation and use names for variables that are meaningful. I hope this does not cause any confusion.

17.4.1 Examples

We saw how to use the continuous running integral in the valuation of Asian options. But what if that integral is replaced by a discrete sum? In practice, the payoff for an Asian option depends on the quantity

$$I_M = \sum_{k=1}^{M} S(t_k),$$

where M is the total number of sampling dates. This is the discretely sampled sum. A more natural quantity to consider is

$$A_M = \frac{I_M}{M} = \frac{1}{M} \sum_{k=1}^{M} S(t_k), \qquad (17.4)$$

because then the payoff for the discretely-sampled arithmetic average strike put is

$$\max(A_M - S, 0).$$

Can we write (17.4) in terms of an updating rule? Yes, easily: if we write

$$A_i = \frac{1}{i} \sum_{k=1}^{i} S(t_k)$$

then we have

$$A_1 = S(t_1), \quad A_2 = \frac{S(t_1) + S(t_2)}{2} = \tfrac{1}{2}A_1 + \tfrac{1}{2}S(t_2),$$

$$A_3 = \frac{S(t_1) + S(t_2) + S(t_3)}{3} = \tfrac{2}{3}A_2 + \tfrac{1}{3}S(t_3), \dots$$

or generally

$$A_i = \frac{1}{i}S(t_i) + \frac{i-1}{i}A_{i-1}.$$

We will see how to use this for pricing in the next section. But first, another example.

The lookback option has a payoff that depends on the maximum or minimum of the realized asset price. If the payoff depends on the maximum sampled at times t_i then we have

$$I_1 = S(t_1), \quad I_2 = \max(S(t_2), I_1), \quad I_3 = \max(S(t_3), I_2) \dots.$$

The updating rule is therefore simply

$$I_i = \max(S(t_i), I_{i-1}).$$

In Chapter 19, where we consider lookbacks in detail, we use the notation M_i for minimum or maximum.

How do we use these updating rules in the pricing of derivatives?

17.5 DISCRETE SAMPLING: THE PRICING EQUATION

Following the continuous-sampling case we can anticipate that the option value will be a function of three variables, $V(S, I, t)$. We derive the pricing equation in a heuristic fashion. The derivation can be made more rigorous, but there is little point since the conclusion is correct and so obvious.

The first step in the derivation is the observation that the stochastic differential equation for I is degenerate:

$$dI = 0.$$

This is because the variable I can only change at the discrete set of dates t_i. This is true if $t \neq t_i$ for any i. So provided we are not *on* a sampling date the quantity I is constant, the stochastic differential equation for I reflects this, and the pricing equation is simply the basic Black–Scholes equation:

$$\frac{\partial V}{\partial t} + \tfrac{1}{2}\sigma^2 S^2 \frac{\partial^2 V}{\partial S^2} + rS\frac{\partial V}{\partial S} - rV = 0.$$

Remember, though, that V is still a function of *three* variables; I is effectively treated as a parameter.

How does the equation know about the path dependency? What happens *at* a sampling date?

The answer to the latter question gives us the answer to the former. Across a sampling date nothing much happens. Across a sampling date the option value stays the same. As we get closer and closer to the sampling date we become more and more sure about the value that I will take according to the updating rule. Since the outcome on the sampling date is known and since *no money changes hands* there cannot be any jump in the value of the option. This is a simple application of the no arbitrage principle.

Across a sampling date the option value is continuous. If we introduce the notation t_i^- to mean the time infinitesimally before the sampling date t_i and t_i^+ to mean infinitesimally after the sampling date, then continuity of the option value is represented mathematically by

$$V(S, I_{i-1}, t_i^-) = V(S, I_i, t_i^+).$$

In terms of the updating rule, we have

A FAIRLY GENERAL JUMP CONDITION FOR PATH-DEPENDENT CONTRACTS

$$V(S, I, t_i^-) = V(S, F(S, I, i), t_i^+)$$

This is called a **jump condition**.

We call this a jump condition even though there is no jump in this case. When money does change hands on a special date there will be a sudden change in the value of the option at that time, as discussed in Chapter 15. If we follow the path of S in time we see that it is continuous. However, the path for I is discontinuous. There is a deterministic jump in I across the sampling date. If we were to plot V as a function of S and I just before and just after the sampling date we would see that *for fixed S and I* the option price would be discontinuous. But this plot would have to be interpreted correctly; $V(S, I, t)$ may be discontinuous as a function of S and I but V is continuous along each *realized* path of S and I.

In Figure 17.1 I show the relationship between before and after option values.

17.5.1 Examples

To price an arithmetic Asian option with the average sampled at times t_i solve the Black–Scholes equation for $V(S, A, t)$ with

$$V(S, A, t_i^-) = V\left(S, \frac{i-1}{i}A + \frac{1}{i}S, t_i^+\right),$$

and a suitable final condition representing the payoff.

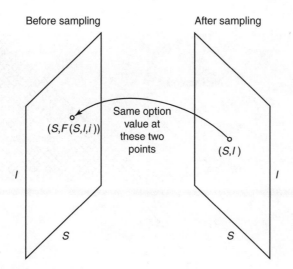

Before sampling

After sampling

$(S, F(S, I, i))$

Same option
value at
these two
points

(S, I)

I

I

S

S

Figure 17.1 Representation of the jump condition.

To price a lookback depending on the maximum sampled at times t_i solve the Black–Scholes equation for $V(S, M, t)$ with

$$V(S, M, t_i^-) = V(S, \max(S, M), t_i^+). \tag{17.5}$$

How this particular jump condition works is shown in Figure 17.2. The top right-hand plot is the S, M plane just after the sample of the maximum has been taken. Because the sample has just been taken the region $S > M$ cannot be reached, it is the region labeled 'Unattainable'. When we come to solve the Black–Scholes equation numerically in Part Six we will see how we work backwards in time, so that we will find the option value for time t_i^+ *before* the value for time t_i^-. So we will have found the option value $V(S, M, t_i^+)$ for all $S < M$. To find the option value just before the sampling we must apply the jump condition (17.5). Pictorially, this means that the option value at time t_i^- for $S < M$ is the same as the t_i^+ value–just follow the left-hand arrow in the figure. *However*, for $S > M$ (which is attainable before the sample is taken) the option value comes from the $S = M$ line at time t_i^+ for the same S value; after all, S is continuous. Now, just follow the right-hand arrow.

17.5.2 The Algorithm for Discrete Sampling

Because the path-dependent quantity, I, is updated discretely and is therefore constant between sampling dates, the partial differential equation for the option value between sampling dates is the Black–Scholes equation with I treated as a parameter. The algorithm for valuing an option on a discretely-sampled quantity is as follows:

- Working backwards from expiry, solve

$$\frac{\partial V}{\partial t} + \tfrac{1}{2}\sigma^2 S^2 \frac{\partial^2 V}{\partial S^2} + rS \frac{\partial V}{\partial S} - rV = 0$$

between sampling dates. (How to do this is the subject of Part Seven.) Stop when you get to the timestep on which the sampling takes place.

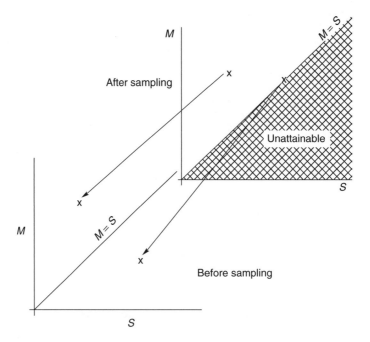

Figure 17.2 The jump condition for a lookback option.

- Then apply the appropriate jump condition across the current sampling date to deduce the option value immediately before the present sampling date using the calculated value of the option just after. Use this as your final condition for further timestepping of the Black–Scholes equation.
- Repeat this process as necessary to arrive at the current value of the option.

17.6 HIGHER DIMENSIONS

The methods outlined above are not restricted to a single path-dependent quantity. Any finite number of path-dependent variables can be accommodated, theoretically. Imagine a contract that pays off the difference between a continuous geometric and a continuous arithmetic average. To price this one would need to introduce I_g and I_a, defined by

$$I_g = \int_0^t \log(S) \, d\tau \quad \text{and} \quad I_a = \int_0^t S \, d\tau.$$

The solution would then be a function of four variables, $V(S, I_g, I_a, t)$. However, this is at the limit of practicality for a numerical solution of a partial differential equation. Unless there is a similarity solution, reducing the dimensionality of the problem, it may be better to consider Monte Carlo simulation.

The same thoughts apply to discrete sampling or a combination of discrete and continuous.

17.7 **PRICING VIA EXPECTATIONS**

In Chapter 10 I showed how we can value options in the Black–Scholes world by taking the present value of the expected payoff under a risk-neutral random walk. This approach applies perfectly well to all of the path-dependent options we have described or are going to describe. Simply simulate the random walk

$$dS = rS\,dt + \sigma S\,dX,$$

as will be discussed in Chapter 66, for many paths, calculate the payoff for each path—and this means calculating the value of the path-dependent quantity which is usually very simple to do—take the average payoff over all the paths and then take the present value of that average. That is the option fair value. Note that there is no μ in this, it is the risk-neutral random walk that must be simulated.

This is a very general and powerful technique, useful for path-dependent contracts for which either a partial differential equation approach is impossible or too high dimensional. The only disadvantage, and it's a big one, is that it is hard to value American options in this framework.

17.8 **EARLY EXERCISE**

If you have found a partial differential equation formulation of the option problem then it is simple to incorporate the early exercise feature of American and Bermudan options. Simply apply the constraint

$$V(S, I, t) \geq P(S, I),$$

together with continuity of the delta of the option, where $P(S, I)$ is the payoff function (and it can also be time-dependent). This condition is to be applied at any time that early exercise is allowed. If you have found a partial differential equation formulation of the problem and it is in sufficiently low dimension then incorporating early exercise in the numerical scheme is a matter of adding a couple of lines of code; see Chapter 64.

17.9 **SUMMARY**

The basic theory has been explained above for the pricing of many path-dependent contracts in the partial differential equation framework. We have examined both continuously-sampled and discretely-sampled path-dependent quantities. In the next three chapters we discuss these matters in more detail, first with Asian and lookback options and then for a wider range of exotic contracts. The practical implementation of these models is described in Chapter 64.

FURTHER READING

- See Bergman (1985) for the early work on a unified partial differential equation framework for path-dependent contracts.
- The excellent book by Ingersoll (1987) also discusses the partial differential equation approach to pricing exotics.
- The general framework for exotics is described in a *Risk* magazine article by Dewynne & Wilmott (1994a) and also in Dewynne & Wilmott (1996).

CHAPTER 18
Asian options

In this Chapter...

- many types of Asian option with a payoff that depends on an average
- the different types of averaging of asset prices that are used in defining the payoff
- how to price these contracts in the partial differential equation framework

18.1 INTRODUCTION

Asian options give the holder a payoff that depends on the average price of the underlying over some prescribed period. This averaging of the underlying can significantly reduce the price of an Asian option compared with a similar vanilla contract. Anything that reduces the up-front premium in an option contract tends to make them more popular.

Asian options might be bought by someone with a stream of cashflows in a foreign currency, due to sales abroad for example, and who wants to hedge against fluctuations in the exchange rate. The Asian tail that we see later is designed to reduce exposure to sudden movements in the underlying just before expiry; some pension schemes have such a feature.

In this chapter we find differential equations for the value of a wide variety of such Asian options. There are many ways to define an 'average' of the price, and we begin with a discussion of the obvious possibilities. We will see how to write the price as the solution of a partial differential equation in *three* variables: the underlying asset, time, and a new state variable representing the evolution of the average.

18.2 PAYOFF TYPES

Assuming for the moment that we have defined our average A, what sort of payoffs are common? As well as calls, puts etc. there is also the classification of **strike** and **rate**. These classifications work as follows. Take the payoff for a vanilla option, a vanilla call, say,

HERE ARE SOME DEFINITIONS OF BASIC ASIAN CONTRACTS

$$\max(S - E, 0).$$

Replace the strike price E with an average and you have an **average strike call**. This has payoff

$$\max(S - A, 0).$$

An **average strike put** thus has payoff

$$\max(A - S, 0).$$

Now take the vanilla payoff and instead replace the asset with its average; what you get is a rate option. For example, an **average rate call** has payoff

$$\max(A - E, 0)$$

and an **average rate put** has payoff

$$\max(E - A, 0).$$

The average rate options can be used to lock in the price of a commodity or an exchange rate for those who have a continual and fairly predictable exposure to one of these over extended periods.

The difference between calls and puts is simple from a pricing point of view, but the strike/rate distinction can make a big difference. Strike options are easier to value numerically.

THE TYPE OF AVERAGING WILL AFFECT THE PRICING OF THE CONTRACT

18.3 **TYPES OF AVERAGING**

The precise definition of the average used in an Asian contract depends on two elements: how the data points are combined to form an average and which data points are used. The former means whether we have an arithmetic or geometric average or something more complicated. The latter means how many data points we use in the average, all quoted prices, or just a subset, and over what time period.

18.3.1 Arithmetic or Geometric

The two simplest and obvious types of average are the **arithmetic average** and the **geometric average**. The arithmetic average of the price is the sum of all the constituent prices, equally weighted, divided by the total number of prices used. The geometric average is the *exponential* of the sum of all the *logarithms* of the constituent prices, equally weighted, divided by the total number of prices used. Another popular choice is the exponentially weighted average, meaning instead of having an equal weighting to each price in the average, the recent prices are weighted more than past prices in an exponentially decreasing fashion.

18.3.2 Discrete or Continuous

How much data do we use in the calculation of the average? Do we take every traded price or just a subset? If we take closely-spaced prices over a finite time then the sums that we calculate in the average become integrals of the asset (or some function of it) over the averaging period. This would give us a **continuously-sampled average**. More commonly, we only take data points that are reliable, using closing prices, a smaller set of data. This is called **discrete sampling**. This issue of continuous or discrete sampling was discussed in the previous chapter.

18.4 EXTENDING THE BLACK–SCHOLES EQUATION

18.4.1 Continuously-sampled Averages

Figure 18.1 shows a realization of the random walk followed by an asset, in this case YPF, an Argentine oil company, together with a continuously-sampled running arithmetic average. This average is defined as

$$\frac{1}{t} \int_0^t S(\tau)\,d\tau.$$

If we introduce the new state variable

$$I = \int_0^t S(\tau)\,d\tau$$

then, following the analysis of Chapter 17, the partial differential equation for the value of an option contingent on this average is

$$\frac{\partial V}{\partial t} + S\frac{\partial V}{\partial I} + \tfrac{1}{2}\sigma^2 S^2 \frac{\partial^2 V}{\partial S^2} + rS\frac{\partial V}{\partial S} - rV = 0.$$

Figure 18.2 shows a realization of an asset price random walk with the continuously-sampled geometric running average.

The continuously-sampled geometric average is defined to be

$$\exp\left(\frac{1}{t} \int_0^t \log S(\tau)\,d\tau\right).$$

To value an option contingent on this average we define

$$I = \int_0^t \log S(\tau)\,d\tau$$

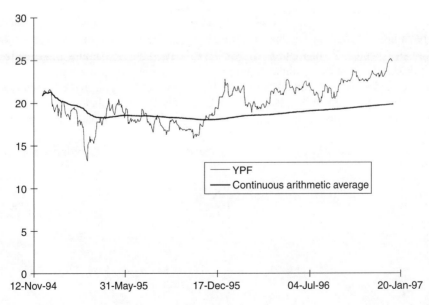

Figure 18.1 An asset price random walk and its continuously measured arithmetic running average.

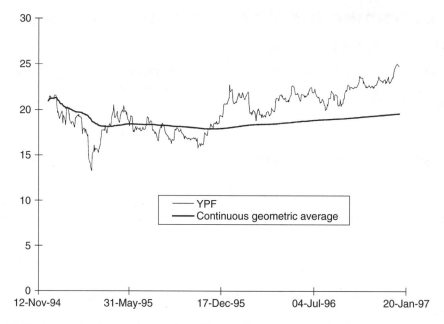

Figure 18.2 An asset price random walk and its continuous geometric running average.

and, following again the analysis of Chapter 17, the partial differential equation for the value of the option is

$$\frac{\partial V}{\partial t} + \log S \frac{\partial V}{\partial I} + \tfrac{1}{2}\sigma^2 S^2 \frac{\partial^2 V}{\partial S^2} + rS \frac{\partial V}{\partial S} - rV = 0.$$

18.4.2 Discretely-sampled Averages

Discretely-sampled averages, whether arithmetic or geometric, fit easily into the framework established in Chapter 17. In Figures 18.3 and 18.4 are shown examples of a realized asset price and discretely-sampled arithmetic and geometric averages respectively.

Above, we modeled the continuously-sampled average as an integral. By a discretely-sampled average we mean the sum, rather than the integral, of a finite number of values of the asset during the life of the option. Such a definition of average is easily included within the framework of our model.

If the sampling dates are t_i, $i = 1, \ldots$ then the discretely-sampled arithmetic averages are defined by

$$A_i = \frac{1}{i} \sum_{k=1}^{i} S(t_k).$$

In particular

$$A_1 = S(t_1), \quad A_2 = \frac{S(t_1) + S(t_2)}{2} = \tfrac{1}{2}A_1 + \tfrac{1}{2}S(t_2),$$

$$A_3 = \frac{S(t_1) + S(t_2) + S(t_3)}{3} = \tfrac{2}{3}A_2 + \tfrac{1}{3}S(t_3), \ldots.$$

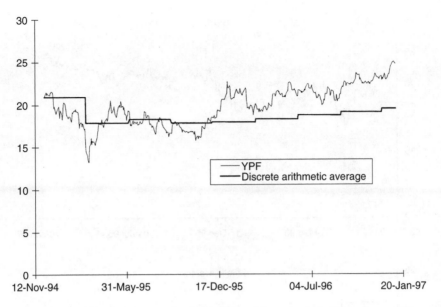

Figure 18.3 An asset price random walk and its discretely-sampled arithmetic running average.

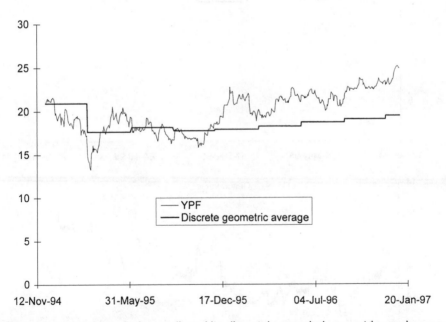

Figure 18.4 An asset price random walk and its discretely-sampled geometric running average.

It is easy to see that these are equivalent to

$$A_i = \frac{i-1}{i} A_{i-1} + \frac{1}{i} S(t_i).$$

At the top of Figure 18.5 is shown a realized asset price path. Below that is the discretely-sampled average. This is necessarily piecewise constant. At the bottom of the figure is the value

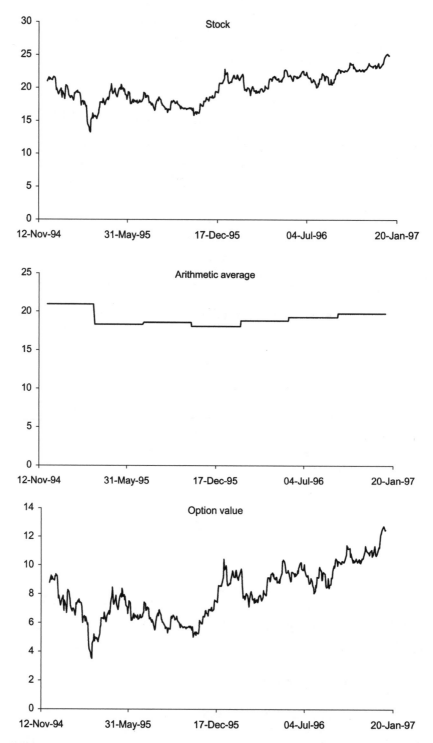

Figure 18.5   Top: An asset price random walk. Middle: Its discretely-sampled arithmetic running average. Bottom: The option value.

of some option (it doesn't matter which). The option value must be continuous to eliminate arbitrage opportunities.

Using the results of Chapter 17, the jump condition for an Asian option with discrete arithmetic averaging is then simply

$$V(S, A, t_i^-) = V\left(S, \frac{i-1}{i}A + \frac{1}{i}S, t_i^+\right).$$

This is a result of the continuity of the option price across a sampling date i.e. no arbitrage.

Similarly the discretely-sampled geometric average has the jump condition

$$V(S, A, t_i^-) = V\left(S, \exp\left(\frac{i-1}{i}\log(A) + \frac{1}{i}\log(S)\right), t_i^+\right)$$

where

$$A_i = \exp\left(\frac{i-1}{i}\log(A_{i-1}) + \frac{1}{i}\log(S(t_i))\right).$$

Figures 18.6 and 18.7 show details of an arithmetic average rate call with discrete sampling.

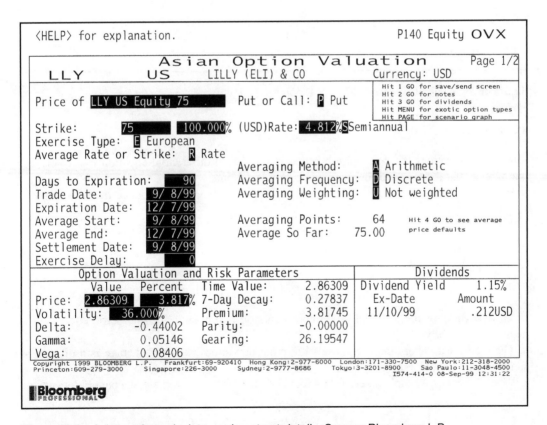

Figure 18.6 Asian option calculator and contract details. Source: Bloomberg L.P.

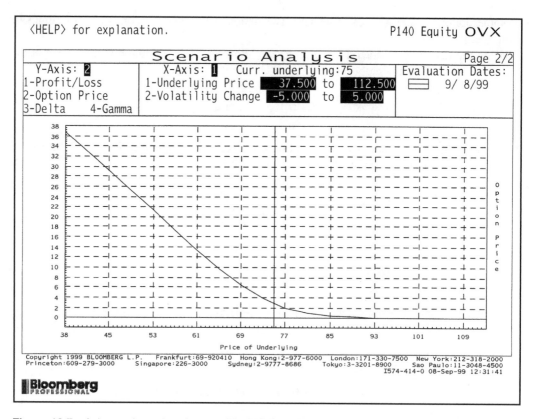

<HELP> for explanation.   P140 Equity OVX
```
                         Scenario Analysis              Page 2/2
   Y-Axis: 2              X-Axis: 1   Curr. underlying:75      Evaluation Dates:
  1-Profit/Loss          1-Underlying Price  37.500 to 112.500    ▭  9/ 8/99
  2-Option Price         2-Volatility Change -5.000 to  5.000
  3-Delta    4-Gamma
```

Figure 18.7 Asian option price. Source: Bloomberg L.P.

18.4.3 Exponentially-weighted and Other Averages

Simple modifications that are easily handled in the partial differential equation framework are the exponential average and the average up to a fixed time.

In the exponential continuously-sampled arithmetic average just introduce the new variable

$$I = \lambda \int_{-\infty}^{t} e^{-\lambda(t-\tau)} S(\tau) \, d\tau$$

which satisfies

$$dI = \lambda(S - I) \, dt.$$

From this, the governing partial differential equation is obvious. The geometric equivalent is dealt with similarly.

The reader can determine for himself the jump condition when the average is a discretely-sampled exponential average.

When the average is only taken up to a fixed point, so that, for example, the payoff depends on

$$I = \int_{0}^{T_0} S(\tau) \, d\tau \quad \text{with} \quad T_0 < T,$$

then the new term in the partial differential equation (the derivative with respect to I) disappears for times greater that T_0. That is,

$$\frac{\partial V}{\partial t} + S\mathcal{H}(T_0 - t)\frac{\partial V}{\partial I} + \frac{1}{2}\sigma^2 S^2 \frac{\partial^2 V}{\partial S^2} + rS\frac{\partial V}{\partial S} - rV = 0,$$

where $\mathcal{H}$ is the Heaviside function.

One type of contract that is *not* easily put into a partial differential equation framework with a finite number of underlyings is the moving window option. In this option, the holder can exercise early for an amount that depends on the average over the previous three months, say. The key point about this contract that makes it difficult is that the starting point of the averaging period is not fixed in time. As a result the stochastic differential equation for the path-dependent quantity cannot be written in terms of *local* values of the independent variables: all details of the path need to be known and recorded.

18.4.4 The Asian Tail

Often the averaging is confined to only a part of the life of the option. For example, if the averaging of the underlying is only over the final part of the option's life it is referred to as an **Asian tail**. Such a contract would reduce the exposure of the option to sudden moves in the underlying just before the payoff is received. A feature like this is also common in pension awards.

18.5 EARLY EXERCISE

There is not much to be said about early exercise that has not already been said elsewhere in this book. The only point to mention is that the details of the payoff on early exercise have to be well defined. The payoff at expiry depends on the value of the average up to expiry, which will, of course, not be known until expiry. Typically, on early exercise it is the average to date that is used. For example, in an American arithmetic average strike put the early payoff would be

$$\max\left(\frac{1}{t}\int_0^t S(\tau)\,d\tau - S, 0\right).$$

18.6 ASIAN OPTIONS IN HIGHER DIMENSIONS

We are not restricted to an average of a single underlying. The **anteater option**[1], so called for obvious reasons, has a payoff defined in terms of the average of the ratio of two underlying, S_1 and S_2:

$$I = \int_0^t \frac{S_1}{S_2}\,d\tau.$$

This contract will be in four dimensions, S_1, S_2, I and t.

---

[1] Gunner Wilkins' favorite contract.

IF A SIMILARITY REDUCTION IS POSSIBLE, USE IT. COMPUTATION WILL BE EASIER AND FASTER

18.7 **SIMILARITY REDUCTIONS**

As long as the stochastic differential equation or updating rule for the path-dependent quantity only contains references to S, t and the path-dependent quantity itself then the value of the option depends on three variables. Unless we are very lucky, the value of the option must be calculated numerically. Some options have a particular structure that permits a reduction in the dimensionality of the problem by use of a similarity variable. I will illustrate the idea with an example. The dimensionality of the continuously-sampled arithmetic average strike option can be reduced from three to two.

The payoff for the call option is

$$\max\left(S - \frac{1}{T}\int_0^T S(\tau)\,d\tau, 0\right).$$

We can write the running payoff for the call option as

$$I\max\left(R - \frac{1}{t}, 0\right),$$

where

$$I = \int_0^t S(\tau)\,d\tau$$

and

$$R = \frac{S}{\displaystyle\int_0^t S(\tau)\,d\tau}. \tag{18.1}$$

The payoff at expiry may then be written as

$$I\max\left(R - \frac{1}{T}, 0\right).$$

In view of the form of the payoff function, it seems plausible that the option value takes the form

$$V(S, R, t) = IW(R, t), \quad \text{with} \quad R = \frac{S}{I}.$$

We find that W satisfies

$$\frac{\partial W}{\partial t} + \tfrac{1}{2}\sigma^2 R^2 \frac{\partial^2 W}{\partial R^2} + R(r - R)\frac{\partial W}{\partial R} - (r - R)W \leq 0. \tag{18.2}$$

If the option is European we have strict equality in (18.2). If it is American we may have inequality in (18.2) but the constraint

$$W(R, t) \geq \max\left(R - \frac{1}{t}, 0\right)$$

must be satisfied. Moreover, if the option price ever meets the early exercise payoff it must do so smoothly. That is, the function $W(R, t)$ and its first R-derivative must be continuous everywhere.

For the European option we must impose boundary conditions at both $R = 0$ and as $R \to \infty$:

$$W(0, t) = 0,$$

and

$$W(R, t) \sim R \quad \text{as} \quad R \to \infty.$$

The solution of the European problem can be written as an infinite sum of confluent hypergeometric functions. I do not give this exact solution because it is easier (and certainly a more flexible approach) to obtain values by applying numerical methods directly to the partial differential equation.

In Figure 18.8 we see W against R at three months before expiry and with three months' averaging completed; $\sigma = 0.4$ and $r = 0.1$.

In the case of an American option, we have to solve the partial differential inequality (18.2) subject to the constraint, the final condition and the boundary conditions. We cannot do this analytically and we must find the solution numerically.

18.7.1 Put-call Parity for the European Average Strike

The payoff at expiry for a portfolio of one European average strike call held long and one put held short is

$$I \max\left(R - \frac{1}{T}, 0\right) - I \max\left(\frac{1}{T} - R, 0\right).$$

Whether R is greater or less than T at expiry, this payoff is simply

$$S - \frac{I}{T}.$$

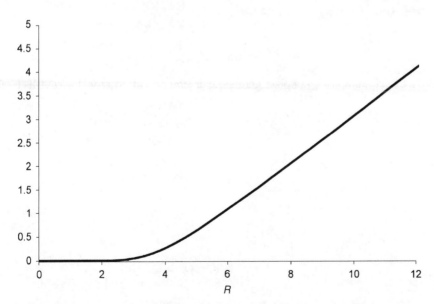

Figure 18.8 The European average strike call option; similarity variable W versus similarity variable R with $\sigma = 0.4$ and $r = 0.1$ at three months before expiry; there has already been three months' averaging.

The value of this portfolio is identical to one consisting of one asset and a financial product whose payoff is

$$-\frac{I}{T}.$$

In order to value this product find a solution of the average strike equation of the form

$$W(R, t) = b(t) + a(t)R \tag{18.3}$$

and with $a(T) = 0$ and $b(T) = -1/T$; such a solution would have the required payoff of $-I/T$. Substituting (18.3) into (18.2) and satisfying the boundary conditions, we find that

$$a(t) = -\frac{1}{rT}(1 - e^{-r(T-t)}), \quad b(t) = -\frac{1}{T}e^{-r(T-t)}.$$

We conclude that

$$V_C - V_P = S - \frac{S}{rT}(1 - e^{-r(T-t)}) - \frac{1}{T}e^{-r(T-t)}\int_0^t S(\tau)\,d\tau,$$

where V_C and V_P are the values of the European arithmetic average strike call and put. This is put-call parity for the European average strike option.

18.8 SOME FORMULAE

There are very few nice formulae for the values of Asian options. The best known are for average rate calls and puts when the average is a continuously-sampled, geometrical average.

The Geometric Average Rate Call

This option has payoff

$$\max(A - E, 0),$$

where A is the continuously-sampled geometric average. This option has a Black–Scholes value of

$$e^{-r(T-t)}\left[G\exp\left(\frac{(r - D - \sigma^2/2)(T - t)^2}{2T} + \frac{\sigma^2(T - t)^3}{6T^2}\right)N(d_1) - EN(d_2)\right]$$

where $\quad I = \int_0^t \log(S(\tau))\,d\tau,$

$\qquad G = e^{I/T}S^{(T-t)/T},$

where

$$d_1 = \frac{T\log(G/E) + (r - D - \sigma^2/2)(T - t)^2/2 + \sigma^2(T - t)^3/3T}{\sigma\sqrt{(T - t)^3/3}}$$

and

$$d_2 = \frac{T\log(G/E) + (r - D - \sigma^2/2)(T - t)^2/2}{\sigma\sqrt{(T - t)^3/3}}.$$

The geometric average of a lognormal random walk is itself lognormally distributed, but with a reduced volatility.

The Geometric Average Rate Put

This option has payoff

$$\max(E - A, 0),$$

where A is the continuously-sampled geometric average. This option has a Black–Scholes value of

$$e^{-r(T-t)} \left[EN(-d_2) - G \exp\left(\frac{(r - D - \sigma^2/2)(T - t)^2}{2T} + \frac{\sigma^2(T - t)^3}{6T^2} \right) N(-d_1) \right]$$

with the same G, d_1 and d_2 as before.

18.9 SUMMARY

I applied the general theory of Chapter 17 to the problem of pricing Asian options, options with a payoff depending on an average. The partial differential equation approach is very powerful for these types of options, and hard to beat if your option has a similarity reduction or is American style. The approach can be generalized much further. We will see this in Chapter 21.

FURTHER READING

- Some exact solutions can be found in Boyle (1991) and Angus (1999).
- Bergman (1985) and Ingersoll (1987) present the partial differential equation formulation of some average strike options and demonstrate the similarity reduction.
- For another, rather abstract, method for valuing Asian options see Geman & Yor (1993).
- The application of the numerical Monte Carlo method is described by Kemna & Vorst (1990). They also derive some exact formulae.
- More examples of the methods described here can be found in Dewynne & Wilmott (1995b, c).
- For an approximate valuation of arithmetic Asian options see Levy (1990) who replaces the density function for the distribution of the average by a lognormal function.

CHAPTER 19
lookback options

In this Chapter...

- the features that make up a lookback option
- how to put the lookback option into the Black–Scholes framework for both continuous- and discrete-sampling cases

19.1 INTRODUCTION

The dream contract has to be one that pays the difference between the highest and the lowest asset prices realized by an asset over some period. Any speculator is trying to achieve such a trade. The contract that pays this is an example of a **lookback option**, an option that pays off some function of the realized maximum and/or minimum of the underlying asset over some prescribed period. Since lookback options have such an extreme payoff they tend to be expensive.

We can price these contracts in the Black–Scholes environment quite easily, theoretically. There are two cases to consider; whether the maximum/minimum is measured continuously or discretely. Here I will always talk about the 'maximum' of the asset. The treatment of the 'minimum' should be obvious.

19.2 TYPES OF PAYOFF

For the basic lookback contracts, the payoff comes in two varieties, like the Asian option. These are the *rate* and the *strike* option, also called the **fixed strike** and the **floating strike** respectively. These have payoffs that are the same as vanilla options except that in the strike option the vanilla exercise price is replaced by the maximum. In the rate option it is the asset value in the vanilla option that is replaced by the maximum.

19.3 CONTINUOUS MEASUREMENT OF THE MAXIMUM

In Figure 19.1 is shown the termsheet for a foreign exchange lookback swap. One counterparty pays a floating interest rate every six months, while the other counterparty pays on maturity a linear function of the maximum realized level of the exchange rate. The floating interest

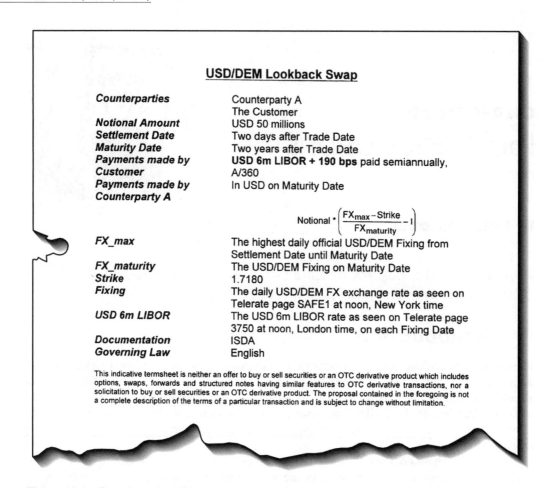

Figure 19.1 Termsheet for USD/DEM lookback swap.

payments of six-month LIBOR can be represented as a much simpler cashflow, to be explained in Chapter 39. Then this contract becomes a straightforward lookback option.

Introduce the new variable M as the realized maximum of the asset from the start of the sampling period $t = 0$, say, until the current time t:

$$M = \max_{0 \leq \tau \leq t} S(\tau).$$

In Figure 19.2 is shown a realization of an asset price path and the continuously-sampled maximum. An obvious point about this plot, but one that is worth mentioning, is that the asset price is always below the maximum. (This will not be the case when we come to examine the discretely-sampled case.) The value of our lookback option is a function of three variables, $V(S, M, t)$ but now we have the restriction

$$0 \leq S \leq M.$$

This observation will also lead us to the correct partial differential equation for the option's value, and the boundary conditions. We derive the equation in a heuristic fashion that can be made rigorous.

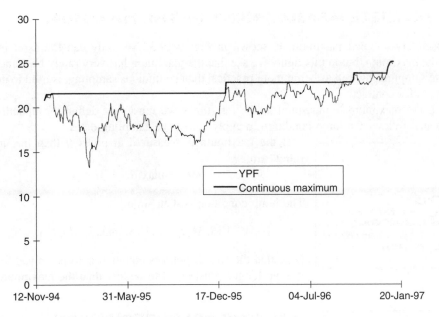

Figure 19.2 An asset price path and the continuously-sampled maximum.

From Figure 19.2 we can see that most of the time the asset price is below the maximum. But there are times when they coincide. When $S < M$ the maximum cannot change and so the variable M satisfies the stochastic differential equation

$$dM = 0.$$

Hold that thought. While $0 \leq S < M$ the governing equation must be Black–Scholes

$$\frac{\partial V}{\partial t} + \tfrac{1}{2}\sigma^2 S^2 \frac{\partial^2 V}{\partial S^2} + rS \frac{\partial V}{\partial S} - rV = 0,$$

with M as a 'parameter' and only for $S < M$.

The behavior of the option value when $S = M$ tells us the *boundary condition* to apply there. The boundary condition is

$$\frac{\partial V}{\partial M} = 0 \quad \text{on} \quad S = M.$$

The reason for this boundary condition is that the option value is insensitive to the level of the maximum when the asset price is *at* the maximum. This is because the probability of the present maximum still being the maximum at expiry is zero. The rigorous derivation of this boundary condition is rather messy; see the original paper by Goldman, Sosin & Gatto (1979) for the details.

Finally, we must impose a condition at expiry to reflect the payoff. As an example, consider the lookback rate call option. This has a payoff given by

$$\max(M - E, 0).$$

The lookback strike put has a payoff given by

$$\max(M - S, 0).$$

19.4 **DISCRETE MEASUREMENT OF THE MAXIMUM**

The discretely-sampled maximum is shown in Figure 19.3. Not only can the asset price go above the maximum, but in this figure we see that the maximum has very rarely been increased. Discrete sampling, as well as being more practical than continuous sampling, is used to decrease the value of a contract.

When the maximum is measured at discrete times we must first define the updating rule, from which follows the jump condition to apply across the sampling dates.

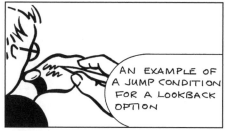

If the maximum is measured at times t_i then the updating rule is simply

$$M_i = \max(S(t_i), M_{i-1}).$$

The jump condition is then simply

$$V(S, M, t_i^-) = V(S, \max(S, M), t_i^+).$$

Note that the Black–Scholes equation is to be solved for all S; it is no longer constrained to be less than the maximum.

19.5 **SIMILARITY REDUCTION**

The general lookback option with a payoff depending on one path-dependent quantity is a three-dimensional problem. The three dimensions are asset price, the maximum and time. The numerical solution of this problem is more time-consuming than a two-dimensional problem. However, there are some special, and important, cases when the dimensionality of the problem can be reduced.

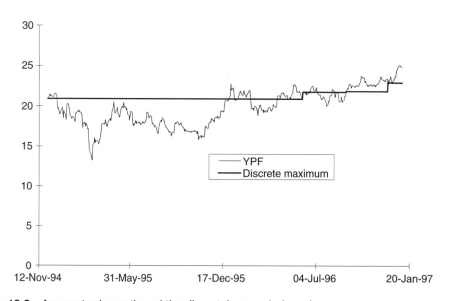

Figure 19.3 An asset price path and the discretely-sampled maximum.

This reduction relies on some symmetry properties in the equation and is not something that can be applied to all or, indeed, many, lookback contracts. It is certainly possible if the payoff takes the form

$$M^\alpha P\left(\frac{S}{M}\right).$$ (19.1)

For example, this is true for the lookback strike put which has payoff

$$\max(M - S, 0) = M \max\left(1 - \frac{S}{M}, 0\right).$$

Generally, if the payoff takes the form (19.1), then the substitution

$$\xi = \frac{S}{M}$$

leads to a problem for $W(\xi, t)$ where

$$V(S, M, t) = M^\alpha W(\xi, t)$$

where W satisfies the governing equation

$$\frac{\partial W}{\partial t} + \frac{1}{2}\sigma^2 \xi^2 \frac{\partial^2 W}{\partial \xi^2} + r\xi \frac{\partial W}{\partial \xi} - rW = 0,$$

the final condition

$$W(\xi, T) = P(\xi)$$

and the boundary condition

$$\frac{\partial W}{\partial \xi} - \alpha W = 0 \quad \text{on} \quad \xi = 1.$$

19.6 SOME FORMULAE

Floating Strike Lookback Call

The continuously-sampled version of this option has a payoff

$$\max(S - M, 0) = S - M,$$

where M is the realized minimum of the asset price. In the Black–Scholes world the value is

$$Se^{-D(T-t)}N(d_1) - Me^{-r(T-t)}N(d_2) + Se^{-r(T-t)}\frac{\sigma^2}{2(r - D)}$$

$$\times \left(\left(\frac{S}{M}\right)^{-(2(r-D)/\sigma^2)} N\left(-d_1 + \frac{2(r - D)\sqrt{T - t}}{\sigma}\right) - e^{(r-D)(T-t)}N(-d_1)\right),$$

where

$$d_1 = \frac{\log\left(\frac{S}{M}\right) + \left(r - D + \frac{1}{2}\sigma^2\right)(T - t)}{\sigma\sqrt{T - t}}$$

and

$$d_2 = d_1 - \sigma\sqrt{T - t}.$$

Floating Strike Lookback Put

The continuously-sampled version of this option has a payoff

$$\max(M - S, 0) = M - S,$$

where M is the realized maximum of the asset price. The value is

$$Me^{-r(T-t)}N(-d_2) - Se^{-D(T-t)}N(-d_1) + Se^{-r(T-t)}\frac{\sigma^2}{2(r-D)}$$

$$\times \left(-\left(\frac{S}{M}\right)^{-(2(r-D)/\sigma^2)} N\left(d_1 - \frac{2(r-D)\sqrt{T-t}}{\sigma}\right) + e^{(r-D)(T-t)}N(d_1)\right),$$

where

$$d_1 = \frac{\log\left(\frac{S}{M}\right) + \left(r - D + \frac{1}{2}\sigma^2\right)(T-t)}{\sigma\sqrt{T-t}}$$

and

$$d_2 = d_1 - \sigma\sqrt{T-t}.$$

Fixed Strike Lookback Call

This option has a payoff given by

$$\max(M - E, 0)$$

where M is the realized maximum. For $E > M$ the fair value is

$$Se^{-D(T-t)}N(d_1) - Ee^{-r(T-t)}N(d_2) + Se^{-r(T-t)}\frac{\sigma^2}{2(r-D)}$$

$$\times \left(-\left(\frac{S}{E}\right)^{-(2(r-D)/\sigma^2)} N\left(d_1 - \frac{2(r-D)\sqrt{T-t}}{\sigma}\right) + e^{(r-D)(T-t)}N(d_1)\right),$$

where

$$d_1 = \frac{\log\left(\frac{S}{E}\right) + \left(r - D + \frac{1}{2}\sigma^2\right)(T-t)}{\sigma\sqrt{T-t}}$$

and

$$d_2 = d_1 - \sigma\sqrt{T-t}.$$

When $E < M$ the value is

$$(M - E)e^{-r(T-t)} + Se^{-D(T-t)}N(d_1) - Me^{-r(T-t)}N(d_2) + Se^{-r(T-t)}\frac{\sigma^2}{2(r-D)}$$

$$\times \left(-\left(\frac{S}{M}\right)^{-(2(r-D)/\sigma^2)} N\left(d_1 - \frac{2(r-D)\sqrt{T-t}}{\sigma}\right) + e^{(r-D)(T-t)}N(d_1)\right),$$

where

$$d_1 = \frac{\log\left(\frac{S}{M}\right) + \left(r - D + \frac{1}{2}\sigma^2\right)(T-t)}{\sigma\sqrt{T-t}}$$

and

$$d_2 = d_1 - \sigma\sqrt{T-t}.$$

Fixed Strike Lookback Put

This option has a payoff given by

$$\max(E - M, 0)$$

where M is the realized minimum. For $E < M$ the fair value is

$$Ee^{-r(T-t)}N(-d_2) - Se^{-D(T-t)}N(-d_1) + Se^{-r(T-t)}\frac{\sigma^2}{2(r-D)}$$

$$\times \left(\left(\frac{S}{E}\right)^{-(2(r-D)/\sigma^2)} N\left(-d_1 + \frac{2(r-D)\sqrt{T-t}}{\sigma}\right) - e^{(r-D)(T-t)}N(-d_1)\right),$$

where

$$d_1 = \frac{\log\left(\frac{S}{E}\right) + \left(r - D + \frac{1}{2}\sigma^2\right)(T-t)}{\sigma\sqrt{T-t}}$$

and

$$d_2 = d_1 - \sigma\sqrt{T-t}.$$

When $E > M$ the value is

$$(E - M)e^{-r(T-t)} - Se^{-D(T-t)}N(-d_1) + Me^{-r(T-t)}N(-d_2) + Se^{-r(T-t)}\frac{\sigma^2}{2(r-D)}$$

$$\times \left(\left(\frac{S}{M}\right)^{-(2(r-D)/\sigma^2)} N\left(-d_1 + \frac{2(r-D)\sqrt{T-t}}{\sigma}\right) - e^{(r-D)(T-t)}N(-d_1)\right),$$

where

$$d_1 = \frac{\log\left(\frac{S}{M}\right) + \left(r - D + \frac{1}{2}\sigma^2\right)(T-t)}{\sigma\sqrt{T-t}}$$

and

$$d_2 = d_1 - \sigma\sqrt{T-t}.$$

19.7 SUMMARY

Lookback options, and lookback features generally, are seen in many types of contract. They are quite common in fixed-income products where an interest payment may depend on the maximum level that rates have reached over some previous period. The same partial differential equation framework that we have seen in the equity world carries over in principle to the more complicated stochastic interest rate world.

FURTHER READING

- See Goldman, Sosin & Gatto (1979) for the first academic description of lookback options. They show how to rigorously derive the crucial boundary condition.
- Conze & Viswanathan (1991) give the derivation of formulae for several types of lookback option.
- Babbs (1992) puts the lookback option into a binomial stteing.
- See Dewynne & Wilmott (1994b) for a derivation of the governing equation and boundary conditions.
- Heynen & Kat (1995) discuss the discrete and partial monitoring of the maximum.
- Two contracts that are related to lookback options are the stop-loss option, described by Fitt, Dewynne & Wilmott (1994), and the Russian option, see Duffie & Harrison (1992).

CHAPTER 20
derivatives and stochastic control

In this Chapter...

- an intro to the subject of stochastic control
- valuing a rather complex exotic

20.1 INTRODUCTION

Some options give the holder an element of control during the life of the contract. The commonest of these is the American option which the holder can exercise whenever he wants to and doesn't have to wait until expiry. Some contracts give the holder even greater flexibility and require him to make many and complex decisions.

In this chapter I want to describe an option that sits outside the framework we have developed so far. It is not too difficult to analyze but it introduces some new ideas, and in particular it is a gentle introduction to the subject of **stochastic control**.

20.2 PERFECT TRADER AND PASSPORT OPTIONS

Suppose that you invest in a particular stock, keeping track of its movements and buying or selling according to your view of its future direction. The amount of money that you accumulate due to trading in this stock is called the **trading account**. If you are a good trader or lucky then the amount in the account will grow, and if you are a bad trader or unlucky then the amount will be negative. How much would you pay to be insured against losing money over a given time horizon? A **perfect trader** or **passport option** is a call option on the trading account, giving the holder the amount in his account at the end of the horizon if it is positive, or zero if it is negative.

To value this contract we must introduce a new variable π which is the value of the trading account, meaning the value of the stocks held together with any cash accumulated. This quantity satisfies the following stochastic differential equation

$$d\pi = r(\pi - qS)\,dt + q\,dS, \tag{20.1}$$

Classification	Passport Option
Time dependence	No
Cashflow	No
Decisions	Yes
Path dependence	Weak
Dimension	3
Order	First

Classification option table for Passport Option.

where S is still the lognormally distributed stock price and q is the amount of stock held at time t. It's quite easy to see why this should be. The trading account accumulates value for two reasons. First, the account changes because of the change in the asset holding, $q\,dS$. Second, the holding in cash, $\pi - qS$, earns the risk-free rate of interest.

The quantity q is of our choosing, will vary as time and the stock price change, and is called the **strategy**; it will be a function of S, π and t. I will restrict the size of the position in the stock by insisting that $|q| \leq 1$. Equation (20.1) contains a deterministic and a random term. The first term says that there is growth in the cash holding, $\pi - qS$, due to the addition of interest at a rate r, and the second term is due to the change in value of the stock holding.

The contract will pay off an amount

$$\max(\pi, 0)$$

at time T. This will be the final condition for our option value $V(S, \pi, t)$. Note that the option value is a function of three variables.

Now let us hedge this option:

$$\Pi = V - \Delta S.$$

We find that

$$d\Pi = \left(\frac{\partial V}{\partial t} + \tfrac{1}{2}\sigma^2 S^2 \frac{\partial^2 V}{\partial S^2} + q\sigma^2 S^2 \frac{\partial^2 V}{\partial S\,\partial \pi} + \tfrac{1}{2}q^2\sigma^2 S^2 \frac{\partial^2 V}{\partial \pi^2} \right) dt + \frac{\partial V}{\partial S}\,dS + \frac{\partial V}{\partial \pi}\,d\pi - \Delta\,dS.$$

Since $d\pi$ contains a dS term the correct hedge ratio is

$$\Delta = \frac{\partial V}{\partial S} + q\frac{\partial V}{\partial \pi}.$$

From the no-arbitrage principle follows the pricing equation

$$\frac{\partial V}{\partial t} + \tfrac{1}{2}\sigma^2 S^2 \frac{\partial^2 V}{\partial S^2} + q\sigma^2 S^2 \frac{\partial^2 V}{\partial S\,\partial \pi} + \tfrac{1}{2}q^2\sigma^2 S^2 \frac{\partial^2 V}{\partial \pi^2} + rS\frac{\partial V}{\partial S} + r\pi\frac{\partial V}{\partial \pi} - rV = 0. \qquad (20.2)$$

This is not a diffusion equation in two space-like variables because S and π are perfectly correlated; the equation really has one space-like and two time-like variables.

We come to the stochastic control part of the problem in choosing q. If we are selling this contract then we should assume that the holder acts optimally, making the contract's value as high as possible. That doesn't mean the holder will follow such a strategy since he will have other priorities, he will have a view on the market and will not be hedging. The highest value for the contract occurs when q is chosen to maximize the q terms in (20.2):

$$\max_{|q|\leq 1}\left(q\sigma^2 S^2 \frac{\partial^2 V}{\partial S\,\partial\pi} + \tfrac{1}{2}q^2\sigma^2 S^2 \frac{\partial^2 V}{\partial\pi^2} \right).$$

This is the only term containing q.

20.2.1 Similarity Solution

If the payoff is simply

$$V(S, \pi, T) = \max(\pi, 0)$$

then we can find a similarity solution of the form

$$V(S, \pi, t) = SH(\xi, t), \quad \xi = \frac{\pi}{S}.$$

In this case Equation (20.2) becomes

$$\frac{\partial H}{\partial t} + \tfrac{1}{2}\sigma^2(\xi - q)^2 \frac{\partial^2 H}{\partial\xi^2} = 0, \tag{20.3}$$

the payoff is

$$H(\xi, T) = \max(\xi, 0)$$

and the optimal strategy is

$$\max_{|q|\leq 1}\left((\xi - q)^2 \frac{\partial^2 H}{\partial\xi^2} \right).$$

Assuming that $\partial^2 H/\partial\xi^2 > 0$ (and this can be verified *a posteriori*) because of the nature of the equation and its final condition, the optimal strategy is

$$q = \begin{cases} -1 & \text{when } \xi > 0 \\ 1 & \text{when } \xi < 0. \end{cases}$$

For a more general payoff the strategy would depend on the sign of $\partial^2 H/\partial\xi^2$.

The option value, assuming the optimal strategy, thus satisfies

$$\frac{\partial H}{\partial t} + \tfrac{1}{2}\sigma^2(|\xi| + 1)^2 \frac{\partial^2 H}{\partial\xi^2} = 0.$$

The termsheet for a perfect trader option is shown in Figure 20.1. The holder of the option is allowed to make a series of hypothetical trades on the USD/DEM exchange rate. Two trades are allowed per day with a maximum allowed position long and short. The holder then receives the positive part of his transactions. If his trades result in a negative net balance this is written off.

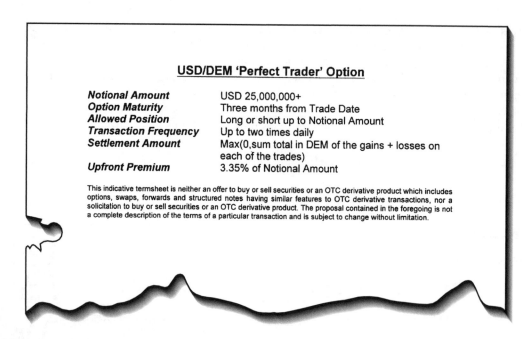

Figure 20.1 Termsheet for a 'perfect trader' option.

20.3 LIMITING THE NUMBER OF TRADES

The passport option described above allows the trader to buy and sell the underlying asset as often as he wishes. How valuable is the right to trade at will? Will it make much difference if the number of, or time between, trades is restricted? In practice, the trader can only make a finite number of purchases or sales.

Let's start by examining the case of limited number of trades.

Introduce the notation $V^{n+}(S, \pi, t)$ and $V^{n-}(S, \pi, t)$ to mean the value of the passport option when there are still n trades allowed and the $+/-$ refers to whether the trader is currently long or short the (maximum permitted quantity of the) underlying.

These two functions satisfy the inequalities

$$\frac{\partial V^{n+}}{\partial t} + \frac{1}{2}\sigma^2 S^2 \frac{\partial^2 V^{n+}}{\partial S^2} + \sigma^2 S^2 \frac{\partial^2 V^{n+}}{\partial S\, \partial \pi} + \frac{1}{2}\sigma^2 S^2 \frac{\partial^2 V^{n+}}{\partial \pi^2} + rS\frac{\partial V^{n+}}{\partial S} + r\pi \frac{\partial V^{n+}}{\partial \pi} - rV^{n+} \leq 0$$

and

$$\frac{\partial V^{n-}}{\partial t} + \frac{1}{2}\sigma^2 S^2 \frac{\partial^2 V^{n-}}{\partial S^2} - \sigma^2 S^2 \frac{\partial^2 V^{n-}}{\partial S\, \partial \pi} + \frac{1}{2}\sigma^2 S^2 \frac{\partial^2 V^{n-}}{\partial \pi^2} + rS\frac{\partial V^{n-}}{\partial S} + r\pi \frac{\partial V^{n-}}{\partial \pi} - rV^{n-} \leq 0.$$

Observe the sign difference between these.

Final data are

$$V^{0\pm}(S, \pi, T) = \max(\pi, 0).$$

A trade is optimal when the value with n trades left (and currently long/short) is the same as with $n - 1$ trades left and the opposite position (short/long):

$$V^{n+} \geq V^{(n-1)-} \quad \text{and} \quad V^{n-} \geq V^{(n-1)+}.$$

This completes the formulation of the problem.

A simple modification is to include a fixed penalty P to be paid on each trade. This is modeled by

$$V^{n+} \geq V^{(n-1)-} + P \quad \text{and} \quad V^{n-} \geq V^{(n-1)+} + P.$$

20.4 LIMITING THE TIME BETWEEN TRADES

Instead of restricting the number of trades we could restrict the minimum time interval between trades. If a trade has been made then another is not allowed until a time ω has passed. We must introduce a 'clock' that keeps track of the time since the last trade. This clock is reset to zero as soon as a trade is made. The passport option value is now given by $V^+(S, \pi, t, \tau)$ and $V^-(S, \pi, t, \tau)$ where τ is the time on the clock.

These two functions satisfy the inequalities

$$\frac{\partial V^+}{\partial t} + \frac{\partial V^+}{\partial \tau} + \tfrac{1}{2}\sigma^2 S^2 \frac{\partial^2 V^+}{\partial S^2} + \sigma^2 S^2 \frac{\partial^2 V^+}{\partial S\, \partial \pi} + \tfrac{1}{2}\sigma^2 S^2 \frac{\partial^2 V^+}{\partial \pi^2} + rS\frac{\partial V^+}{\partial S} + r\pi \frac{\partial V^+}{\partial \pi} - rV^+ \leq 0$$

and

$$\frac{\partial V^-}{\partial t} + \frac{\partial V^-}{\partial \tau} + \tfrac{1}{2}\sigma^2 S^2 \frac{\partial^2 V^-}{\partial S^2} - \sigma^2 S^2 \frac{\partial^2 V^-}{\partial S\, \partial \pi} + \tfrac{1}{2}\sigma^2 S^2 \frac{\partial^2 V^-}{\partial \pi^2} + rS\frac{\partial V^-}{\partial S} + r\pi \frac{\partial V^-}{\partial \pi} - rV^- \leq 0.$$

Final data is

$$V^{\pm}(S, \pi, T, \tau) = \max(\pi, 0).$$

A trade is optimal when the value of the option is the same as with the opposite position in the underlying and with a trade permitted (i.e. $\tau = \omega$):

$$V^+(S, \pi, t, \omega) \geq V^-(S, \pi, t, 0) \quad \text{and} \quad V^-(S, \pi, t, \omega) \geq V^+(S, \pi, t, 0).$$

In the absence of any penalties, there is a similarity solution for the passport option price of the form

$$V = S\Phi\left(\frac{\pi}{S}, t\right),$$

for some function Φ. Φ is a function also of the remaining variables. For the option with limited trades it is a function of n and for the time-limited option it is a function of τ.

Example

Volatility of the underlying asset is 20%, six months to expiry. The plain passport option value is $0.0589 \times S$. In Figure 20.2 is shown the effect on the price of limiting the number of trades. In the first plot is shown the option value (when $S = 100$) as a function of the number of permitted trades. The solid line is the option value with unlimited number of trades, the plain passport option. Note that even with only three or four trades allowed the option value is close to the unlimited trades case. The second plot shows the option value against T/ω when the time between trades is restricted. The structure of the price is similar to the restricted number of trades case.

Allowing a trader to trade more frequently has little effect on the theoretical price of the contract. It does make hedging of the contract slightly harder since the option writer will be hedging at least as often as the owner trades. The more frequent the hedging the greater the effect of transaction costs.

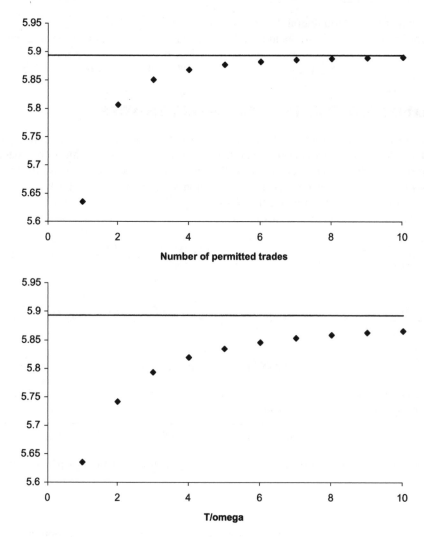

Figure 20.2 Values of different types of passport option.

20.5 NON-OPTIMAL TRADING AND THE BENEFITS TO THE WRITER

In this section I want to discuss issues to do with what is the best trading strategy for the option holder.

For the sake of argument, let's assume that we are in a Black–Scholes world. It's reasonable to also assume that the holder is not delta hedging. If he is then he's following the same strategy as the option writer. This would be a pointless strategy unless the holder thinks he has bought the option very cheaply. So let's suppose that he has another strategy in mind; maybe he sees himself as a hot trader with insight into the market's direction. Is he right to trade in a 'non-optimal' fashion?

We've used 'optimal' to mean something very specific; the strategy is optimal if it gives the option the highest value to someone who is delta hedging. So what is optimal to a delta hedger will generally not be optimal to our punter. Yet that is his concern. If he wants to take a chance, then I hope he gets a better return. If he trades well he would expect to make a greater return than the risk-free rate by taking more risk.

How does the option writer feel about this? He valued the contract on the assumption that the trader would trade optimally, with the hedger's definition of optimal. But the trader is following some other strategy. By definition, the 'value' of the contract to the holder is less than he paid for it *if he is following a non-optimal strategy*. In other words, the seller of the option is pleased that the holder is acting in a non-optimal fashion; he will probably make more money that way. Note that I am not saying that the trader is wrong to trade according to his own priorities, nor am I saying that the passport option is incorrectly valued. We'll discuss this again in Chapter 35.

20.6 **SUMMARY**

This new breed of derivative is particularly exciting. It is a challenge from the pricing and hedging points of view, especially because complex and continuous decisions must be made by the holder/trader. As with most quantitative finance, the math is quite straightforward once you've seen how it's done.

FURTHER READING

- Hyer, Lipton-Lifschitz & Pugachevsky (1997) describe the passport option.
- Ahn, Penaud & Wilmott (1998) and Penaud, Wilmott & Ahn (1998) discuss many, many extensions to the passport option concept.

CHAPTER 21
miscellaneous exotics

In this Chapter...

- contract specifications for many more exotic derivatives
- more 'tricks of the trade' for valuing exotic contracts in the partial differential equation framework

21.1 INTRODUCTION

The universe of exotic derivatives is large, and becoming larger all the time. I have tried to bring together, and even classify, many of these contracts. For example, a whole chapter was devoted entirely to Asian options, options depending on an average of the realized asset path. Nevertheless, because of the complexity of the instruments available and the increase in their number, the classification exercise can become a labor of Sisyphus. In this chapter I give up on this exercise and introduce a miscellany of exotics, with the aim of expanding the techniques available to the reader for pricing new contracts. Typically, I introduce the pricing and hedging concepts in an equity framework, but by the end of the book the astute reader will appreciate their applicability to other worlds, such as fixed income.

21.2 FORWARD START OPTIONS

As its name suggests, a **forward start** is an option that comes into being some time in the future. Let us consider an example: a forward start call option is bought now, at time $t = 0$, but with a strike price that is not known until time T_1, when the strike is set at the asset price on that date, say. The option expires later at time T. There are two ways to solve this problem in a Black–Scholes world, the simple and the complicated. We begin with the former.

The simple way to price this contract is to ask what happens at time T_1. At that time we get an at-the-money option with a time $T - T_1$ left to expiry. If the stock price at time T_1 is S_1 then the value of the contract is simply the Black–Scholes value with $S = S_1$, $t = T - T_1$, $E = S_1$ and with given values for r and σ. For a call option this value, as a function of S_1, is

$$S_1 N(d_1) - S_1 e^{-r(T-T_1)} N(d_2),$$

where

$$d_1 = \frac{r + \frac{1}{2}\sigma^2}{\sigma}\sqrt{T - T_1}$$

and

$$d_2 = \frac{r - \frac{1}{2}\sigma^2}{\sigma}\sqrt{T - T_1}.$$

The value is proportional to S. Thus, at time T_1 we will hold an asset worth

$$S_1 f(T - T_1).$$

Since this is a constant multiplied by the asset price at time T_1 the value today must be

$$Sf(T - T_1)$$

where S is today's asset price.

The other way of valuing this option, within our general path-dependent framework, is to introduce a new state variable $\mathcal{S}$ which is defined for $t \geq T_1$ as being the asset price at time T_1,

$$\mathcal{S} = S(T_1). \tag{21.1}$$

For times before that, we set $\mathcal{S} = 0$, although it does not actually matter what it is. The result of this pricing method is, as we now see, identical to the above simple method but the technique of introducing a new variable has a very wide applicability.

The option has a value that depends on three variables: $V(S, \mathcal{S}, t)$. This function satisfies the Black–Scholes equation in S and t since $\mathcal{S}$ is not stochastic and is constant after the date T_1. At expiry we have

$$V(S, \mathcal{S}, T) = \max(S - \mathcal{S}, 0).$$

At the start date, T_1, the strike price is set to the current asset price, this is Equation (21.1). The jump condition across T_1 is simply

$$V(S, \mathcal{S}, T_1^-) = V(S, S, T_1^+).$$

And that's all there is to it.

For this, the simplest of forward start options, we can take the analysis considerably further. We can either observe that the option value after time T_1 is that of a vanilla call, and therefore we have a formula for it as above, or we can use the similarity variable, $\xi = S/\mathcal{S}$ to transform the problem to

$$\frac{\partial H}{\partial t} + \frac{1}{2}\sigma^2 \xi^2 \frac{\partial^2 H}{\partial \xi^2} + r\frac{\partial H}{\partial \xi} - rH = 0$$

with

$$H(\xi, T) = \max(\xi - 1, 0)$$

and where $V(S, \mathcal{S}, t) = \mathcal{S}H(\xi, t)$.

Across time T_1 we have

$$V(S, \mathcal{S}, T_1^-) = V(S, S, T_1^+) = SH(1, T_1^+).$$

If we use this as the final condition (at time T_1) for the value for the option up to time T_1 then we see that for such times the option value is simply proportional to S. The unique solution is therefore

$$V(S, \mathcal{S}, t) = SH(1, T_1) \quad \text{for } t < T_1.$$

Of course, $H(1, T_1)$ is just the value of an at-the-money call option with a strike of 1 at a time $T - T_1$ before expiry. This takes us back to the result of the simple approach.

I have stressed the path-dependent approach, although this is unnecessarily complicated for the simple forward start option, because of its applicability to many other contracts.

Classification	Forward Start
Time dependence	No
Cashflow	No
Decisions	No
Path dependence	Weak
Dimension	2
Order	Second

Classification option table for Forward Start.

21.3 **SHOUT OPTIONS**

A **shout call** option is a vanilla call option but with the extra feature that the holder can at any time reset the strike price of the option to the current level of the asset (if it is higher than the original strike). There is simultaneously a payment, usually of the difference between the old and the new strike prices. The action of resetting is called 'shouting'.

Since there is clearly an element of optimization in the matter of shouting, one would expect to see a free boundary problem occur quite naturally as with American options.

To value this contract introduce the two functions: $V_a(S, X, t)$ and $V_b(S, X, t)$. The former is the value of the option after shouting and the latter, before. S is the underlying asset value, X the strike level. Because the variable X is updated discretely, the relevant equation to solve is the basic Black–Scholes equation. The final conditions are

$$V_a(S, X, T) = V_b(S, X, T) = \max(S - X, 0).$$

The function $V_b(S, X, t)$ must satisfy the constraint

$$V_b(S, X, t) \geq V_a(S, \max(S, X), t) - R(S, X),$$

with gradient continuity. Here $R(S, X)$ is the amount of money that must be paid on shouting. This represents the optimization of the shouting policy; when the two sides of this expression are equal it is optimal to shout.

This problem must be solved numerically; depending on the form of R there may be a similarity reduction to two dimensions. The option value is then $V(S_0, X_0, t_0)$ where the subscripts denote the initial values of the variables.

The definition of this simple shout option can be easily extended to allow for other rules about how the strike is reset, what the payment is on shouting, and to allow for multiple shouts.

Classification	Shout Option
Time dependence	No
Cashflow	Yes
Decisions	Yes
Path dependence	Strong
Dimension	3
Order	First

Classification option table for Shout Option.

21.4 CAPPED LOOKBACKS AND ASIANS

In **capped lookbacks** and **capped Asians** there is some limit or guarantee placed on the size of the maximum, minimum or average. A typical example of a capped Asian would have the path-dependent quantity being the average of the lesser of the underlying asset and some other prescribed level. This is represented by

$$A = \frac{I}{t}$$

with

$$I = \int_0^t \min(S, S_u) \, d\tau.$$

The stochastic differential equation for I from which follows the governing partial differential equation is

$$dI = \min(S, S_u) \, dt.$$

21.5 COMBINING PATH-DEPENDENT QUANTITIES: THE LOOKBACK-ASIAN ETC.

We have seen in Chapters 17–19 how to value options whose payoff, and therefore value, depend on various path-dependent quantities. There is no reason why we cannot price a contract that depends on more than one path-dependent quantity. Often, all that this requires is the use of one state-variable for each quantity.

As an example, let us consider the pricing of an option that we could call a **lookback-Asian**. By this, we mean a contract that depends on both a maximum (or minimum) and an average. But what do we mean by the 'maximum', is it the realized maximum of the underlying asset or, perhaps, the maximum of the average? Clearly, there are a great many possible meanings for such a contract. We consider three of them here, although the reader can (and should) doubtless think of many more.

Classification	Lookback - Asian
Time dependence	Yes, if discrete Sampling
Cashflow	No.
Decisions	No.
Path dependence	Strong
Dimension	4 (poss. sim. reduction)
Order	First

Classification option table for Lookback-Asian.

21.5.1   The Maximum of the Asset and the Average of the Asset

The simplest example, and the one closest to the problems we have encountered so far, is that of a contract depending on both the realized maximum of the asset and the realized average of the asset. Suppose that the average is arithmetic and that both path-dependent quantities are sampled discretely, and on the same dates. These assumptions can easily be generalized.

First of all, we observe that the option value is a function of *four* variables, S, t, M, the maximum, and A, the average. The variables M and A are measured discretely according to the definitions

$$M_i = \max(S(t_1), S(t_2), \ldots, S(t_i))$$

and

$$A_i = \frac{1}{i} \sum_{j=1}^{i} S(t_j).$$

Now we need to find the jump condition across a sampling date. This follows directly from the updating rule across sampling dates. The updating rule across a sampling date for the maximum is, as we have seen,

$$M_i = \max(M_{i-1}, S(t_i)).$$

The updating rule for the average is

$$A_i = \frac{(i-1)A_{i-1} + S(t_i)}{i}.$$

Thus the jump condition across a sampling date for this type of lookback-Asian is therefore given by

$$V(S, M, A, t_i^-) = V\left(S, \max(M, S), \frac{(i-1)A + S}{i}, t_i^+\right).$$

This, together with the Black–Scholes equation and a suitable final condition, is the full specification of this lookback-Asian.

21.5.2 The Average of the Asset and the Maximum of the Average

In this contract, the payoff depends on the average of the underlying asset and on the maximum *of that average*. We still have an option value that is a function of the four variables but now the updating rules are

$$A_i = \frac{(i-1)A_{i-1} + S(t_i)}{i},$$

and

$$M_i = \max(M_{i-1}, A_i).$$

Observe how we have taken the average first and then used that new average in the definition of the maximum. Again, there is plenty of scope for generalization.

Thus the jump condition across a sampling date for this type of lookback-Asian is given by

$$V(S, M, A, t_i^-) = V\left(S, \max\left(M, \frac{(i-1)A + S}{i}\right), \frac{(i-1)A + S}{i}, t_i^+\right).$$

21.5.3 The Maximum of the Asset and the Average of the Maximum

In this final example, we just swap the role of maximum and average in the previous case. We take the maximum of the asset first and then the average of that maximum, thus we have the following updating rules

$$M_i = \max(M_{i-1}, S(t_i))$$

and

$$A_i = \frac{(i-1)A_{i-1} + M_i}{i},$$

and consequently we have the jump condition

$$V(S, M, A, t_i^-) = V\left(S, \max(M, S), \frac{(i-1)A + \max(M, S)}{i}, t_i^+\right).$$

Finally, let us comment that for many of the obvious and natural payoffs, there is a similarity reduction, so that we need only solve a three-dimensional problem. I leave this issue to the reader to follow up.

21.6 **THE VOLATILITY OPTION**

A particularly interesting path-dependent quantity that can be the payoff for an exotic option is the realized historical volatility. By this, I mean a statistical quantity such as

$$
\sqrt{\frac{1}{\delta t}\frac{1}{M-1}\sum_{j=1}^{M}\log\left(\frac{S(t_j)}{S(t_{j-1})}\right)^2},
$$

where δt is the time interval between samples of the asset price. (Note that I have not taken off the drift of the asset. This becomes increasingly less significant as we let δt tend to zero.) The data points used in this expression are shown in Figure 21.1.

The reader might ask 'Won't this quantity simply be the volatility that we put into the model?' If so, what is the point of having a model for the historical volatility at all? The answer is that either we do not take δt sufficiently small for the above quantity to be necessarily close to the input volatility, or we assume a more complicated model for the volatility (such as stochastic or uncertain volatility, see Chapters 26 and 27, or assume an implied volatility surface, see Chapter 25).

Let us begin by valuing a contract with the above payoff in a Black–Scholes, constant volatility world, and then briefly discuss improvements.

The trick is to introduce *two* path-dependent quantities, the running volatility and the last sampled asset price:

$$
I_i = \sqrt{\frac{1}{\delta t(i-1)}\sum_{j=1}^{i}\log\left(\frac{S(t_j)}{S(t_{j-1})}\right)^2};
$$

$$
\mathcal{S}_i = S(t_{i-1}).
$$

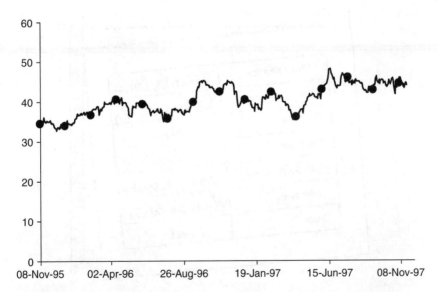

Figure 21.1　A schematic representation of the calculation of the historical volatility.

The option value is a function of four variables: $V(S, \mathcal{S}, I, t)$. The updating rules at time t_i for the two path-dependent quantities are

$$\mathcal{S}_i = S(t_{i-1})$$

and

$$I_i = \sqrt{\frac{1}{\delta t(i-1)} \sum_{j=1}^{i} \log\left(\frac{S(t_j)}{S(t_{j-1})}\right)^2}$$

$$= \sqrt{\frac{i-2}{i-1} I_{i-1}^2 + \frac{1}{\delta t(i-1)} \log(S(t_i)) - \log(\mathcal{S}_i))^2}.$$

We can see from the second updating rule why we had to keep track of the old sampled asset price: it is used in the updating rule for the running volatility.

The jump condition across a sampling date is therefore

$$V(S, \mathcal{S}, I, t_i^-) = V\left(S, S, \sqrt{\frac{i-2}{i-1} I^2 + \frac{1}{\delta t(i-1)} (\log(S) - \log(\mathcal{S}))^2}, t_i^+\right).$$

If the option pays off the realized volatility at expiry, T, then

$$V(S, \mathcal{S}, I, T) = I.$$

The dimensionality of this problem can be reduced to three by the use of the similarity variable $S/\mathcal{S}$.

If we do not believe in constant volatility then we could introduce a stochastic volatility model. This does not change the specification of our model in any way other than to introduce a new variable σ which satisfies some stochastic differential equation. The problem to solve is

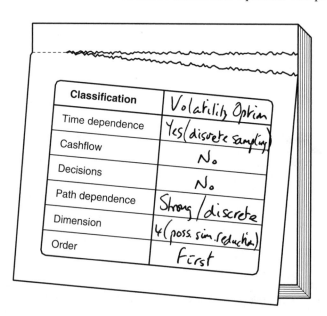

Classification	Volatility Option
Time dependence	Yes (discrete sampling)
Cashflow	No.
Decisions	No.
Path dependence	Strong / discrete
Dimension	4 (poss. sim. reduction)
Order	First

Classification option table for Volatility Option.

then in five dimensions, becoming four with the use of a similarity reduction, with the same path-dependent quantities and the same updating rules and jump condition. The high (four) dimensionality makes this problem computationally intensive and it may well be a candidate for a Monte Carlo simulation (see Chapter 66).

21.6.1 The Continuous-time Limit

We can get to the continuous-sampling limit for this problem by letting $\delta t \to 0$. The analysis is rather messy and is more easily derived directly, as we'll now see.

For the problem to make sense we must allow the volatility to be non-deterministic. For example, we could model it via a stochastic differential equation. However, the simplest model that is internally consistent is to just allow the volatility to be a function of the asset price and time $\sigma(S, t)$. Introduce the new state variable I defined by

$$I = \sqrt{\frac{1}{t} \int_0^t \sigma(S, t)^2 \, dt}.$$

This is a continuous-time version of the above discrete definition for I since

$$\log\left(\frac{S(t_j)}{S(t_{j-1})}\right) \approx \sigma(S_j)^2 \, dX^2.$$

From this we have

$$dI = \frac{(\sigma(S, t)^2 - I^2)}{2tI} dt$$

from which we get

$$\frac{\partial V}{\partial t} + \frac{1}{2}\sigma(S, t)^2 S^2 \frac{\partial^2 V}{\partial S^2} + rS\frac{\partial V}{\partial S} - rV + \frac{(\sigma(S, t)^2 - I^2)}{2tI}\frac{\partial V}{\partial I} = 0.$$

The option value is a function of three variables $V(S, I, t)$ and we are on familiar territory.

21.7 LADDERS

The **ladder option** is a lookback option that is discretely sampled, but this time discretely sampled in asset price rather than time. Thus the option receives a payoff that is a function of the highest asset price achieved out of a given set. For example, the ladder is set at multiples of $5: ..., 50, 55, 60, 65, ... If during the life of the contract the asset reached a maximum of 58, then the maximum registered would be 55. Such an option would clearly be cheaper than the continuous version.

This contract can be decomposed into a series of barrier-type options triggered at each of the rungs of the ladder. Alternatively, using the framework of Chapter 19, we simply have a payoff that is a step function of the maximum, M; see Figure 21.2.

21.8 PARISIAN OPTIONS

Parisian options are barrier options for which the barrier feature (knock-in or knock-out) is only triggered after the underlying has spent a certain prescribed time beyond the barrier. The effect of this more rigorous triggering criterion is to 'smooth' the option value (and delta and

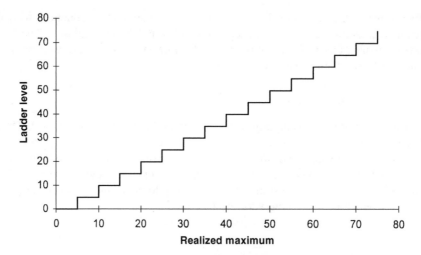

Figure 21.2 The payoff for a ladder option as a function of the realized maximum.

gamma) near the barrier to make hedging somewhat easier. It also makes manipulation of the triggering, by manipulation of the underlying asset, much harder. In the classical Parisian contract the 'clock' measuring the time outside the barrier is reset when the asset returns to within the barrier. In the **Parisian** contract the clock is not reset. We only consider the former here, the latter is a simple modification.

In Figure 21.3 is shown a representation of a Parisian contract. The bottom curve is the (scaled) time that the stock price has been above the barrier level. Once it has reached ten days (here scaled to one) the barrier is triggered.

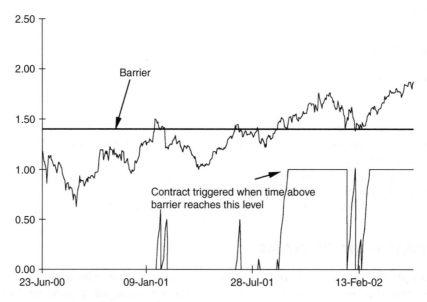

Figure 21.3 Representation of a Parisian contract.

Parisian options are clearly strongly path-dependent, but in a way that is perfectly manageable in the differential equation framework. We do not need all the details of the path taken; the only extra information we need to know is the value of the variable τ, defined as the length of time the asset has been beyond the barrier:

$$\tau = t - \sup\{t' \leq t | S(t') \leq S_u\}$$

for an up barrier at S_u, and there is a similar expression for a down barrier. The stochastic differential equation for τ is given by

$$d\tau = \begin{cases} dt & S > S_u \\ -\tau^- & S = S_u \\ 0 & S < S_u \end{cases}$$

where τ^- is the value of τ before it jumps to zero on resetting. We use this equation to derive the partial differential equation for the option value. Notice how, outside the barrier, real time t and the trigger time τ are increasing at the same rate. The barrier is triggered when τ reaches the value ω.

We must solve for V as a function of three variables S, t and τ, $V(S, t, \tau)$. We solve for V in two regions: inside the barrier and outside. Inside the barrier the clock is switched off and we solve

$$\frac{\partial V}{\partial t} + \tfrac{1}{2}\sigma^2 S^2 \frac{\partial^2 V}{\partial S^2} + rS\frac{\partial V}{\partial S} - rV = 0.$$

In this region there is no τ dependence (this is not true for Parasians). Outside the barrier the clock is ticking and we have

$$\frac{\partial V}{\partial t} + \frac{\partial V}{\partial \tau} + \tfrac{1}{2}\sigma^2 S^2 \frac{\partial^2 V}{\partial S^2} + rS\frac{\partial V}{\partial S} - rV = 0.$$

At $S = S_u$, where τ is reset, we must impose continuity of the option value:

$$V(S_u, t, \tau) = V(S_u, t, 0).$$

Now we come to the exact specification of the payoff in the event of triggering (or the event of not-triggering). If the barrier has not been triggered by expiry T then the option has the payoff $F(S, \tau)$. If the barrier has been triggered before expiry the option pays off $G(S)$ at expiry. For example an up-and-in Parisian put would have $F = 0$ and $G = \max(E - S, 0)$. An up-and-out call would have $F = \max(S - E, 0)$ and $G = 0$. In this framework ins and outs are treated the same. The boundary conditions are applied as follows:

$$V(S, T, \omega) = G(S)$$

and

$$V(S, T, \tau) = F(S, \tau).$$

American-style Parisians (Henry Miller options?) have the additional constraint that

$$V(S, t, \tau) \geq A(S, t, \tau),$$

with continuity of the delta, where the function A defines the payoff in the event of early exercise.

Classification	*Parisian*
Time dependence	No
Cashflow	No
Decisions	No
Path dependence	Strong/Continuous
Dimension	3
Order	First(out), second?(in)

Classification option table for Parisian Option.

21.8.1 Examples

First consider the case of a Parisian, European, down-and-in put on an asset with no dividends, an expiration time of $T = 0.25$ years, a volatility of $\sigma = 0.2$, an interest rate of $r = 0.08$; strike $E = 10$, barrier $S = 8$, and barrier trigger time $\omega = 0.05$. Figure 21.4 depicts the option

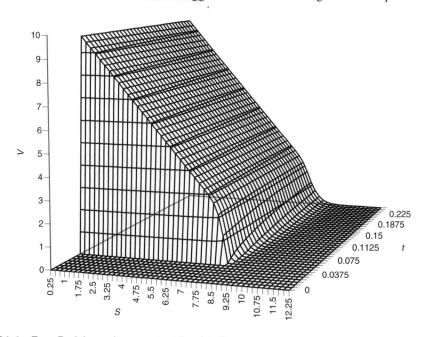

Figure 21.4 Euro Parisian value, see text for details.

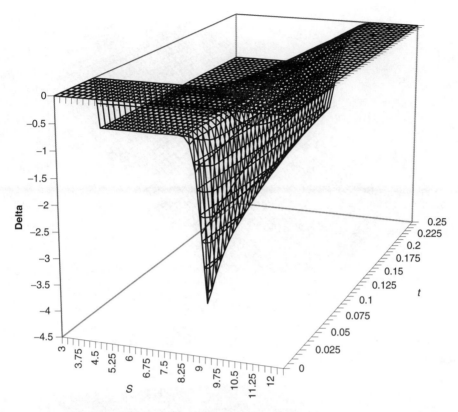

Figure 21.5 Euro Parisian delta, see text for details.

value V versus price S and time t. As expected, the option is worthless for values of time less than the barrier time ω. However for time greater than the barrier time the function appears quite smooth. This is validated by examining the hedge value versus price and time, as in Figure 21.5. The diffusion, prior to the barrier time, acts to smooth the data such that the hedge ratio remains reasonably manageable, compared to a traditional knock-out barrier.

Next consider the more sophisticated example of a Parisian, American, up-and-out call with dividend rate $D = 0.04$, an expiration time of $T = 0.25$ years, a volatility of $\sigma = 0.2$, an interest rate of $r = 0.08$, strike $E = 8$, barrier $S = 10$, and barrier time $\omega = 0.05$. The resulting plots of option value V and hedge ratio of this more complicated example are shown in Figures 21.6 and 21.7, respectively.

21.9 **SUMMARY**

I hope that after reading the chapters in this part of the book the reader will feel confident to find partial differential equation formulations for many other types of derivative contract. In Part Four, on interest rates and products, it is assumed that the reader can transfer ideas from the lognormal asset world to the more complicated world of fixed income.

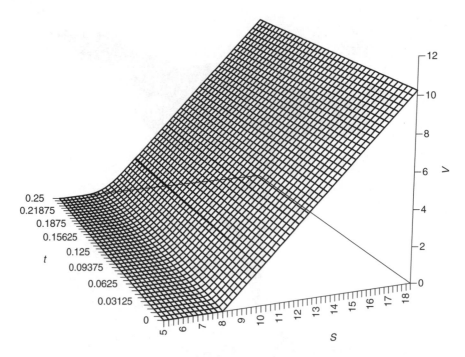

Figure 21.6 American Parisian value, see text for details.

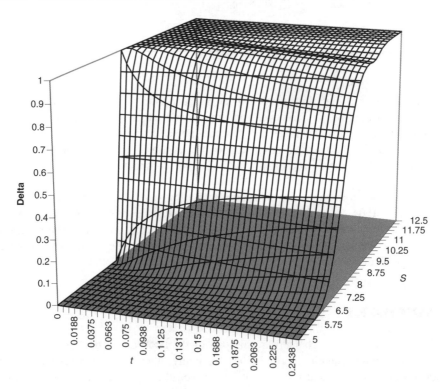

Figure 21.7 American Parisian delta, see text for details.

FURTHER READING

- See Dewynne & Wilmott (1994c) for a mixed bag of exotics and their modeling.
- Chesney, Cornwall, Jeanblanc-Picqué, Kentwell & Yor (1997) price Parisians via Laplace transforms.
- See Haber, Schönbucher & Wilmott (1997) for a more detailed description of partial differential equation approach to pricing Parisian options. A fully-functional standalone Parisian option pricer may be downloaded from www.wilmott.com.
- See Whaley (1993) for a description of his implied volatility index and options on it.

PART THREE
extending Black–Scholes

The third part of the book concerns extensions to the Black–Scholes world. All of the assumptions in the Black–Scholes model are incorrect to a greater or lesser degree. I will show how to incorporate ideas aimed at improving the modeling of the underlying asset. Some of these ideas are relatively standard, fitting easily into a classical Black–Scholes-type framework. Others take us far from the well-trodden path into new and uncharted territories. One of the most important points to watch out for is whether or not a model is linear or nonlinear; is the value of a portfolio of options the same as the sum of the values of the individual components? I believe that nonlinear models capture some important features that are neglected by the more common linear models. I will expand on this point at various times throughout the rest of the book.

Chapter 22: Defects in the Black–Scholes Model Many deficiencies in the simple Black–Scholes model are brought together in this chapter for a general discussion. The issues raised here will be expanded upon in later chapters.

Chapter 23: Discrete Hedging Even if all of the other assumptions that go into the Black–Scholes model were correct, the impossibility of delta hedging continuously in time would have an enormous effect on the actual profit and loss experienced during the business of hedging.

Chapter 24: Transaction Costs This was one of the first areas of research for me when I discovered mathematical finance. Unfortunately, most of the advanced work in this subject cannot easily be condensed into a chapter of a book. For that reason I will summarize many of the ideas and results, giving details only for the Leland model for vanilla options and its Hoggard–Whalley–Wilmott extension to arbitrary contracts.

Chapter 25: Volatility Smiles and Surfaces Volatility is the most important parameter in the valuation of options. It is difficult to measure statistically from time series data; however, we can tell from the prices of traded instruments what the market thinks is the correct value or structure for volatility.

Chapter 26: Stochastic Volatility It is natural to model volatility as a stochastic variable because it is so clearly not constant and not predictable.

Chapter 27: Uncertain Parameters The difficulties associated with observing, estimating and predicting important parameters means that we should try to reduce the dependence of a price on such quantities. One way of doing this is to allow parameters to lie in a prescribed range; we don't even need to know their value at any point in time if we price in a worst-case scenario. This is the Avellaneda, Levy, Parás and Lyons model for uncertain volatility.

Chapter 28: Empirical Analysis of Volatility   It is not too difficult to find a decent stochastic model for volatility if one knows how to examine the data. I make a few suggestions in that direction.

Chapter 29: Jump Diffusion   One of the real-life features not captured by the Black–Scholes world is that of discontinuous asset price paths. I describe the Merton model for jump-diffusive random walks.

Chapter 30: Crash Modeling   The Merton model for jumps requires quite a few parameters to be input, as well as an estimate of the distribution of jump sizes. This, together with the impossibility of hedging in the Merton world, makes the model somewhat unsatisfactory. If we price assuming the worst-case scenario then most of these difficulties vanish.

Chapter 31: Speculating with Options   Option theory is built on the idea of hedging. But what about those people who use options for speculation? Very little is ever said about how they can make investment decisions. I will try to remedy that in this chapter.

Chapter 32: Static Hedging   The nonlinearity in many of the preceding models means that the value of a portfolio of contracts is not the same as the sum of the values of the individual components. Thus the value of a contract depends on what you hedge it with. The beautiful consequences of this are discussed in this chapter.

Chapter 33: The Feedback Effect of Hedging in Illiquid Markets   This chapter describes a highly non-classical detour away from the Black–Scholes world. What if there is so much trade in the underlying asset for the hedging of derivatives that the normal causal link between the underlying and the option gets confused?

Chapter 34: Utility Theory   What if you can't hedge or don't want to? You'll be left with some uncertainty/randomness/risk. Utility theory is a way of assigning a value to a random outcome. This topic is popular with economists but not with practitioners. I also used to not like the subject but now I feel that sometimes you don't have a choice but to *carpe urticum*.

Chapter 35: More About American Options and Related Matters   There is some confusion concerning when it is best to exercise an American option. I try to clear up this confusion in this chapter. You've seen the theory from the point of view of the option writer in Chapter 9; now read about the option holder's strategy.

Chapter 36: Stochastic Volatility and Mean-variance Analysis   Can you really hedge away volatility risk?

Chapter 37: Advanced Dividend Modeling   Stock dividends can have a big impact on the value of an option. In this chapter we look at sophisticated models for dividends to try and better capture the effect.

CHAPTER 22
defects in the Black–Scholes model

In this Chapter...

* why the Black–Scholes assumptions are wrong
* how to improve the Black–Scholes model

22.1 INTRODUCTION

Before pointing out some of the flaws in the assumptions of the Black–Scholes world, I must emphasize how well the model has done in practice, how widespread is its use and how much impact it has had on financial markets, not to mention a Nobel Prize for two of its three creators. The model is used by everyone working in derivatives, whether they are salesmen, traders or quants. It is used confidently in situations for which it was not designed, usually successfully. The value of vanilla options are often not quoted in monetary terms, but in volatility terms, with the understanding that the price of a contract is its Black–Scholes value using the quoted volatility. You use a model other than Black–Scholes only with extreme caution, and you will have to be pretty convincing to persuade your colleagues that you know what you are doing. The ideas of delta hedging and risk-neutral pricing have taken a formidable grip on the minds of academics and practitioners alike. In many ways, especially with regards to commercial success, the Black–Scholes model is remarkably robust.

Nevertheless, there is room for improvement. Certainly, we can find models that better describe the underlying; what is not clear is whether these models make us more money. In the next few sections I introduce the ideas and models that will be expanded upon in the rest of the book. Some of these ideas are classical, in the sense that they are academically respectable, require no great leaps of imagination and are thoroughly harmless. Other ideas are too new to say what the future has in store for them. Yet there is at least a grain of truth in all that follows.

Each section corresponds to a chapter. First I give the relevant Black–Scholes assumption, and then I say why it is wrong and how. Later, I try to relax the assumption. All the details will be found in the relevant chapters.

IT IS QUITE FRIGHTENING HOW INACCURATE BS IS BECAUSE YOU CAN'T HEDGE CONTINUOUSLY

22.2 **DISCRETE HEDGING**

Black–Scholes assumes

* Delta hedging is continuous

Consider first, the continuous-time world of stochastic calculus. When we derived the Black–Scholes equation we used the continuous-time Itô's lemma. The delta hedging that was necessary for risk elimination also had to take place continuously. If there is a finite time between rehedges then there is risk that has not been eliminated.

In Chapter 23 we consider the effect of hedging at discrete intervals of time, taking real expectations over finite time periods rather than applying continuous-time calculus.

YOU CAN MAKE PRICING WITH TRANSACTION COSTS AS SIMPLE OR AS COMPLEX AS YOU WANT

22.3 **TRANSACTION COSTS**

Black–Scholes assumes

* Delta hedging is continuous
* There are no costs in delta hedging

But not only must we worry about hedging discretely, we must also worry about how much it costs us to rehedge. The buying and selling of assets exposes us to bid-offer spreads. In some markets this is insignificant, then we rehedge as often as we can. In other markets, the cost can be so great that we cannot afford to hedge as often as we would like. These issues, and several models, are described in Chapter 24.

I DON'T LIKE THE ASSUMPTIONS THAT GO INTO THIS 'IMPROVEMENT' OF BS

22.4 **VOLATILITY SMILES AND SURFACES**

Black–Scholes assumes

* Volatility is a known constant (or a known deterministic function)

If volatility is not a simple constant then perhaps it is a more complicated function of time and/or the underlying. If it is a function of time alone then we can find explicit solutions as described in Chapter 8. But what if it is a function of both time and the underlying? Perhaps the market even knows what this function is. Perhaps the market is so clever that it prices all traded contracts consistently with this volatility function and all we have to do is deduce from these traded prices what the volatility function is.

The link between the prices of vanilla options in the market and a deterministic volatility is the subject of Chapter 25. We see how to back out from market prices the 'implied volatility surface'.

This technique is popular for pricing exotic options, yielding prices that are 'consistent' with traded prices of similar contracts.

22.5 **STOCHASTIC VOLATILITY**

Black–Scholes assumes

- Volatility is a known constant (or a known deterministic function)

The Black–Scholes *formulae* require the volatility of the underlying to be a known deterministic function of time. The Black–Scholes *equation* requires the volatility to be a known function of time and the asset value. Neither of these is true. All volatility time series show volatility to be a highly unstable quantity. It is very variable and unpredictable. It is therefore natural to represent volatility itself as a random variable. I show the theory behind this in Chapter 26.

Stochastic volatility models are currently popular for the pricing of contracts that are very sensitive to the behavior of volatility. Barrier options are the most obvious example.

22.6 **UNCERTAIN PARAMETERS**

Black–Scholes assumes

- Volatility, interest rates and dividends are known constants (or known deterministic functions)

So volatility is not constant. Nor, actually, is it a deterministic function of time and the underlying. It's definitely unpredictable. Worse still it may not even be measurable.

Volatility cannot be directly observed and its measurement is very difficult. How then can we hope to model it? Maybe we should not attempt to model something we can't even observe. What we should do is to make as few statements about its behavior as possible, so we will not say what volatility currently is or even what probability distribution it has. We shall content ourselves with placing a bound on its value, restricting it to lie within a given range. The probability distribution of the volatility within this range will not be prescribed. If it so desires, the volatility can jump from one extreme to the other as often as it wishes. This 'model' is then used to price contracts in a 'worst-case scenario'.

The idea of ranges for parameters is extended to allow the short-term interest rate and dividends to be uncertain, and to lie within specified ranges.

22.7 **EMPIRICAL ANALYSIS OF VOLATILITY**

Black–Scholes assumes

- Volatility is a known constant (or a known deterministic function)

In Chapter 28 I show how to estimate such quantities as the volatility of volatility, and how to deduce the drift rate of the volatility by analyzing its distribution.

Having determined a plausible stochastic differential equation model for the volatility from data, I suggest ways in which it can be used. It can be used directly in a stochastic volatility model, or indirectly in an uncertain volatility model to determine the likelihood of volatility ranges being breached.

I THINK THERE ARE BETTER WAYS OF TREATING MARKET JUMP/CRASHES, SEE BELOW

22.8 JUMP DIFFUSION

Black–Scholes assumes

● The underlying asset path is continuous

It is common experience that markets are discontinuous: from time to time they 'jump', usually downwards. This is not incorporated in the lognormal asset price model, for which all paths are continuous.

When I say 'jump' I mean two things. First, that the sudden moves are not contained in the lognormal model; they are too large, occurring too frequently, to be from a Normally distributed returns model. Second, they are unhedgeable; the moves are too sudden for continuous hedging through to the bottom of the jump.

The jump-diffusion model described in Chapter 29 is an attempt to incorporate discontinuities into the price path. These discontinuities are not modeled by the lognormal random walk that we have been using so far. Jump-diffusion is an improvement on the model of the underlying but introduces some unsatisfactory elements: risk elimination is no longer possible and we must price in an 'expected' sense.

A NON-PROBABILISTIC TREATMENT OF MARKET CRASHES, WORST-CASE SCENARIOS

22.9 CRASH MODELING

Black–Scholes assumes

● The underlying asset path is continuous

If risk elimination is not possible can we consider worst-case scenarios? That is, assume that the worst does happen and then allow for it in the pricing. But what exactly is 'the worst'? The worst outcome will be different for different portfolios.

In Chapter 30 I show how to model the worst-case scenario and price options accordingly.

22.10 SPECULATING WITH OPTIONS

Black–Scholes assumes

● Options are delta hedged

In Chapter 31 I show how to 'value' options when one is *not* hedging, rather when one is speculating with derivatives because one has a good idea where the underlying is going. If one has a view on the underlying, then it is natural to invest in options because of their gearing. But how can this view be quantified? One way is to estimate real expected returns from an unhedged position. This, together with an estimate of the risk in a position, enables one to choose which option gives the best risk/reward profile for the given market view.

This idea can be extended to consider many types of model for the underlying, each one representing a different view of the behavior of the market. Furthermore, many trading strategies can be modeled. For example, how can one model the optimal closure of an option position? When should one sell back an option? Should you at times hedge, and at others speculate? Is there a best way to choose between the two?

22.11 **OPTIMAL STATIC HEDGING**

NONLINEAR MODELS HAVE PROPERTIES THAT MEAN THEY ALWAYS GET TRADED PRICES RIGHT

Many of the non-Black–Scholes models of this part of the book are nonlinear. This includes the models of Chapters 24, 27, 30 and 31 (and 48). If the governing equation for pricing is nonlinear then the value of a portfolio of contracts is *not* the same as the sum of the values of each component on its own. One aspect of this is that the value of a contract depends on what it is hedged with. As an extreme example, consider the contract whose cashflows can be hedged exactly with traded instruments — to price this contract we do not even need a model. In fact, to use a model would be suicidal.

The beauty of the nonlinear equation is that fitting parameters to traded prices (as in the implied volatility surfaces in Chapter 25 or in 'yield curve fitting' of Chapter 41) becomes redundant. Traded prices may be right or wrong, we don't much care which. All we care about is that if we want to put them into our portfolio then we know how much they will cost.

Imagine that we have a contract called 'contract', with a value that we can write as

$$V_{NL}(\text{contract}),$$

where V_{NL} means the value of the contract using whatever is our nonlinear pricing equation (together with relevant boundary and final conditions). Now imagine we want to hedge 'contract' with another contract called 'hedge'. And suppose that it costs 'cost' to buy or sell this second contract in the market. Suppose that we buy λ of these hedging contracts and put them in our portfolio, then the *marginal value* of our original 'contract' is

$$V_{NL}(\text{contract} + \lambda \text{ hedge}) - \lambda \text{ cost}. \tag{22.1}$$

In this expression 'contract $+ \lambda$ hedge' should be read as the portfolio made up of the 'union' of the original contract and λ of the hedging contract. Since V_{NL} is nonlinear, this marginal value is not the value of the contract on its own. We have hedged 'contract' statically, we may hold 'hedge' until expiry of 'contract'. We can go one step further and hedge *optimally*. Since the quantity λ can be chosen, let us choose it to maximize the marginal value of 'contract'. That is, choose λ to maximize (22.1). This is optimal static hedging. We can, of course, have as many traded contracts for hedging as we want, and we can easily incorporate bid-offer spread.

In the event that 'contract' can have all its cashflows hedged away by one or more 'hedge' contracts, we find that we are using our nonlinear equation to value an empty portfolio and that the contract value is model independent.

22.12 **THE FEEDBACK EFFECT OF HEDGING IN ILLIQUID MARKETS**

Black–Scholes assumes

- The underlying asset is unaffected by trade in the option

The buying and selling of assets moves their prices. A large trade will move prices more than a small trade. In the Black–Scholes model it is assumed that moves in the underlying are exogenous, that some cosmic random number generator tells us the prices of all 'underlyings'. In reality, a large trade in the underlying will move the price in a fairly predictable fashion.

For example, it is not unheard of for unscrupulous people to deliberately move prices to ensure that a barrier option is triggered. In that case, a small move in the underlying could have a very big effect on the payoff of an option.

In Chapter 33 I will try to quantify this effect, introducing the idea that a trade in the underlying initiated by the need to delta hedge can move the price of the underlying. Thus it is no longer the case that the underlying moves and the option price follows; now it is more of a chicken-and-egg scenario. We will see that close to expiry of an option, when the gamma is large, the underlying can move in a very dramatic way.

22.13 UTILITY THEORY

Black–Scholes assumes

* Delta hedging eliminates all risk

If, for any reason, we cannot perfectly delta hedge then we are left with some residual risk. We will therefore need a framework for valuing this risk. In Chapter 34 I introduce some ideas in this direction, which we will build on in later chapters.

THIS CONCEPT IS A HARD SELL, BUT PROFOUNDLY IMPORTANT

22.14 MORE ABOUT AMERICAN OPTIONS AND RELATED MATTERS

Black–Scholes assumes

* The American option should/will be exercised at optimal time

But what does optimal mean? Does it mean the same thing to both the writer and the holder of the American option?

In Chapter 35 we examine the exercise strategy from the position of the option holder and I explain why he might want to exercise at an unexpected time. Such apparently 'non-optimal' exercise may actually be quite rational and will have a large impact on the writer of the option.

SOMETIMES YOU CAN'T HEDGE PERFECTLY, SO WHAT IS YOUR LEFT-OVER RISK?

22.15 STOCHASTIC VOLATILITY AND MEAN-VARIANCE ANALYSIS

Black–Scholes assumes

* Volatility is a known deterministic function of asset value and time

When volatility is itself stochastic we can derive a theory (Chapter 26) that is consistent but requires knowledge of a new function, the market price of risk. This function is only observable via option prices themselves and so we find ourselves with a circular argument; we can price options if we know their market values. This is not entirely satisfactory and so in Chapter 36 we explore the possibility of valuing options when we know that there is some unhedged volatility risk.

22.16 **ADVANCED DIVIDEND MODELING**

Black–Scholes assumes

- The underlying asset has deterministic dividends

Dividend payment is a subtle subject for modeling, and can have a significant effect on the prices of derivatives. In Chapter 37 we look at various forms of dividend models, including uncertain amount and payment dates, and stochastic dividends.

22.17 **SUMMARY**

There are many faults with the Black–Scholes assumptions. Some of these are addressed in the next few chapters. Although it is easy to come up with any number of models that improve on Black–Scholes from a technical and mathematical point of view, it is nearly impossible to improve on its commercial success.

FURTHER READING

- For a discussion of some real-world modeling issues see the collection of papers edited by Kelly, Howison & Wilmott (1995).
- There are many other ways to improve on the lognormal model for the underlying. One which I don't have space for is a model of asset returns using hyperbolic, instead of Normal, distributions: see Eberlein & Keller (1995).

CHAPTER 23
discrete hedging

In this Chapter...

- the effect of hedging at discrete times
- hedging error
- the real distribution of profit and loss

23.1 INTRODUCTION

In this chapter we concentrate on one of the erroneous assumptions in the Black–Scholes model, that of continuous hedging. The Black–Scholes analysis requires the continuous rebalancing of a hedged portfolio according to a delta neutral strategy. This strategy is, in practice, impossible.

The structure of this chapter is as follows. We begin by examining the concept of delta hedging in a discrete-time framework. We will see that taking expectations leads to the Black–Scholes equation without any need for stochastic calculus. I then show how this can be extended to a higher-order approximation, valid when the hedging period is not infinitesimal. We then discuss the nature of the hedging error, the error between the expected change in portfolio and the actual. This quantity is commonly ignored (perhaps because it averages out to zero) but is important, especially when one examines the real distribution of returns.

23.2 A MODEL FOR A DISCRETELY-HEDGED POSITION

The Black–Scholes analysis requires *continuous* hedging, which is possible in theory but impossible, and even undesirable, in practice. Hence one hedges in some discrete way.

Our first step in analyzing the discrete hedging problem is to choose a hedging strategy. If there are no transaction costs then there is no penalty for continuous rehedging and so the 'optimal' strategy is simply the Black–Scholes strategy, and the option value is the Black–Scholes value. This is the 'mathematical solution'. However, this strategy is clearly impractical. Unfortunately, it may be difficult to associate a 'cost' to the inconvenience of continuous rehedging. (This contrasts with the later problems when we do include transaction costs and optimal strategies are found.) For this reason a common assumption is that rehedging takes place regularly at times separated by a constant interval, the hedging period, here denoted by δt. This is a strategy commonly used in practice with δt ranging from half a day to a couple of weeks. Note the use of $\delta \cdot$ to denote a discrete change in a quantity; this is to make a distinction with $d\cdot$, the earlier continuous changes.

	A	B		C	D	E
1	Spot price	100	Strike	100		
2	Volatility	0.2	Expiry	1		
3	Return	0.12				
4	Int. rate	0.05	Timestep	0.01		

Formula annotations

- $H1$ (Total H. E. = 0.104310869): `=F107+C107*E107-H107`
- Time column: `=B7+$E$4`
- Asset column: `=C8*(1+$B$3*$E$4+$B$2*SQRT($E$4)*(RAND()+RAND()+RAND()+RAND()+RAND()+RAND()+RAND()+RAND()+RAND()+RAND()+RAND()+RAND()-6))`
- d1 column: `=(LN(C13/$E$1)+($B$4+0.5*$B$2*$B$2)*($E$2-B13))/$B$2/SQRT($E$2-B13)`
- Delta column: `=NORMSDIST(D13)`
- Option column: `=C14*NORMSDIST(D14)-$E$1*EXP(-$B$4*($E$2-B14))*NORMSDIST(D14-$B$2*SQRT($E$2-B14))`
- Balance column: `=H9*EXP($B$4*$E$4)+G10`
- Cashflow column: `=(E13-E12)*C13`

Main simulation table (Total H. E. = 0.104310869)

Time	Asset	d1	Delta	Option	Cashflow	Balance
0	100	0.35	0.63683059	10.45057563		53.23248337
0.01	100.1123086	0.353886158	0.638287839	10.45789692	0.145888571	53.40499483
0.02	99.59672193	0.326072522	0.627815192	10.06713079	-1.04304128	52.38866273
0.03	97.9454227	0.239321328	0.594571739	8.993910811	-3.256046101	49.15881751
0.04	96.19617638	0.145027267	0.557655352	7.92345508	-3.55121526	45.6327878
0.05	97.67591118	0.220507287	0.587261915	8.708093487	2.891848029	48.546
0.06	96.35566426	0.147885155	0.558703319	7.88867366	-2.744074033	45.82
0.07	97.93453976	0.229317242	0.590688782	8.7330524	3.1246468	48.97462917
0.08	97.60898883	0.209553868	0.582992027	8.478029115	-0.751272453	48.24785016
0.09	97.57099263			8.391889969	-0.173782294	48.09819782
0.1	100.7329397			10.26526582	6.37582014	54.4980307
0.11	101.8190694	0.425733427	0.664848916	10.9090656	2.071374769	56.59670369
0.12	104.1163135	0.543333961	0.706550086	12.41653882	4.341772093	60.96678121
0.13	103.2636408	0.498				59.38884711
0.16						55.69144464
	101.6795342	948179	0.658906895	10.5001587	0.47846141	56.19775873
0.16	101.6645465		0.659700309	10.47498377	0.080673944	56.30653858
0.17	100.5185021	0.347248158	0.635797491	9.654028136	-2.402675453	53.93202344
0.18	103.1987043	0.490791447	0.688213021	11.359784993	5.409214799	59.36821099
0.19	103.3275011	0.496852115	0.690353334	11.378266858	0.221153139	59.61905566
0.2	104.9466342	0.582952925	0.720037551	12.444949039	3.115258687	62.76413133
0.21	107.244597	0.704541252	0.759452189	14.078885413	4.227007006	67.02252825
0.22	105.2790073	0.600355094	0.725865248	12.54707454	-3.535999785	63.5200481
0.23	105.8326889	0.630140759	0.735698815	12.87990023	1.040712796	64.59252887

Figure 23.1 Spreadsheet for simulating hedging during the life of an option.

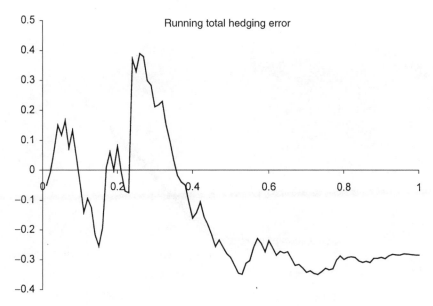

Figure 23.2 The running total hedging error.

The first work in this area was by Boyle and Emanuel who examined the discrepancy, the hedging error, between the Black–Scholes strategy of continuous rehedging and discrete hedging. They found that in each interval the hedging error was a random variable, from a chi-squared distribution, and proportional to the option gamma. We will see why this is shortly.

The spreadsheet on page 320 shows how to simulate delta hedging in the Black–Scholes fashion, but in discrete time. In this example the option is a call, there are no dividends on the underlying and there are no transaction costs. All of these can easily be incorporated. The spreadsheet uses a simple approximation to the Normal distribution. Note that the real drift of the asset is used in the simulation.

Results of this simulation are shown in Figures 23.2 and 23.3. In the first figure is the time series of the running **total hedging error** for a single realization of the underlying asset. This is the difference between the actual and theoretical values of the net option position as time evolves from the start to the end of the contract's life. The final value of -0.282 means that if you sold the option for the Black–Scholes value and hedged in the Black–Scholes manner until expiry to eliminate risk, you would have lost 0.282.

In Figure 23.3 are the results of option replication over many realizations of the underlying asset. Each dot represents the final asset price and accumulated profit and loss after hedging for the life of the contract. If hedging had been perfect each dot would lie on the payoff function, in this case $\max(S - 60, 0)$.

23.3 **A HIGHER-ORDER ANALYSIS**

Now we take this analysis one stage further. We use simple Taylor series expansions first to find a *better* hedge than the Black–Scholes and second to find an adjusted value for the option. The better hedge comes from hedging the option with the underlying using the number of shares that minimizes the variance of the hedged portfolio over the next timestep.

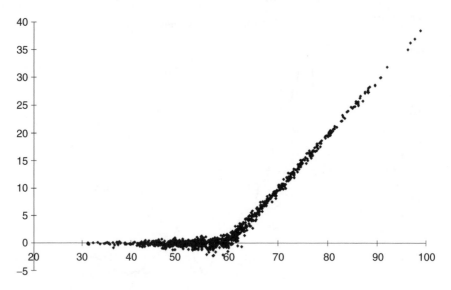

Figure 23.3 The results of discrete hedging.

The first step towards valuing an option is to choose a good model for the underlying in discrete time. A sensible choice is

$$S = e^x \tag{23.1}$$

where

$$\delta x = \left(\mu - \frac{\sigma^2}{2} \right) \delta t + \sigma \phi \, \delta t^{1/2}. \tag{23.2}$$

This is a discrete-time version of the earlier continuous-time stochastic differential equation for S.[1] Here ϕ is a random variable drawn from a standardized Normal distribution and the term $\phi \, \delta t^{1/2}$ replaces the earlier Wiener process. In principle, the ideas that we describe do not depend on the random walk being lognormal and many models for S could be examined. In particular, ϕ need not be Normal but could even be measured empirically. If historic volatility is to be used then it should be measured at the same frequency as the rehedging takes place, i.e., using data at intervals of δt.

As in the Black–Scholes analysis we construct a hedged portfolio

$$\Pi = V - \Delta S, \tag{23.3}$$

with Δ to be chosen. As in Black–Scholes we first choose the hedge and then use it to derive an equation for the option value. (I have called this a 'hedged portfolio' but we will see that it is not perfectly hedged.)

We no longer have Itô's lemma, since we are in discrete time, but we still have the Taylor series expansion. Thus, it is a very simple matter to derive $\delta \Pi$ as a power series in δt and δx. On substituting for δx from (23.2) this expression becomes

$$\delta \Pi = \delta t^{1/2} A_1(\phi, \Delta) + \delta t A_2(\phi, \Delta) + \delta t^{3/2} A_3(\phi, \Delta) + \delta t^2 A_4(\phi, \Delta) + \cdots. \tag{23.4}$$

---

[1] Why have I not chosen $\delta S = \mu S \, \delta t + \sigma S \phi \, \delta t^{1/2}$? Because that would only be an approximation to the continuous-time stochastic differential equation. We are going to look at high order terms in the Taylor series expansion of V and the above is exact.

Actually, V and its derivatives appear in each A_i, but only the dependence on ϕ and Δ is shown. For example, the first term A_1 is given by

$$A_1(\phi, \Delta) = \sigma \phi S \left(\frac{\partial V}{\partial S} - \Delta \right),$$

and the second term by

$$A_2(\phi, \Delta) = \frac{\partial V}{\partial t} + S \left(\frac{\partial V}{\partial S} - \Delta \right) (\mu + \tfrac{1}{2}\sigma^2(\phi^2 - 1)) + \tfrac{1}{2}\sigma^2\phi^2 S^2 \frac{\partial^2 V}{\partial S^2}.$$

This expansion in powers of $\delta t^{1/2}$ can be continued indefinitely; we stop at the order shown above because it is at this order that we find results that differ from Black–Scholes. Since time is measured in units of one year, δt is small but not zero. I have not put all of the algebraic details here; this is definitely big-picture time. They are not hard to derive, just take up a lot of space. Then treat the derivation of A_1, A_2, A_3 and A_4 as an exercise.

Now I can state the very simple hedging strategy and valuation policy.

- Choose Δ to minimize the variance of $\delta\Pi$
- Value the option by setting the *expected* return on Π equal to the risk-free rate

The first of these, the hedging strategy, is easy to justify: the portfolio is, after all, hedged so as to reduce risk. But how can we justify the second, the valuation policy, since the portfolio is not riskless? The argument for the latter is that since options are in practice valued according to Black–Scholes yet necessarily discretely hedged, the second assumption is already being used by the market, but with an inferior choice for Δ.

Because we cannot totally eliminate risk I could also argue for a pricing equation that depends on risk preferences. I won't pursue that here, just note that even though we'll get an option 'value' that is different from the Black–Scholes value this does not mean that there is an arbitrage opportunity.

23.3.1 Choosing the Best Δ

The variance of $\delta\Pi$ is easily calculated from (23.4) since

$$\text{var } [\delta\Pi] = E[\delta\Pi^2] - (E[\delta\Pi])^2. \tag{23.5}$$

In taking the expectations of $\delta\Pi$ and $\delta\Pi^2$ to calculate (23.5), all of the ϕ terms are integrated out leaving the variance of $\delta\Pi$ as a function of V, its derivatives, and, most importantly, Δ. Then to minimize the variance we find the value of Δ for which

$$\frac{\partial}{\partial\Delta} \text{ var } [\delta\Pi] = 0.$$

The result is that the optimal Δ is given by

$$\Delta = \frac{\partial V}{\partial S} + \delta t(\cdots). \tag{23.6}$$

The first term will be recognized as the Black–Scholes delta. The second term, which I give explicitly in a moment, is the correction to the Black–Scholes delta that gives a better reduction in the variance of $\delta\Pi$, and thus a reduction in risk. This term contains V and its derivatives.

23.3.2 The Hedging Error

The leading-order random term in the 'hedged' portfolio is, with this choice for Δ,

$$\tfrac{1}{2}\sigma^2\phi^2 S^2 \frac{\partial^2 V}{\partial S^2}.$$

We can write this as

$$\tfrac{1}{2}\sigma^2 S^2 \frac{\partial^2 V}{\partial S^2} + \tfrac{1}{2}(\phi^2 - 1)\sigma^2 S^2 \frac{\partial^2 V}{\partial S^2}.$$

The first term will, in the next section, be part of our pricing equation (and is part of the Black–Scholes equation). The second, which has a mean value of zero because ϕ is drawn from a standardized Normal distribution, is the **hedging error**. This term is random. The distribution of the square of a standardized Normal variable is called the **chi-squared distribution** (with a single degree of freedom). This chi-squared distribution is plotted in Figure 23.4. It has a mean of 1.

This distribution is asymmetrical about its mean, to say the least. The average value is 1 but 68% of the time the variable is less than this, and only 32% above. If one is long gamma, most of the time one loses money on the hedge (these are the small moves in the underlying), but 32% of the time you gain (and the size of that gain is on average larger than the small losses). The net position is zero. This result demonstrates the path dependency of hedging errors: ideally, with long gamma, one would like large moves when gamma is large and small moves when gamma is small.

During the life of the hedged option the hedging errors at each rehedge add up to give the total hedging error. This is the final discrepancy between the theoretical profit and loss on the position and the actual. The total hedging error also has a mean of zero, and a standard deviation of $O(\delta t^{1/2})$. The actual outcome for the total hedging error is highly path-dependent. In Figure 23.5 is shown the distribution of the *total* hedging error for a call option hedged

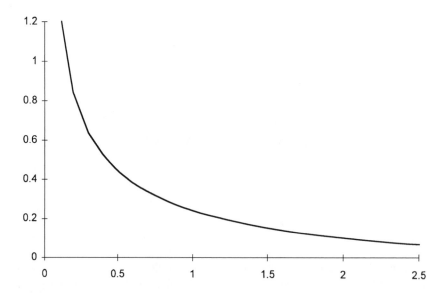

Figure 23.4 The chi-squared distribution.

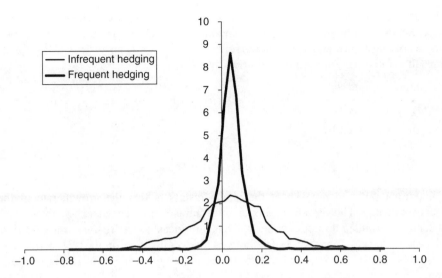

Figure 23.5 The distribution of the total hedging error for a call hedged frequently and infrequently.

at different intervals. Note how the spread of the hedging error is much greater for the more infrequently hedged option.

In passing we note that the commonly-held belief that 'time decay is the expected profit and loss (P&L) from a position' is wrong. The expected P&L from a self-financed and delta-hedged option position is zero.

23.3.3 Pricing the Option

Having chosen the best Δ, we now derive the pricing equation. The option should not be valued at the Black–Scholes value since that assumes perfect hedging and no risk: the fair value to an imperfectly hedged investor may be different. We find that the option value to the investor is equal to the Black–Scholes value plus a correction which prices the hedging risk.

The pricing policy that we have adopted has been stated as equating the expected return on the discretely hedged portfolio with the risk-free rate. This may be written as

$$E[\delta\Pi] = \left(r\,\delta t + \tfrac{1}{2}r^2\,\delta t^2 + \cdots\right)\Pi. \tag{23.7}$$

This is slightly different from the usual right-hand side, but simply represents a consistent higher order correction to exponential growth: $e^{r\,\delta t} = 1 + r\,\delta t + r^2\,\delta t^2/2 + \cdots$. Now substitute (23.3) and (23.4) into (23.7) to get the equation

$$\frac{\partial V}{\partial t} + \tfrac{1}{2}\sigma^2 S^2 \frac{\partial^2 V}{\partial S^2} + rS\frac{\partial V}{\partial S} - rV + \delta t(\cdots) = 0. \tag{23.8}$$

Again the first term is that derived by Black and Scholes and the second term is a correction to allow for the imperfect hedge; it contains V and its derivatives. I give the second term shortly.

23.3.4 The Adjusted Δ and Option Value

The as yet undisclosed terms in parentheses in (23.6) and (23.8) contain V and its derivatives, up to the second derivative with respect to t and up to the fourth with respect to S. However, since the adjusted option price is clearly close in value to the Black–Scholes price we can put the Black–Scholes value into the terms in parentheses without any reduction in accuracy. This amounts to solving (23.6) and (23.8) iteratively.

The result[2] is that the adjusted option price satisfies

$$\frac{\partial V}{\partial t} + \tfrac{1}{2}\sigma^2 S^2 \frac{\partial^2 V}{\partial S^2} + rS\frac{\partial V}{\partial S} - rV + \tfrac{1}{2}\delta t(\mu - r)(r - \mu - \sigma^2)S^2 \frac{\partial^2 V}{\partial S^2} = 0 \qquad (23.9)$$

and the better Δ is given by

$$\Delta = \frac{\partial V}{\partial S} + \delta t \left(\mu - r + \tfrac{1}{2}\sigma^2\right) S \frac{\partial^2 V}{\partial S^2}.$$

The important point to note about the above results is that the growth rate of the asset μ appears explicitly. This is a very important contrast with the Black–Scholes result. The Black–Scholes formulae do not contain μ. Thus any ideas about 'risk-neutral valuation' must be used with great care. There is no such thing as 'perfect hedging' in the real world. In practice the investor is *necessarily* exposed to risk in the underlying, and this manifests itself in the appearance of the drift of the asset price.

Notice how the second derivative terms in (23.9) are both of the form constant $\times S^2$. Therefore the correction to the option price can very easily be achieved by adjusting the volatility and using the value σ^* where

$$\sigma^* = \sigma \left(1 + \frac{\delta t}{2\sigma^2}(\mu - r)(r - \mu - \sigma^2)\right).$$

There is a similar volatility adjustment when there are transaction costs, see Chapter 24. The correction is symmetric for long and short positions.

Is this volatility effect important? Fortunately, in most cases it is not. With typical values for the parameters and daily rehedging there is a volatility correction of one or two percent. In trending markets, however, when large μ can be experienced, this correction can reach five or ten percent, a value that cannot be ignored.

More importantly, in trending markets the corrected Δ will give a better risk reduction since it is in effect an anticipatory hedge: the variance is minimized over the time horizon until the next rehedge. This has been called **hedging with a view**.

Since the option should be valued with a modified volatility, the difference between the adjusted option value and the Black–Scholes value is proportional to the option vega, its derivative with respect to σ.

RETURNS DON'T QUITE MATCH THE THEORY

23.4 THE REAL DISTRIBUTION OF RETURNS AND THE HEDGING ERROR

All of the above assumes that the return on the underlying is Normally distributed with a known volatility. In reality the distribution is close to, but certainly not identical to, Normal. In Figure 23.6 is shown the distribution of daily returns for

---

[2] They were published in *Risk* magazine in 1994 but with an algebraic error. Oops!

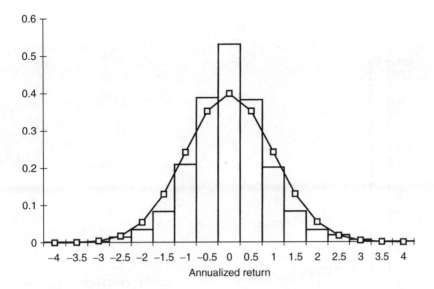

Figure 23.6 The distribution of returns on the Dow Jones index and the Normal distribution.

the Dow Jones index from 1977 until 1996, scaled to have a mean of zero and a standard deviation of one, and the standardized Normal distribution. The empirical distribution has a higher peak and fatter tails than the Normal. This is typical of the randomness in financial variables, whether they are stocks, commodities, currencies or even interest rates. How does this distribution affect our hedging arguments and the hedging error?

Let us assume that the return on the underlying in a time δt is given by

$$\frac{\delta S}{S} = \psi,$$

where ψ is a random variable (with distribution determined empirically). We will examine the change in the value of a hedged portfolio assuming that the option component satisfies the Black–Scholes equation with an implied volatility of σ_i. Our hedged portfolio has a random return in excess of the risk-free rate given by

$$\delta\Pi - r\Pi\,\delta t = S\left(\Delta - \frac{\partial V}{\partial S}\right)(r\,\delta t - \psi) + \tfrac{1}{2}S^2\frac{\partial^2 V}{\partial S^2}(\psi^2 - \sigma_i^2\,\delta t) + \cdots. \tag{23.10}$$

If we delta hedge in the Black–Scholes fashion then we are left with

$$\delta\Pi - r\Pi\,\delta t = \tfrac{1}{2}S^2\frac{\partial^2 V}{\partial S^2}(\psi^2 - \sigma_i^2\,\delta t),$$

to leading order. Obviously, we are interested in the distribution of ψ and in particular its variance. If the variance of ψ is different from σ_i^2 then theoretically we make money on the hedged portfolio (if our gamma is of the right sign; otherwise we lose money). This amounts to saying that the actual and implied volatilities are different. Of course, it is notoriously difficult to measure the actual volatility of the underlying. But what about the *distribution* of the returns? Is the hedging error chi-squared as the theory effectively assumes?

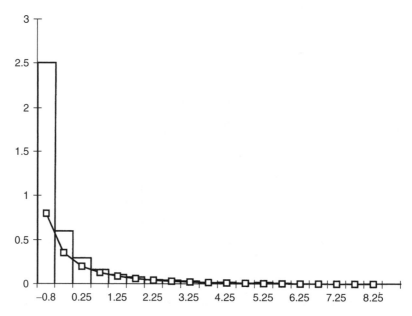

Figure 23.7 The distribution of the square of returns on the Dow Jones index and the translated chi-squared distribution.

In Figure 23.7 is shown the normalized distribution of the square of returns and the chi-squared distribution, translated to have zero means. Both of these distributions have a mean of zero, but the empirical distribution is considerably higher at small values of the square and has a fatter tail (too far out to fit on the figure). Assuming that the standard deviation of the distribution of returns is $\sigma_i \, \delta t^{1/2}$ (i.e. the actual and implied volatilities are the same), then the Black–Scholes value for a discretely-hedged position will be correct in the sense of *expectations*. On *average* the portfolio earns the risk-free rate, but this is achieved via a particularly extreme distribution. Most of the time the stock price moves less than the theory would have, but makes up for this by the occasional large movement.

If the distribution were Normally distributed then 68% of the time the return would be less than the standard deviation and 32% greater. With a long gamma position you would lose a small amount 68% of the time but regain that loss on average due to the rarer but larger moves. However, this situation is exaggerated with the real distribution. From the Dow Jones data the higher peak and fatter tails mean that 78% of the time the move is less than the standard deviation and only 22% of the time is it greater. If you have a long gamma position then approximately one fifth of the asset moves will lose you money. But again this is recovered on average with the rare large asset moves.

23.5 TOTAL HEDGING ERROR

We've looked at the hedging error on each rehedge, but these add up to give a total hedging error by the expiry of the option. This quantity is highly path-dependent since each local

Table 23.1 Total hedging error factors.

Asset	Factor
Ajinomoto	1.9
Bank of Scotland	1.7
Bass	1.3
Dover	1.8
Eastman Kodak	1.5
Guinness	1.6
Kanebo	2.3
Mitsui Mng & Smelt.	2.2
Nepool	7.3
Nippon Yakin Kogyo	1.9
Tesco	1.3

hedging error depends on S. The order of magnitude of the total hedging error is $\delta t^{1/2}$ since the errors should not be serially autocorrelated. This is the size of the standard deviation of the accumulated errors.

The mean of each local error is

$$E[\psi^2 - 1]\tfrac{1}{2}\sigma^2 S^2 \Gamma \, \delta t = 0$$

assuming that the correct volatility is used. The standard deviation of each hedging error is

$$\tfrac{1}{2}\sigma^2 S^2 \Gamma \, \delta t \sqrt{E[(\psi^2 - 1)^2]}.$$

For a Normal distribution the expectation term in this is 2. This need not be true for the actual distribution of returns. In Table 23.1 are shown the square root of the ratio of this expectation to its Normal equivalent for a number of assets. These numbers, which are scaled so that theoretically they should be 1, give a better idea of the size of the total hedging error that can be expected. All of the numbers are significantly greater than 1, and so total hedging error would in practice be larger than the theory says. One of the numbers is enormous; guess what asset that is.

23.6 **SUMMARY**

Hedging error is often overlooked in the pricing of a contract. In each timestep the order of magnitude of the change in P&L is the same order as the growth due to interest, i.e. of the order of the timestep. Just because this averages out to zero does not mean it should be ignored. In this chapter I showed some of the effects of hedging error on an option value and how real returns differ from the theoretical Normal in such a way as to make the hedging error distribution even worse than theoretical. The asset with the worst hedging error factor was Nepool, the price of electricity. We will see why this is such a strange asset in Chapter 62.

FURTHER READING

- Boyle & Emanuel (1980) explain some of the problems associated with the *discrete* re-hedging of an option portfolio.

- Wilmott (1994) derives the 'better' hedge.
- Leland (1985), Henrotte (1993) and Lacoste (1996) derive analytical results for the statistical properties of the tracking error.
- Mercurio & Vorst (1996) discuss the implications of discrete hedging strategies.
- See Hua (1997) for more details and examples of hedging error.

CHAPTER 24
transaction costs

In this Chapter...

- how to allow for transaction costs in option prices
- how economies of scale work
- a variety of hedging strategies and their effects on option prices are discussed

24.1 INTRODUCTION

Transaction costs are the costs incurred in the buying and selling of the underlying, related to the bid-offer spread. The Black–Scholes analysis requires the continuous rebalancing of a hedged portfolio, and no 'friction' such as transaction costs in such rebalancing. In practice, this assumption is incorrect. Depending on the underlying market in question, costs may or may not be important. In a market with high transaction costs, stocks in emerging markets for example, it will be too costly to rehedge frequently. In more liquid markets, government bonds in first-world countries for example, costs are low and portfolios can be hedged often. If costs are important then this will be an important factor in the bid-offer spread in option prices.

The modeling of transaction costs was initiated by Hayne Leland. I will describe his model and then a simple model for transaction costs, the Hoggard–Whalley–Wilmott model for non-vanilla options and option portfolios, which is based on Leland's hedging strategy and model. These models will give us some insight into the effect of costs on the prices of option. Due to the mathematical complexity of this subject, the latter part of this chapter is necessarily just a review of key results with little of the underlying mathematical theory. The interested reader is referred to the original papers for further details.

24.2 THE EFFECT OF COSTS

In the Black–Scholes analysis we assumed that hedging took place continuously. In a sense, we take the limit as the time between rehedges goes to zero, $\delta t \to 0$. We therefore find ourselves rehedging an infinite number of times until expiry. With a bid-offer spread on the underlying the cost of rehedging leads to infinite total transaction costs. Clearly, even with discrete hedging the consequences of these costs associated with rehedging are important. Different people have different levels of transaction costs; there are economies of scale, so that the larger the amount a person trades, the less significant are his costs. In contrast with the basic Black–Scholes model, we may expect that there is no unique option value. Instead, the value of the option depends on the investor.

Not only do we expect different investors to have different values for contracts, we also expect an investor *if they are hedging* to have different values for long and short positions in the same contract. Why is this? It is because transaction costs are always a sink of money for hedgers; they always lose out on the bid-offer spread on the underlying. Thus we expect a hedger to value a long position at less than the Black–Scholes value and a short position at more than the Black–Scholes value; whether the position is long or short, some estimate of hedging costs must be taken away from the value of the option. Since the 'sign' of the payoff is now important, it is natural to think of a long position as having a positive payoff and a short position a negative payoff.

24.3 **THE MODEL OF LELAND (1985)**

The groundwork of modeling the effects of transaction costs was done by Leland (1985). He adopted the hedging strategy of rehedging at *every* timestep. That is, every δt the portfolio is rebalanced, whether or not this is optimal in any sense. He assumes that the cost of trading v assets costs an amount $\kappa v S$ for both buying and selling; this models bid-offer spread, the cost is proportional to the value traded.

In the main the Leland assumptions are those mentioned in Chapter 5 for the Black–Scholes model but with the following exceptions:

- The portfolio is revised every δt where δt is a finite and fixed, small timestep.

- The random walk is given in discrete time by[1]

$$\delta S = \mu S \, \delta t + \sigma S \phi \, \delta t^{1/2}$$

  where ϕ is drawn from a standardized Normal distribution

- Transaction costs are proportional to the *value* of the transaction in the underlying. Thus if v shares are bought ($v > 0$) or sold ($v < 0$) at a price S, then the cost incurred is $\kappa|v|S$, where κ is a constant. The value of κ will depend on the individual investor. A more complex cost structure can be incorporated into the model with only a small amount of effort; see later. We will then also see economies of scale appearing.

- The hedged portfolio has an *expected* return equal to that from a risk-free bank deposit. This is exactly the same valuation policy as earlier on discrete hedging with no transaction costs.

He allows for the cost of trading in valuing his hedged portfolio and by equating the *expected* return on the portfolio with the risk-free rate, and he finds that long call and put positions should be valued with an adjusted volatility of $\breve{\sigma}$ where

$$\breve{\sigma} = \sigma \left(1 - \sqrt{\left(\frac{8}{\pi \, \delta t} \right) \frac{\kappa}{\sigma}} \right)^{1/2}.$$

---

[1] We don't need the more accurate discrete lognormal model of (23.1) and (23.2) since we are not going to a high order of accuracy.

Similarly short positions should be valued using $\hat{\sigma}$ where

$$\hat{\sigma} = \sigma \left(1 + \sqrt{\left(\frac{8}{\pi \, \delta t} \right) \frac{\kappa}{\sigma}} \right)^{1/2}.$$

REMEMBER THESE
ADJUSTMENTS....
BUT THEY ONLY WORK
FOR VANILLAS

As I have mentioned, long and short positions have different values.

Although the Leland concept is sound it is, in this form, only applicable to vanilla calls and puts, or any contract having a gamma of the same sign for all S and t. In the next section the Leland idea is extended to arbitrary option payoffs or portfolios, and the Leland result is derived along the way.

24.4 THE MODEL OF HOGGARD, WHALLEY & WILMOTT (1992)

The Leland strategy can be applied to arbitrary payoffs and to portfolios of options but the final result is not as simple as an adjustment to the volatility in a Black–Scholes formula. Instead, we will arrive at a nonlinear equation for the value of an option, derived by Hoggard, Whalley & Wilmott, and one of the first nonlinear models in derivatives theory.

Let us suppose we are going to hedge and value a portfolio of European options and allow for transaction costs. We can still follow the Black–Scholes analysis but we must allow for the cost of the transaction. If Π denotes the value of the hedged portfolio and $\delta\Pi$ the change in the portfolio over the timestep δt, then we must subtract the cost of any transaction from the equation for $\delta\Pi$ at each timestep. Note that we are not going to the limit $\delta t = 0$. Full details can be found in the Appendix at the end of this chapter.

After a timestep the change in the value of the hedged portfolio is now given by

$$\delta\Pi = \sigma S \left(\frac{\partial V}{\partial S} - \Delta \right) \phi \, \delta t^{1/2}$$

$$+ \left(\tfrac{1}{2}\sigma^2 S^2 \frac{\partial^2 V}{\partial S^2} \phi^2 + \mu S \frac{\partial V}{\partial S} + \frac{\partial V}{\partial t} - \mu \Delta S \right) \delta t - \kappa S |v|. \tag{24.1}$$

This is similar to the Black–Scholes expression but now contains a transaction cost term. Transaction costs have been subtracted from the change in the value of the portfolio. Because these costs are always positive, there is a modulus sign, $| \cdot |$, in the above.

We will follow the same hedging strategy as before, choosing $\Delta = \partial V/\partial S$. However, the portfolio is only rehedged at discrete intervals. The number of assets held short is therefore

$$\Delta = \frac{\partial V}{\partial S}(S, t)$$

where this has been evaluated at time t and asset value S. I have not given the details, but this choice minimizes the risk of the portfolio, as measured by the variance, to leading order. After a timestep δt and rehedging, the number of assets we hold is

$$\frac{\partial V}{\partial S}(S + \delta S, t + \delta t).$$

Note that this is evaluated at the new time and asset price. We can subtract the former from the latter to find the number of assets we have traded to maintain a 'hedged' position. The number of assets traded is therefore

$$\nu = \frac{\partial V}{\partial S}(S + \delta S, t + \delta t) - \frac{\partial V}{\partial S}(S, t).$$

Since the timestep and the asset move are both small we can apply Taylor's theorem to expand the first term on the right-hand side:

$$\frac{\partial V}{\partial S}(S + \delta S, t + \delta t) = \frac{\partial V}{\partial S}(S, t) + \delta S \frac{\partial^2 V}{\partial S^2}(S, t) + \delta t \frac{\partial^2 V}{\partial S \, \partial t}(S, t) + \cdots.$$

Since $\delta S = \sigma S \phi \, \delta t^{1/2} + O(\delta t)$, the dominant term is that which is proportional to δS; this term is $O(\delta t^{1/2})$ and the other terms are $O(\delta t)$. To leading order the number of assets bought (sold) is

$$\nu \approx \frac{\partial^2 V}{\partial S^2}(S, t)\delta S \approx \frac{\partial^2 V}{\partial S^2}\sigma S \phi \, \delta t^{1/2}.$$

We don't know beforehand how many shares will be traded, but we can calculate the *expected* number, and hence the expected transaction costs. The expected transaction cost over a timestep is

$$E[\kappa S|\nu|] = \sqrt{\frac{2}{\pi}}\kappa\sigma S^2 \left|\frac{\partial^2 V}{\partial S^2}\right| \delta t^{1/2}, \tag{24.2}$$

where the factor $\sqrt{2/\pi}$ is the expected value of $|\phi|$. We can now calculate the expected change in the value of our portfolio from (24.1), including the usual Black–Scholes terms and also the new cost term:

$$E[\delta\Pi] = \left(\frac{\partial V}{\partial t} + \tfrac{1}{2}\sigma^2 S^2 \frac{\partial^2 V}{\partial S^2} - \kappa\sigma S^2 \sqrt{\frac{2}{\pi \, \delta t}} \left|\frac{\partial^2 V}{\partial S^2}\right|\right) \delta t. \tag{24.3}$$

Except for the modulus sign, the new term above is of the same form as the second S derivative that has appeared before; it is a gamma term, multiplied by the square of the asset price, multiplied by a constant.

Now assuming that the holder of the option *expects* to make as much from his portfolio as if he had put the money in the bank, then we can replace the $E[\delta\Pi]$ in (24.3) with $r(V - S(\partial V/\partial S))\delta t$ as before to yield an equation for the value of the option:

A NONLINEAR PRICING EQUATION... NONLINEARITY HAS IMPORTANT CONSEQUENCES

$$\frac{\partial V}{\partial t} + \tfrac{1}{2}\sigma^2 S^2 \frac{\partial^2 V}{\partial S^2} - \kappa\sigma S^2 \sqrt{\frac{2}{\pi \, \delta t}} \left|\frac{\partial^2 V}{\partial S^2}\right|$$
$$+ rS \frac{\partial V}{\partial S} - rV = 0. \tag{24.4}$$

There is a nice financial interpretation of the term that is not present in the usual Black–Scholes equation. The second derivative of the option price with respect to the asset price, the gamma, $\Gamma = \partial^2 V/\partial S^2$, is a measure of the degree of mishedging of the hedged portfolio. The leading-order component of randomness is proportional to δS and this has been eliminated by delta hedging. But this delta hedging leaves behind a small component of risk proportional to the gamma. The gamma of an option or portfolio of options is related to the amount of rehedging that is expected to take place at the next rehedge and hence to the expected transaction costs.

The equation is a *nonlinear* parabolic partial differential equation, one of the first such in finance. It is obviously also valid for a portfolio of derivative products. This is the first time in this book that we distinguish between single options and a portfolio of options. But for much of the rest of the book this distinction will be important. In the presence of transaction costs, the value of a portfolio is not the same as the sum of the values of the individual components. We can best see this by taking a very extreme example.

We have positions in two European call options with the same strike price and the same expiry date and on the same underlying asset. One of these options is held long and the other short. Our net position is therefore exactly zero because the two positions exactly cancel each other out. But suppose that we do not notice the cancellation effect of the two opposite positions and decide to hedge each of the options separately. Because of transaction costs we lose money at each rehedge on both options. At expiry the two payoffs will still cancel, but we have a negative net balance due to the accumulated costs of all the rehedges in the meantime. This contrasts greatly with our net balance at expiry if we realize beforehand that our positions cancel. In the latter case we never bother to rehedge, leaving us with no transaction costs and a net balance of zero at expiry.

Now consider the effect of costs on a single vanilla option held long. We know that

$$\frac{\partial^2 V}{\partial S^2} > 0$$

for a single call or put held long in the absence of transaction costs. Postulate that this is true for a single call or put when transaction costs are included. If this is the case then we can drop the modulus sign from (24.4). Using the notation

$$\check{\sigma}^2 = \sigma^2 - 2\kappa\sigma\sqrt{\frac{2}{\pi\,\delta t}} \tag{24.5}$$

the equation for the value of the option is identical to the Black–Scholes value with the exception that the actual variance σ^2 is replaced by the modified variance $\check{\sigma}^2$. Thus our assumption that $\partial^2 V/\partial S^2 > 0$ is true for a single vanilla option even in the presence of transaction costs. The modified volatility will be recognized as the Leland volatility correction mentioned at the start of this chapter.

For a short call or put option position we simply change all the signs in the above analysis with the exception of the transaction cost term, which must always be a drain on the portfolio. We then find that the call or put is valued using the new variance

$$\hat{\sigma}^2 = \sigma^2 + 2\kappa\sigma\sqrt{\frac{2}{\pi\,\delta t}}. \tag{24.6}$$

Again this is the Leland volatility correction.

The results (24.5) and (24.6) show that a long position in a single call or put with costs incorporated has an apparent volatility that is less than the actual volatility. When the asset price rises the owner of the option must sell some assets to remain delta hedged. But then the effect of the bid-offer spread on the underlying is to reduce the price at which the asset is sold. The effective increase in the asset price is therefore less than the actual increase, being seen as a reduced volatility. The converse is true for a short option position.

The above volatility adjustments are applicable when you have an option or a portfolio of options having a gamma of one sign. If the gamma is always and everywhere positive use the lower volatility value for a long position, and the higher value for a short position. If gamma is always and everywhere negative, swap these values around.

For a single vanilla call or put, we can get some idea of the total transaction costs associated with the above strategy by examining the difference between the value of an option with the cost-modified volatility and one with the usual volatility; that is, the difference between the Black–Scholes value and the value of the option taking into account the costs. Consider

$$V(S, t) - \hat{V}(S, t),$$

where the hatted function is Black–Scholes with the modified volatility. Expanding this expression for small κ we find that it becomes

$$(\sigma - \hat{\sigma})\frac{\partial V}{\partial \sigma} + \cdots.$$

This is proportional to the vega of the option. We know the formula for a European call option and therefore we find the expected spread to be

$$\frac{2\kappa S N'(d_1)\sqrt{(T - t)}}{\sqrt{2\pi \, \delta t}},$$

where $N(d_1)$ has its usual meaning.

The most important quantity appearing in the model is

$$K = \frac{\kappa}{\sigma \, \delta t^{1/2}}. \tag{24.7}$$

K is a non-dimensional quantity, meaning that it takes the same value whatever units are used for measuring the parameters. If this parameter is very large, we write $K \gg 1$, then the transaction costs term is much greater than the underlying volatility. This means that costs are high and that the chosen δt is too small. The portfolio is being rehedged too frequently. If the transaction costs are very large or the portfolio is rehedged very often then it is possible to have $\kappa > \sigma/\sqrt{8\pi \, \delta t}$. In this case the equation becomes forward parabolic for a long option position. Since we are still prescribing final data, the equation is ill-posed. Although the asset price may have risen, its effective value due to the addition of the costs will actually have dropped. I discuss such ill posedness later.

If the parameter K is very small, we write $K \ll 1$, then the costs have only a small effect on the option value. Hence δt could be decreased to minimize risk. The portfolio is being rehedged too infrequently.

We can see how to use this result in practice if we have data for the bid-offer spread, volatility and time between rehedges for a variety of stocks. Plot the parameter κ/σ against the quantity $1/\delta t^{1/2}$ for each stock. An example of this for a real portfolio is shown in Figure 24.1. In this figure are also shown lines on which K is constant. To be consistent in our attitude towards transaction costs across all stocks we might decide that a value of $K = K'$ is ideal. If this is the bold line in the figure then options on those stocks above the line, such as A, are too infrequently hedged, while those below, such as R, are hedged too often. Of course, this is a very simple approach to optimizing a hedging strategy. A more sophisticated approach would also take into account the advantage of increased hedging: the reduction of risk.

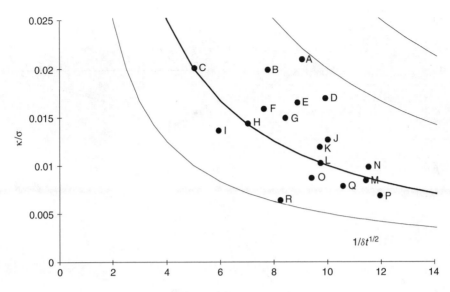

Figure 24.1 The parameter κ/σ against $1/\delta t^{1/2}$ for a selection of stocks. On each curve the transaction cost parameter K is the same.

24.5 NON-SINGLE-SIGNED GAMMA

For an arbitrary portfolio of options, the gamma, $\partial^2 V/\partial S^2$, is not of one sign. If this is the case then we cannot drop the modulus sign. Since the Hoggard–Whalley–Wilmott equation is nonlinear we must solve it numerically.

In Figures 24.2 and 24.3 is shown the value of a long bull spread consisting of one long call with $E = 45$ and one short call with $E = 55$ and the delta at six months before expiry for the two cases, with and without transaction costs. The volatility is 20% and the interest rate 10%. In this example $K = 0.25$. The bold curve shows the values in the presence of transaction costs and the other curve in the absence of transaction costs. The latter is simply the Black–Scholes value for the combination of the two options. The bold line approaches the other line as the transaction costs decrease.

In Figures 24.4 and 24.5 is shown the value of a long butterfly spread and its delta, before and at expiry. In this example the portfolio contains one long call with $E = 45$, two short calls with $E = 55$ and another long call with $E = 65$. The results are shown for one month until expiry for the two cases, with and without transaction costs. The volatility, the interest rate and K are as in the previous example.

24.6 THE MARGINAL EFFECT OF TRANSACTION COSTS

Suppose we hold a portfolio of options, let's call its value $P(S, t)$, and we want to add another option to this portfolio. What will be the effect of transaction costs? Call the value of the new, larger, portfolio $P + V$: what equation is satisfied by the marginal value V? This is not just the cost equation applied to V because of the nonlinearity of the problem. We can write

$$\frac{\partial P}{\partial t} + \frac{1}{2}\sigma^2 S^2 \frac{\partial^2 P}{\partial S^2} - \kappa \sigma S^2 \sqrt{\frac{2}{\pi \, \delta t}} \left| \frac{\partial^2 P}{\partial S^2} \right| + rS \frac{\partial P}{\partial S} - rP = 0$$

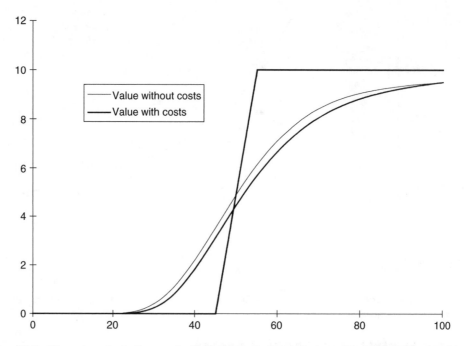

Figure 24.2   The value of a bull spread with (bold) and without transaction costs. The payoff is also shown.

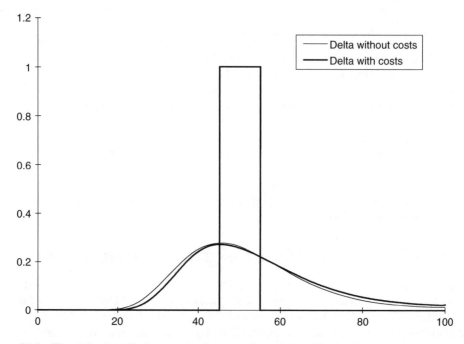

Figure 24.3   The delta for a bull spread prior to and at expiry with (bold) and without transaction costs.

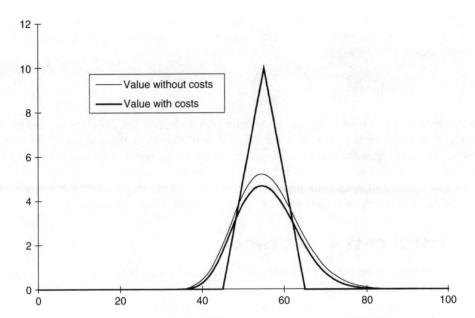

Figure 24.4 The value of a butterfly spread with (bold) and without transaction costs.

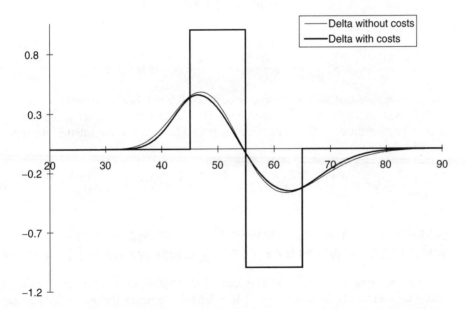

Figure 24.5 The delta for a butterfly spread with (bold) and without transaction costs.

since the original portfolio must satisfy the costs equation. But now the new portfolio also satisfies this equation:

$$\frac{\partial(P+V)}{\partial t} + \tfrac{1}{2}\sigma^2 S^2 \frac{\partial^2(P+V)}{\partial S^2} - \kappa\sigma S^2 \sqrt{\frac{2}{\pi\,\delta t}} \left|\frac{\partial^2(P+V)}{\partial S^2}\right| + rS\frac{\partial(P+V)}{\partial S} - r(P+V) = 0.$$

Both of these equations are nonlinear. But if the size of the new option is much less than the original portfolio we can linearize the latter equation to examine the equation satisfied by the marginal value V. We find that

$$\frac{\partial V}{\partial t} + \frac{1}{2}\sigma^2 S^2 \frac{\partial^2 V}{\partial S^2} - \text{sgn}\left(\frac{\partial^2 P}{\partial S^2}\right)\kappa\sigma S^2 \sqrt{\frac{2}{\pi\,\delta t}}\frac{\partial^2 V}{\partial S^2} + rS\frac{\partial V}{\partial S} - rV = 0.$$

This equation is now linear. The important point to note is that the volatility correction only depends on the sign of the gamma of the original portfolio, P. If the gamma of the original portfolio is positive then the addition of another contract with positive gamma only makes the cost situation worse. However, the addition of a new contract with negative gamma will reduce the level of transaction costs. The benefits of reducing gamma may even make it worthwhile to buy/sell a contract for more/less than the Black–Scholes value, theoretically.

24.7 OTHER COST STRUCTURES

The above model can be extended to accommodate other transaction cost structures. Suppose that the cost of buying or selling v assets is $\kappa(v, S)$. We follow the analysis of Section 24.4 up to the point where we take expectations of the transaction cost of hedging. The number of assets traded is still proportional to the gamma, and the expected cost of trading is

$$E\left[\kappa\left(\sigma S\,\delta t^{1/2}\Gamma\phi, S\right)\right].$$

The option pricing equation then becomes

$$\frac{\partial V}{\partial t} + \frac{1}{2}\sigma^2 S^2 \frac{\partial^2 V}{\partial S^2} + rS\frac{\partial V}{\partial S} - rV = \frac{1}{\delta t}E\left[\kappa\left(\sigma S\,\delta t^{1/2}\Gamma\phi, S\right)\right].$$

For example, suppose that trading in shares costs $k_1 + k_2 v + k_3 vS$, where k_1, k_2 and k_3 are constants. This cost structure contains fixed costs (k_1), a cost proportional to volume traded ($k_2 v$), and a cost proportional to the value traded ($k_3 vS$). The option value satisfies the nonlinear diffusion equation

$$\frac{\partial V}{\partial t} + \frac{1}{2}\sigma^2 S^2 \frac{\partial^2 V}{\partial S^2} + rS\frac{\partial V}{\partial S} - rV = \frac{k_1}{\delta t} + \sqrt{\frac{2}{\pi}}\delta t\sigma S(k_2 + k_3 S)\left|\frac{\partial^2 V}{\partial S^2}\right|. \tag{24.8}$$

24.8 HEDGING TO A BANDWIDTH: THE MODEL OF WHALLEY & WILMOTT (1993) AND HENROTTE (1993)

We have so far seen how to model option prices when hedging takes place at fixed intervals of time. Another commonly used strategy is to rehedge whenever the position becomes too far out of line with the perfect hedge position. Prices are therefore monitored continuously but hedging still has to take place discretely.

Due to the complexity of this problem and those that follow, I only give a brief sketch of the ideas and results.

With $V(S, t)$ as the option value, the perfect Black–Scholes hedge is given by

$$\Delta = \frac{\partial V}{\partial S}.$$

Suppose, however, that we are not perfectly hedged, that we hold $-D$ of the underlying asset but do not want to accept the extra cost of buying or selling to reposition our hedge. The risk, as measured by the variance over a timestep δt of this imperfectly hedged position, is, to leading order,

$$\sigma^2 S^2 \left(D - \frac{\partial V}{\partial S} \right)^2 \delta t.$$

I can make two observations about this expression. The first is simply to confirm that when $D = \partial V / \partial S$ this variance is zero. The second observation is that a natural hedging strategy is to bound the variance within a given tolerance and that this strategy is equivalent to restricting D so that

$$\sigma S \left| D - \frac{\partial V}{\partial S} \right| \leq H_0. \tag{24.9}$$

The parameter H_0 is now a measure of the maximum expected risk in the portfolio. When the perfect hedge $\partial V / \partial S$ and the current hedge (D) move out of line so that (24.9) is violated, then the position should be rebalanced. Equation (24.9) defines the **bandwidth** of the hedging position.

The model of Whalley & Wilmott (1993) and Henrotte (1993) takes this as the hedging strategy: the investor prescribes H_0 and on rehedging rebalances to $D = \partial V / \partial S$.

We find that the option value satisfies the nonlinear diffusion equation

$$\frac{\partial V}{\partial t} + \tfrac{1}{2}\sigma^2 S^2 \frac{\partial^2 V}{\partial S^2} + rS \frac{\partial V}{\partial S} - rV = \frac{\sigma^2 S^4 \Gamma^2}{H_0} \left(k_1 + (k_2 + k_3 S) \frac{H_0^{1/2}}{S} \right), \tag{24.10}$$

where Γ is the option's gamma and the parameters k_1, k_2 and k_3 are the cost parameters for the cost structure introduced in Section 24.7. Note that again there is a nonlinear correction to the Black–Scholes equation that depends on the gamma.

24.9 UTILITY-BASED MODELS

24.9.1 The model of Hodges & Neuberger (1989)

All of the above models for transaction costs take the hedging strategy as exogenously given; that is, the investor chooses his strategy and then prices his option afterwards. Strategies like this have been called **local in time** because they only worry about the state of an option at the present moment in time. An alternative, first examined by Hodges & Neuberger (1989), is to find a strategy that is in some sense *optimal*. These have been called **global-in-time models** because they are concerned with what may happen over the rest of the life of the option.

The seminal work in this area, combining both utility theory and transaction costs, was by Hodges & Neuberger (HN), with Davis, Panas & Zariphopoulou (DPZ) making improvements to the underlying philosophy. HN explain that they assume that a financial agent holds a portfolio that is already optimal in some sense but then has the opportunity to issue an option and hedge the risk using the underlying. However, since rehedging is costly, he must define his strategy in terms of a 'loss function'. He thus aims to maximize expected utility. This entails the investor specifying a 'utility function'. The case considered in most detail by HN and DPZ is of the exponential utility function. This has the nice property of constant risk aversion.

Mathematically, such a problem is one of stochastic control and the differential equations involved are very similar to the Black–Scholes equation.

24.9.2 The Model of Davis, Panas & Zariphopoulou (1993)

The ideas of HN were modified by DPZ. Instead of valuing an option on its own, they embed the option valuation problem within a more general portfolio management approach. They then consider the effect on a portfolio of adding the constraint that at a certain date, expiry, the portfolio has an element of obligation due to the option contract. They introduce the investor's utility function; in particular, they assume it to be exponential. They only consider costs proportional to the value of the transaction (κvS), in which case they find that the optimal hedging strategy is not to rehedge until the position moves out of line by a certain amount. Then, the position is rehedged as little as possible to keep the delta at the edge of this hedging bandwidth. This result is shown schematically in Figure 24.6. Here we see the Black–Scholes delta position and the hedging bandwidth.

In HN and DPZ the value of the option and, most importantly, the hedging strategy are given in terms of the solution of a three-dimensional free boundary problem. The variables in the problem are asset price S, time t, as always, and also D, the number of shares held in the hedged portfolio.

24.9.3 The Asymptotic Analysis of Whalley & Wilmott (1993)

The models of HN and DPZ are unwieldy because they are time-consuming to compute. As such it is difficult to gain any insight into the optimal hedging strategy. Whalley & Wilmott did an asymptotic analysis of the DPZ model assuming that transaction costs are small, which is, of course, the case in practice. This analysis shows that the option price is given by the solution of an inhomogeneous diffusion equation, similar to the Black–Scholes equation.

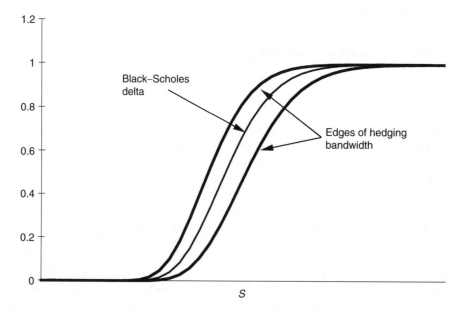

Figure 24.6 The optimal hedging strategy with proportional costs.

This asymptotic analysis also shows that the HN optimal hedging bandwidth is symmetric about the Black–Scholes delta so that

$$\left| D - \frac{\partial V}{\partial S} \right| \leq \left(\frac{3k_3 Se^{-r(T-t)}F(S,t,\Gamma)^2}{2\gamma} \right)^{1/3},$$

where

$$F(S,t,\Gamma) = \left| \Gamma - \frac{e^{-r(T-t)}(\mu - r)}{\gamma S^2 \sigma^2} \right|.$$

The parameter γ is the index of risk aversion in the utility function.

These results are important in that they bring together all the local-in-time models mentioned above and the global-in-time models of HN and DPZ into the same diffusion equation framework.

This hedging bandwidth has been tested using Monte Carlo simulations by Mohamed (1994) and found to be the most successful strategy that he tested. The model has been extended by Whalley & Wilmott (1994) to an arbitrary cost structure, described below.

24.9.4 Arbitrary Cost Structure

The above description concentrates on the proportional cost case. If there is a fixed cost component then shares are traded to position the number of shares to be at some **optimal rebalance point**. This is illustrated schematically in Figure 24.7.

I do not give any of the details but note that the algorithm for finding the optimal rebalance point and the hedging bandwidth is as follows.

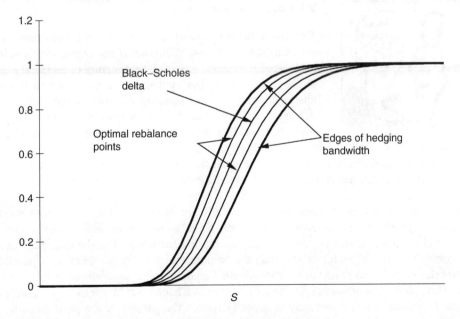

Figure 24.7 The optimal hedging strategy with arbitrary cost structure.

Assume that costs take the form $K(S, v)$, and that this is symmetric for buying and selling. The bandwidth is given by

$$\Delta - A(S, t) \leq D \leq \Delta + A(S, t)$$

where Δ is the Black–Scholes delta. The optimal rebalance points are given by

$$D = \Delta \pm B(S, t).$$

A and B come from solving

$$\gamma AB(A + B) = 3\delta\Gamma^2 \frac{\partial K}{\partial v}\bigg|_{v=A-B}$$

and

$$\gamma(A + B)^3(A - B) = 12\delta\Gamma^2 K(S, A - B),$$

where

$$\delta = e^{-r(T-t)},$$

Γ is the Black–Scholes gamma and γ is the index of risk aversion.

24.10   INTERPRETATION OF THE MODELS

Nonlinear and inhomogeneous diffusion equations appear throughout the physical science literature, thus there is a ready-made source of theoretical results and insights. I describe some of these in this section.

I'LL KEEP COMING BACK TO THIS POINT

24.10.1   Nonlinearity

The effect of nonlinearity on the valuation equations is that the sum of two or more solutions is not necessarily a solution itself. As I have said, a portfolio consisting of an equal number of the same options but held long and short (which has value identically zero), is not equal to the sum of the values of the two sub-portfolios of all the long and short options. This makes sense because in valuing each sub-portfolio separately we are assuming that each would be hedged separately, with attendant transaction costs to be taken into account. Upon recombining the two, the intrinsic values cancel, but the two sets of costs remain, giving a negative net value. The importance of nonlinearity extends far beyond this however.

Consider the following. Transaction cost models are nonlinear. The value of a portfolio of options is generally not the same as the sum of the values of the individual components. We can add contracts to our portfolio by paying the market prices, but the marginal value of these contracts may be greater or less than the amount that we pay for them. Is it possible to optimize the value of our portfolio by adding traded contracts until we give our portfolio its best value? This question is answered (in the affirmative) in Chapter 32. In a sense, the optimization amounts to finding the cheapest way to globally reduce the gamma of the portfolio, since the costs of hedging are directly related to the gamma.

24.10.2 Negative Option Prices

The transaction cost models above can result in negative option prices for some asset values depending on the hedging strategy implied by the model. So for example in the Hoggard–Whalley–Wilmott model with fixed transaction costs, $k_1 > 0$, option prices can become negative if they are sufficiently far out of the money. This model assumes that we rehedge at the end of *every* timestep, irrespective of the level of risk associated with our position and also irrespective of the option value. Thus there is some element of obligation in our position, and the strategy should be amended so that we do not rehedge if this would make the option value go negative. In the case of a call, therefore, there may be an asset price below which we would cease to rehedge and in this case we would regard the option as worthless.

Note that this is not equivalent to discarding the option; if the asset price were subsequently to rise above the appropriate level (which will change over time), we would begin to hedge again and the option would once more have a positive value. So we introduce the additional conditions for a moving boundary:

$$V(S_s(t), t) = 0$$

and

$$\frac{\partial V}{\partial S}(S_s(t), t) = 0.$$

The value $S_s(t)$ is to be found as part of the solution. This problem is now a 'free boundary problem', similar mathematically to the American option valuation problem. In our transaction cost problem for a call option, we must find the boundary, $S_s(t)$, below which we stop hedging. This is illustrated in Figure 24.8.

In this figure we are valuing a long vanilla call with fixed costs at each rehedge. The top curve is the Black–Scholes option value as a function of S at some time before expiry. The bottom curve allows for the cost of rehedging but with the obligation to hedge at each timestep.

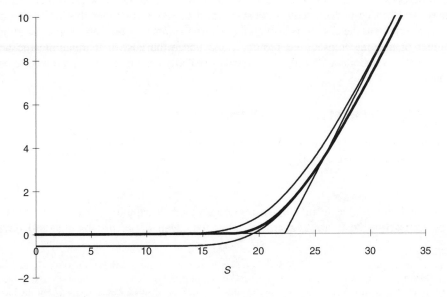

Figure 24.8 When to stop hedging if there are fixed costs.

The option value is thus negative far out of the money. The middle curve also incorporates costs but without the obligation to rehedge. It thus always has a positive value and is, of course, also below the Black–Scholes option value.

24.10.3 Existence of Solutions

Linear diffusion equations have many nice properties, as we discussed in Chapter 6. The solution to a 'sensible' problem exists and is unique. This need not be the case for nonlinear equations. The form of the equation and the final data $V(S, T)$ for the equation (the payoff at expiry) may result in the solution 'blowing up', that is, becoming infinite and thus financially unrealistic.

This can occur in some models even if transaction costs are small because of the effect of the option's gamma, which in those models is raised to some power greater than one in the extra transaction cost term. So wherever the gamma is large this term can dominate. For example, near the exercise price for a vanilla call or put option, $\partial^2 V / \partial S^2(S, 0)$ is infinite. We consider the case for the model of Equation (24.10).

The governing equation in this case has a transaction cost term proportional to Γ^2. Close to expiry and near the exercise price, E, we write $t = T - \tau$ and $S = E + s$ where $|s|/E \ll 1$ and then the equation can be approximated by

$$\frac{\partial V}{\partial \tau} = \beta \left(\frac{\partial^2 V}{\partial s^2} \right)^2, \tag{24.11}$$

where

$$\beta = \frac{\sigma^2 E^4}{H_0} \left(k_1 + H_0^{1/2} \left(\frac{k_2}{E} + k_3 \right) \right). \tag{24.12}$$

Taking H_0 to be a constant, which is equivalent to a fixed bandwidth for the delta, it can be shown that Equation (24.11) is ill-posed, and that it is has no solution, if $\Gamma(S, 0) > 0$.[2]

So a long vanilla call or put hedged under this strategy has no finite value. Note that for short vanilla options, $\Gamma(S, 0) < 0$ and a solution does exist, so they *can* be valued under such a strategy, and can be hedged with a constant level of risk throughout the life of the option. However, returning to the case of payoffs with positive gamma, as the option approaches expiry, the number of hedging transactions required, and hence the cost of maintaining the hedging strategy, increases unboundedly unless the level of risk allowed (H_0) is itself allowed to become unbounded.

24.11 NON-NORMAL RETURNS

We know that returns are not Normally distributed but is this important?

Suppose that returns are given by

$$\frac{\delta S}{S} = \psi$$

for some random variable ψ of empirically determined distribution. What matters as far as expected transaction costs are concerned is not the mean of ψ, nor its standard deviation.

---

[2] We can see this intuitively as follows. If $\Gamma(S, 0) > 0$ then the right-hand side of (24.11) is positive. Thus V increases in time, increasing fastest where the gamma is largest. This in turn further increases the gamma, making the growth in V even faster. The result is a blow-up. The linear diffusion equation also behaves in this way but the increase in V does not get out of control, since the gamma is raised only to the power one.

Table 24.1 Transaction cost factors.

Asset	Factor
Asahi Breweries	0.93
Asda Group	0.93
Cable & Wireless	0.91
Standard Chartered	0.68
Equifax	0.88
Fleetwood Ents	0.92
Ford	0.96
Nepool	0.50
Sumitomo Bank	0.88
Toshiba	0.88

What is most important is the mean of the absolute value of ψ, i.e. the average value of $|\psi|$. We can examine the data to see if this number is greater or less than the theory says, the ratio to the theoretical value giving us a transaction cost factor.

In Table 24.1 are given the transaction cost factors for a selection of stocks, scaled with timestep and volatility so that all numbers would be one if the underlying distribution were Normal. You can see that they are all less than one, but not by an enormous amount, so costs are going to be slightly less important than you might think. As we saw in Chapter 23 electricity prices stand out as being the furthest from the theory.

24.12 **EMPIRICAL TESTING**

In this section we look at empirical results for transaction costs and hedging error using various hedging strategies described above. We will look at the following four strategies:

- Basic Black–Scholes strategy, delta hedging at fixed intervals
- Leland volatility-modified delta, hedging at fixed intervals
- Delta tolerance, hedging to the Black–Scholes delta when the difference between quantity of the underlying held and ideal delta move too far out of line
- The asymptotic version of the utility model

We will use stock price data that are generated randomly, with known and constant volatility, and we will also use real data. Many stock path realizations will be used so that we can examine the statistical properties of the total costs and hedging errors.

Finally, we examine:

- average total transaction costs
- average price (i.e. Black–Scholes value plus average costs)
- standard deviation of price
- price of 95th percentile

The third of these includes the hedging error that would be present even if there were no costs at all. The last simply means the price at which the contract must be sold to ensure that 95% of the time we do not lose money.

When we come to look at real data we also examine for each stock price time series which of the four strategies is the winner. Here 'winner' means the strategy that gives the lowest total cost plus hedging error for that particular realized asset path.

To understand how the random simulations were done, see Chapter 66.

To start with we value and hedge an at-the-money call with a volatility of 20%, risk-free rate of 5%. The underlying is currently at 100 and there is one year to expiry. The values for k_1, k_2 and k_3 are all 0.01.

24.12.1 Black–Scholes and Leland Hedging

This is a straight fight between using the Black–Scholes delta or the Leland volatility-adjusted delta.

Figure 24.9 shows the average amount of costs incurred and the standard deviation of the option price when the time between hedging varies. For each hedging period 5000 simulations were used. The number of days between rehedging varied from 1 up to 25.

Not surprisingly, as the time gap increases the amount of transaction costs paid decreases.

The figure also shows the standard deviation of the price. This includes the hedging error that would still be present in the absence of costs. It is very clear that the level of risk increases when the number of trades decreases. The kinks in the graph between 13 and 20 days reflects the fact that, as the number of days increases, the gap between the last trade and the penultimate trade sometimes decreases, or sometimes increases, which affects the level of risk taken on.

Figure 24.10 shows the average total price of hedging the option. This is a straight average of the 5000 prices accumulated from the simulations. This graph suggests that the fewer trades the better. The Leland strategy produces a lower price than the Black–Scholes.

Figure 24.10 also shows the 95th percentile price. This is the price we must charge for the option for us to make a profit 95% of the time. This is the most informative picture from a risk management point of view.

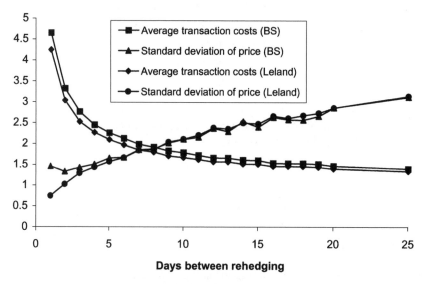

Figure 24.9 Average costs and standard deviation of price under the Black–Scholes and Leland strategies.

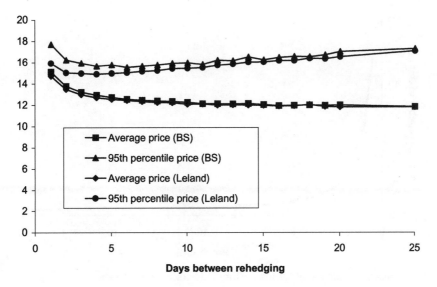

Figure 24.10 Average price and 95th percentile of price under the Black–Scholes and Leland strategies.

Frequent hedging is good for risk control but bad for pricing, infrequent hedging is the opposite. The 95th percentile price is a compromise between taking on risk and incurring transaction costs. With this as our option 'value' the figure shows that the Leland model outperforms the normal Black–Scholes method.

The optimal number of days between rehedging for the Leland method was four, giving a 95th percentile of 14.90.

Now we repeat all of this analysis and plotting for the delta-tolerance strategy.

24.12.2 Market Movement or Delta-tolerance Strategy

In this model the risk in the hedged position is restricted by the parameter H_0, see Equation (24.9).

Figure 24.11 shows the number of trades required on average as H_0 varies.

The next figure, 24.12, shows how much the strategy affects the average total transaction costs. Observe that instead of plotting the costs against the bandwidth I have plotted costs against the inverse of the average number of trades. This is not quite the same as the average number of days between rehedges, hence the inverted commas. As before restricting the number of trades restricts the amount of costs, but increases the standard deviation of price, the risk.

The average contract price is shown in Figure 24.13 along with the 95th percentile price. As in the Leland strategy, there is a compromise point. With the 95th percentile 'value' determining this point we find that we get an option 'value' of 15.15. This is slightly worse than the Leland strategy but better than normal Black–Scholes hedging.

24.12.3 The Utility Strategy

In this strategy the parameter to be varied is γ, the level of risk aversion.

The first figure shows the average number of trades versus the risk aversion parameter. Again, this is used to convert from risk aversion to a measure of the number of days between rehedges, so that all plots can be better compared across strategies.

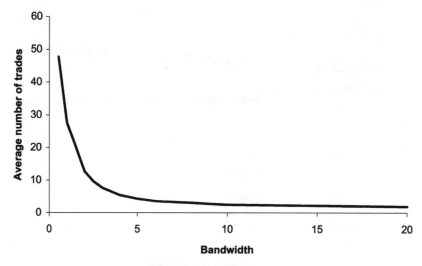

Figure 24.11   Average number of trades for the delta-tolerance strategy.

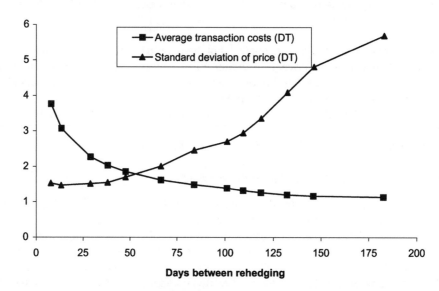

Figure 24.12   Average costs and standard deviation of price for the delta-tolerance strategy.

From the 95th percentile plot the best option 'value' is 15.03. Thus this method turns out to be better than the delta-tolerance method, but still not as good as the Leland fixed-timestep hedging strategy. However over the whole range of values considered for γ this utility method produced a far lower 95th percentile price than the ranges produced from the other strategies; the 95th percentile seems to be quite insensitive to the risk aversion parameter.

There are a couple of points to note about the use of the utility strategy. First, we have only used the asymptotic version since the computational time necessary for the solution of the full partial differential equation would be prohibitively long. Second, we looked at a fairly general cost model, not just proportional costs. The addition of the extra transaction costs, fixed and

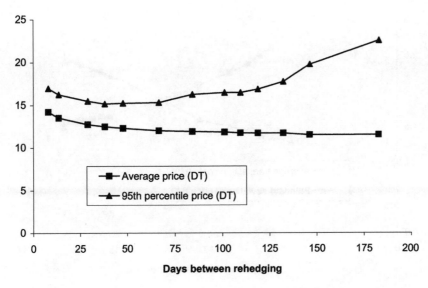

Figure 24.13   Average price and 95th percentile price for the delta-tolerance strategy.

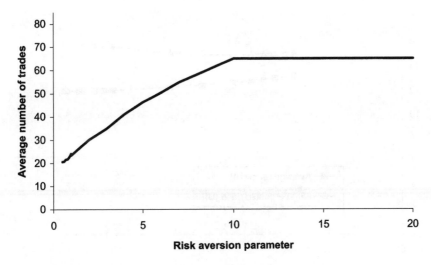

Figure 24.14   Average number of trades for the utility strategy.

proportional to volume, made the simulations much more general and more realistic. However, solving for the edges of the bandwidth and the rebalance point did take quite a lot of time. This wouldn't matter during the real hedging of a position, but slows down the back-testing.

24.12.4   Using the Real Data

In addition to incorporating the full cost structure into the testing framework, the strategies have been used on real data. For nearly 40 stocks, using daily prices from 1985 to 1996, one-year options were set up, commencing on 1st January and expiring on the 31st December. The options were at-the-money when struck.

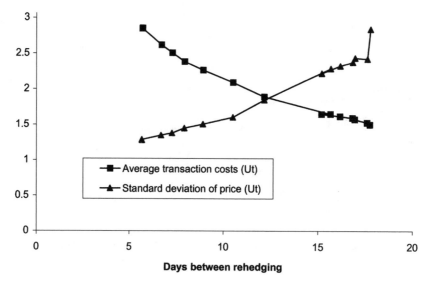

Figure 24.15 Average costs and standard deviation of price for the utility strategy.

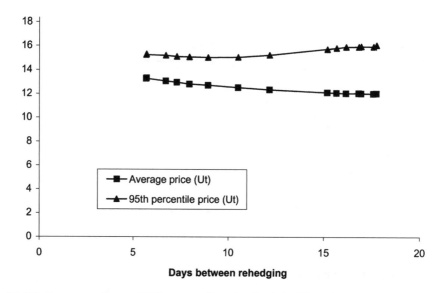

Figure 24.16 Average price and 95th percentile price for the utility strategy.

This is a more realistic test than the basic Monte Carlo simulations used above, since it involves a large range of different volatilities and starting stock prices.

24.12.5 And the Winner is. . .

Although the Leland strategy seemed to be marginally better than the others for the simulations, the results were quite different when real data were used. Admittedly, a slightly different criterion was used when deciding exactly what was meant by 'best'. Because of the relative lack of data, compared with the random simulations, the criterion used was based on which strategy gave the lowest hedging cost for each realized asset path.

The delta-tolerance method proved to be easily the best method. The delta-tolerance method won in nearly 70% of cases. The second place goes to the utility method with 23%. The Leland method produces the lowest price in only 7% of stock price paths.

Given the different results for the simulation and for the real data, it's not at all clear which is the best strategy to use. I've also heard on the grapevine that the utility method is the best, but I haven't got any references to back this up. All in all, despite the empirical work I can't actually give an adequate conclusion. Sorry.

24.13 **SUMMARY**

For equity derivatives and derivatives on emerging market underlyings transaction costs are large and important. In these markets pricing and hedging must take into account costs. In other markets the underlying may be so liquid that costs are irrelevant, then you would hedge as often as you could. When transaction costs are important the problem of pricing and hedging is nonlinear; if you don't price and hedge contracts together then you will miss out on cancellation effects and economies of scale. This is interesting from a mathematical point of view, but cumbersome from a practical point of view.

FURTHER READING

- Some of the material of this chapter is based on the model of Leland (1985) as extended to portfolios of options by Hoggard, Whalley & Wilmott (1994) and Whalley & Wilmott (1993a).

- Gemmill (1992) gives an example taken from practice of the effect of transaction costs on a hedged portfolio.

- The Leland model was put into a binomial setting by Boyle & Vorst (1992). They find a similar volatility adjustment to that of Leland.

- Whalley & Wilmott (1993b, 1994b, 1996) and Henrotte (1993) discuss various hedging strategies and derive more nonlinear equations using ideas similar to those in this chapter.

- For alternative approaches involving 'optimal strategies' see Hodges & Neuberger (1989), Davis & Norman (1990) and Davis, Panas & Zariphopoulou (1993), and the asymptotic analyzes of Whalley & Wilmott (1994a, b, 1995, 1997) for small levels of transaction costs.

- Dewynne, Whalley & Wilmott (1994, 1995) discuss the pricing of exotic options in the presence of costs.

- Avellaneda & Parás (1994) discuss the fixed-timestep equation when transaction costs are large.

- Ahn, Dayal, Grannan & Swindle (1998) discuss the variance of replication error when there are transaction costs.

- For a model of a Poisson process for stocks with transaction costs see Neuberger (1994).

- Jefferies (1999) examines the various hedging strategies using simulated and real data. His results are those presented above.

APPENDIX: DERIVATION OF THE HOGGARD–WHALLEY–WILMOTT EQUATION

In discrete time we approximate the stochastic differential equation

$$dS = \mu S \, dt + \sigma S \, dX$$

by

$$\delta S = \mu S \, \delta t + \sigma S \phi \, \delta t^{1/2}.$$

We still set up a 'hedged' portfolio

$$\Pi = V(S, t) - \Delta S,$$

but we must accept that this cannot be perfect. After a time δt the portfolio value becomes

$$\Pi + \delta \Pi = V(S + \delta S, t + \delta t) - \Delta(S + \delta S).$$

It follows that

$$\delta \Pi = V(S + \delta S, t + \delta t) - \Delta(S + \delta S) - V(S, t) + \Delta S.$$

Expanding this in Taylor series gives

$$\delta \Pi = \frac{\partial V}{\partial t} \delta t + \frac{\partial V}{\partial S} \delta S + \frac{1}{2} \frac{\partial^2 V}{\partial S^2} (\delta S)^2 + \cdots - \Delta \delta S$$

$$= \delta t^{1/2} \sigma S \phi \left(\frac{\partial V}{\partial S} - \Delta \right) + \delta t \left(\frac{\partial V}{\partial t} + \mu S \left(\frac{\partial V}{\partial S} - \Delta \right) + \frac{1}{2} \sigma^2 S^2 \phi^2 \frac{\partial^2 V}{\partial S^2} \right) + \cdots.$$

But this has not accounted for the inevitable transaction costs that will be incurred on rehedging. The costs are

$$\kappa S |\nu|.$$

The quantity ν of the underlying assets that are bought is given by the change in the delta from one timestep to the next:

$$\nu = \frac{\partial V}{\partial S}(S + \delta S, t + \delta t) - \frac{\partial V}{\partial S}(S, t).$$

This can be approximated by

$$\nu = \frac{\partial V}{\partial S} + \frac{\partial^2 V}{\partial S^2} \delta S + \frac{\partial^2 V}{\partial S \partial t} \delta t + \cdots - \frac{\partial V}{\partial S}$$

where all derivatives are now evaluated at (S, t). Two terms in this cancel, leaving the leading-order approximation

$$\nu \approx \frac{\partial^2 V}{\partial S^2} \delta S \approx \frac{\partial^2 V}{\partial S^2} \sigma S \phi \, \delta t^{1/2}.$$

Subtracting the costs from the change in the portfolio value gives a total change of

$$\delta \Pi = \delta t^{1/2} \sigma S \phi \left(\frac{\partial V}{\partial S} - \Delta \right) + \delta t \left(\frac{\partial V}{\partial t} + \mu S \left(\frac{\partial V}{\partial S} - \Delta \right) + \frac{1}{2} \sigma^2 S^2 \phi^2 \frac{\partial^2 V}{\partial S^2} \right)$$

$$- \kappa \sigma S^2 |\phi|; \delta t^{1/2} \left| \frac{\partial^2 V}{\partial S^2} \right| + \cdots.$$

The mean of this is

$$E[\delta\Pi] = \delta t \left(\frac{\partial V}{\partial t} + \mu S \left(\frac{\partial V}{\partial S} - \Delta \right) + \tfrac{1}{2}\sigma^2 S^2 \frac{\partial^2 V}{\partial S^2} \right) - \kappa\sigma S^2 \sqrt{\frac{2}{\pi}} \delta t^{1/2} \left| \frac{\partial^2 V}{\partial S^2} \right| + \cdots.$$

This uses

$$E[\phi] = 0, \quad E[\phi^2] = 1 \quad \text{and} \quad E[|\phi|] = \sqrt{\frac{2}{\pi}}.$$

We also find that

$$E[(\delta\Pi)^2] = \delta t E \left[\sigma^2 S^2 \phi^2 \left(\frac{\partial V}{\partial S} - \Delta \right)^2 - 2\kappa\sigma S^2 \left| \frac{\partial^2 V}{\partial S^2} \right| \sigma S \left(\frac{\partial V}{\partial S} - \Delta \right) \phi|\phi| \right.$$

$$\left. + \kappa^2 \sigma^2 S^4 \left(\frac{\partial^2 V}{\partial S^2} \right)^2 \phi^2 \right] + \cdots$$

$$\doteq \delta t \left(\sigma^2 S^2 \left(\frac{\partial V}{\partial S} - \Delta \right)^2 + \kappa^2 \sigma^2 S^4 \left(\frac{\partial^2 V}{\partial S^2} \right)^2 \right) + \cdots$$

since

$$E[\phi|\phi|] = 0.$$

The variance of the portfolio change is therefore

$$\text{var}\,[\delta\Pi] = E[(\delta\Pi)^2] - (E[\delta\Pi])^2$$

$$= \delta t \left(\sigma^2 S^2 \left(\frac{\partial V}{\partial S} - \Delta \right)^2 + \left(1 - \frac{2}{\pi} \right) \kappa^2 \sigma^2 S^4 \left(\frac{\partial^2 V}{\partial S^2} \right)^2 \right)$$

to leading order. For finite hedging period δt and finite costs κ this cannot generally be made to vanish. However, the variance, or risk, can be minimized by choosing

$$\Delta = \frac{\partial V}{\partial S}.$$

With this choice,

$$E[\delta\Pi] = \delta t \left(\frac{\partial V}{\partial t} + \tfrac{1}{2}\sigma^2 S^2 \frac{\partial^2 V}{\partial S^2} \right) - \kappa\sigma S^2 \sqrt{\frac{2}{\pi}} \delta t^{1/2} \left| \frac{\partial^2 V}{\partial S^2} \right|$$

to leading order. This quantity is an *expectation*, allowing for the *expected* amount of transaction costs. I am now going to set this quantity equal to the amount that would have been earned by a risk-free account:

$$\delta t \left(\frac{\partial V}{\partial t} + \tfrac{1}{2}\sigma^2 S^2 \frac{\partial^2 V}{\partial S^2} \right) - \kappa\sigma S^2 \sqrt{\frac{2}{\pi}} \delta t^{1/2} \left| \frac{\partial^2 V}{\partial S^2} \right| = r\Pi\,\delta t = r \left(V - S\frac{\partial V}{\partial S} \right) \delta t.$$

Dividing by δt and rearranging gives the Hoggard–Whalley–Wilmott equation

$$\frac{\partial V}{\partial t} + \tfrac{1}{2}\sigma^2 S^2 \frac{\partial^2 V}{\partial S^2} - \kappa\sigma S^2 \sqrt{\frac{2}{\pi\delta t}} \left| \frac{\partial^2 V}{\partial S^2} \right| + rS\frac{\partial V}{\partial S} - rV = 0.$$

CHAPTER 25
volatility smiles and surfaces

In this Chapter...

- volatility smiles and skews
- volatility surfaces
- how to determine the volatility surface that gives prices of options that are consistent with the market

25.1 INTRODUCTION

One of the erroneous assumptions of the Black–Scholes world is that the volatility of the underlying is constant. This can be seen in any statistical examination of time-series data for assets, regardless of the sophistication of the analysis. Take a look at Figure 25.1 to see how volatility appears to change with time. This varying volatility is also observed *indirectly* through the market prices of traded contracts. In this chapter we are going to examine the relationship between the volatility of the underlying asset and the prices of derivative products. Since the volatility is not directly observable, and is certainly not predictable, we will try to exploit the relationship between prices and volatility to determine the volatility *from* the market prices. This is the exact inverse of what we have done so far. Previously we modeled the volatility and then found the price, now we take the price and deduce the volatility.

25.2 IMPLIED VOLATILITY

In the Black–Scholes world of constant volatility, the value of a European call option is simply

$$V(S, t; \sigma, r; E, T) = SN(d_1) - Ee^{-r(T-t)}N(d_2),$$

where

$$d_1 = \frac{\log\left(\frac{S}{E}\right) + \left(r + \frac{1}{2}\sigma^2\right)(T - t)}{\sigma\sqrt{T - t}}$$

and

$$d_2 = \frac{\log\left(\frac{S}{E}\right) + \left(r - \frac{1}{2}\sigma^2\right)(T - t)}{\sigma\sqrt{T - t}}.$$

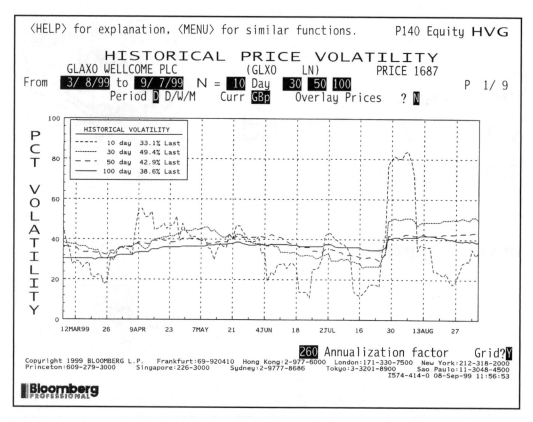

Figure 25.1 Historical volatility using several time periods. Source: Bloomberg L.P.

I have given the function V six arguments: the first two are the independent variables, the second two are parameters of the asset and the financial world, the last two are specific to the contract in question. All but σ are easy to measure (r may be a bit inaccurate but the price is typically not too sensitive to this). If we know σ then we can calculate the option price. Conversely, if we know the option price V then we can calculate σ. We can do this because the value of a call option is monotonic in the volatility. Provided that the market value of the option is greater than $\max(S - E^{-r(T-t)}, 0)$ and less than S there is one, and only one, value for σ that makes the theoretical option value and the market price the same. This is called the **implied volatility**. One is usually taught to think of the implied volatility as the market's view of the future value of volatility. Yes and no. If the 'market' does have a view on the future of volatility then it will be seen in the implied volatility. But the market also has views on the direction of the underlying, and also responds to supply and demand. Let me give examples.

In one month's time there is to be an election; it is not clear who will win. If the right wing party is elected markets will rise, if the left wing is successful, markets will fall. Before the election the market assumes the middle ground, splitting the difference. In fact, little trading occurs and markets have a very low volatility. But option traders know that after the election there will be a lot of movement one way or the other. Prices of both calls and puts are therefore high. If we back out implied volatilities from these option prices we see very high values. Actual and implied volatilities are shown in Figure 25.2 for this scenario.

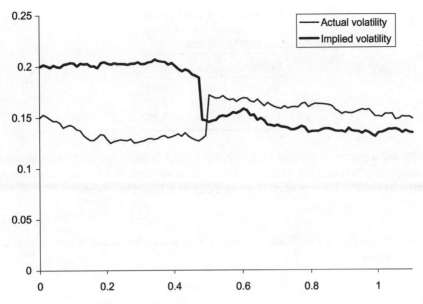

Figure 25.2 Actual and implied volatilities just before and just after an anticipated major news item.

Traders may increase option prices to reflect the expected sudden moves but if we are only observing implied volatilities then we are getting the underlying story very wrong indeed. This illustrates the fact that if you want to play around with prices there is only one parameter you can fudge; the volatility. As long as it is not too out of line compared with implied volatilities of other products no-one will disbelieve it.

Regardless of a market maker's view of future events, he is at the mercy of supply and demand. If everyone wants calls, then it is only natural for him to increase their prices. As long as he doesn't violate put-call parity (either with himself or with another market maker) who's to know that it is supply and demand driving the volatility?

25.3 TIME-DEPENDENT VOLATILITY

In Table 25.1 are the market prices of European call options[1] with one, four and seven months until expiry. All have strike prices of 105 and the underlying asset is currently 106.25. The short-term interest rate over this period is about 5.6%.

As can easily be confirmed by substitution into the Black–Scholes call formula, these prices are consistent with volatilities of 21.2%, 20.5% and 19.4% for the one-, three- and seven-month options respectively. Clearly these prices are cannot be correct if the volatility is constant for the whole seven months. What is to be done?

---

[1] I'm only using call options because put option prices should follow from the prices of the calls by put-call parity, therefore there will not be any more volatility information in the prices of puts than is already present in the prices of calls. Also the following analysis only really works if the options are European. Since most equity options are American we cannot use put prices because the values of American and European puts are different. In the absence of dividends the two call prices are the same.

Table 25.1 Market prices of European call options, see text for details.

Expiry	Value
1 month	3.50
3 months	5.76
7 months	7.97

The simplest adjustment we can make to the Black–Scholes world to accommodate these prices (without any serious effect on the theoretical framework) is to assume a time-dependent, deterministic volatility. Let's assume that volatility is a function of time:

$$\sigma(t).$$

As explained in Chapter 8 the Black–Scholes formulae are still valid when volatility is time-dependent provided we use

$$\sqrt{\frac{1}{T-t} \int_t^T \sigma(\tau)^2 \, d\tau}$$

in place of σ i.e. now use

$$d_1 = \frac{\log\left(\frac{S}{E}\right) + r(T-t) + \frac{1}{2} \int_t^T \sigma(\tau)^2 \, d\tau}{\sqrt{\int_t^T \sigma(\tau)^2 \, d\tau}}$$

and

$$d_2 = \frac{\log\left(\frac{S}{E}\right) + r(T-t) - \frac{1}{2} \int_t^T \sigma(\tau)^2 \, d\tau}{\sqrt{\int_t^T \sigma(\tau)^2 \, d\tau}}.$$

In our example, all we need to do to ensure consistent pricing is to make

$$\sqrt{\frac{1}{T-t} \int_t^T \sigma(\tau)^2 \, d\tau} = \text{implied volatilities.}$$

We do this 'fitting' at time t^*, and if I write $\sigma_{\text{imp}}(t^*, T)$ to mean the implied volatility measured at time t^* of a European option expiring at time T, then

$$\sigma(t) = \sqrt{\sigma_{\text{imp}}(t^*, t)^2 + 2(t - t^*)\sigma_{\text{imp}}(t^*, t)\frac{\partial \sigma_{\text{imp}}(t^*, t)}{\partial t}}. \tag{25.1}$$

Practically speaking, we do not have a continuous (and differentiable) implied volatility curve. We have a discrete set of points (three in the above example). We must therefore make some assumption about the term structure of volatility between the data points. Usually one assumes that the function is piecewise constant or linear. If we have implied volatility for expiries T_i and we assume the volatility curve to be piecewise constant then

$$\sigma(t) = \sqrt{\frac{(T_i - t^*)\sigma_{imp}(t^*, T_i)^2 - (T_{i-1} - t^*)\sigma_{imp}(t^*, T_{i-1})^2}{T_i - T_{i-1}}} \quad \text{for } T_{i-1} < t < T_i$$

25.4 VOLATILITY SMILES AND SKEWS

Now let me throw the cat among the pigeons. Continuing with the example above, suppose that there is also a European call option struck at 100 with an expiry of seven months and a price of 11.48. This corresponds to a volatility of 20.8% in the Black–Scholes equation. Now we have two conflicting volatilities up to the seven-month expiry, 19.4% and 20.8%. Clearly we cannot adjust the time dependence of the volatility in any way that is consistent with *both* of these values. What else can we do? Before I answer this, we'll look at a few more examples. Concentrating on the same example, suppose that there are call options traded with an expiry of seven months and strikes of 90, 92.5, 95, 97.5, 100, 102.5, 105, 107.5 and 110. In Figure 25.3 I plot the implied volatility of these options against the strike price (the actual option prices do not add anything to our insight so I haven't given them).

The shape of this implied volatility versus strike curve is called the **smile**. In some markets it shows considerable asymmetry, a **skew**, and sometimes it is upside down in a **frown**. The general shape tends to persist for a long time in each underlying.

If we managed to accommodate implied volatility that varied with expiry by making the volatility time-dependent perhaps we can accommodate implied volatility that varies with strike by making the volatility asset price-dependent. This is exactly what we'll do. Unfortunately, it's much harder to make analytical progress except in special cases; if we have $\sigma(S)$ then rarely

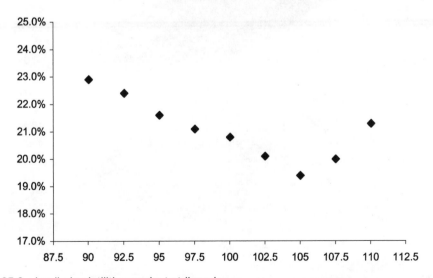

Figure 25.3   Implied volatilities against strike price.

can we solve the Black–Scholes equation to get nice closed-form solutions for the values of derivative products. In fact, we may as well go all the way and assume that volatility is a function of *both* the asset and time, $\sigma(S, t)$.

25.5 VOLATILITY SURFACES

We can show implied volatility against both maturity and strike in a three-dimensional plot. One is shown in Figure 25.4. This implied volatility surface represents the constant value of volatility that gives each traded option a theoretical value equal to the market value.

We saw how the time dependence in implied volatility could be turned into a volatility of the underlying that was time-dependent i.e. we deduced $\sigma(t)$ from $\sigma_{\mathrm{imp}}(t^*, T)$. Can we similarly deduce $\sigma(S, t)$ from $\sigma_{\mathrm{imp}}(t^*, E, T)$, the implied volatility at time t^*? If we could, then we might want to call it the **local volatility surface** $\sigma(S, t)$. This local volatility surface can be thought of as the market's view of the future value of volatility when the asset price is S at time t (which, of course, may not even be realized). The local volatility is also called the **forward volatility** or the **forward forward volatility**.

25.6 BACKING OUT THE LOCAL VOLATILITY SURFACE FROM EUROPEAN CALL OPTION PRICES

Market prices of traded vanilla options are never, in practice, consistent with the constant volatility assumed by Black & Scholes. Nor are they consistent with either a time-dependent or an asset price-dependent local volatility. To match the theoretical prices of traded options to their market prices always requires a volatility structure that is a function of both the asset

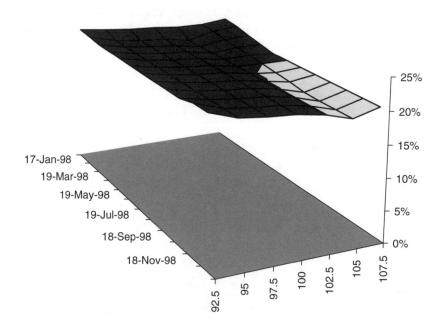

Figure 25.4 Implied volatilities against expiry and strike price.

price, S, and time, t i.e. $\sigma(S, t)$. To back out the local volatility surface from the prices of market traded instruments I am going to assume that we have a distribution of European call prices of all strikes and maturities. This is not a realistic assumption but it gets the ball rolling. These prices will be denoted by $V(E, T)$. I could use puts but these can be converted to call prices by put-call parity. This notation is vastly different from before. Previously, we had the option value as a function of the underlying and time. Now the asset and time are fixed at S^* and t^*, today's values. I will use the dependence of the market prices on strike and expiry to calculate the volatility structure.

I will assume that the risk-neutral random walk for S is

$$dS = rS\,dt + \sigma(S, t)S\,dX.$$

This is our usual one-factor model for which all the building blocks of delta hedging and arbitrage-free pricing hold. The only novelty is that the volatility is dependent on the level of the asset and time.

In the following, I am going to rely heavily on the transition probability density function $p(S^*, t^*; S, T)$ for the risk-neutral random walk. Note that the backward variables are fixed at today's values and the forward time variable is T. Recalling that the value of an option is the present value of the expected payoff, I can write

$$V(E, T) = e^{-r(T-t^*)} \int_0^\infty \max(S - E, 0) p(S^*, t^*; S, T)\,dS$$

$$= e^{-r(T-t^*)} \int_E^\infty (S - E) p(S^*, t^*; S, T)\,dS. \tag{25.2}$$

We are *very* lucky that the payoff is the maximum function so that after differentiating with respect to E we get

$$\frac{\partial V}{\partial E} = -e^{-r(T-t^*)} \int_E^\infty p(S^*, t^*; S, T)\,dS.$$

And after another differentiation, we arrive at

$$p(S^*, t^*; E, T) = e^{r(T-t^*)} \frac{\partial^2 V}{\partial E^2} \tag{25.3}$$

Before even calculating volatilities we can find the transition probability density function. In a sense, this is the market's view of the future distribution. But it's the market view of the risk-neutral distribution and not the real one. An example is plotted in Figure 25.5.

The next step is to use the forward equation for the transition probability density function, the Fokker–Planck equation,

$$\frac{\partial p}{\partial T} = \frac{1}{2} \frac{\partial^2}{\partial S^2}(\sigma^2 S^2 p) - \frac{\partial}{\partial S}(rSp). \tag{25.4}$$

Here σ is our, still unknown, function of S and t. However, *in this equation $\sigma(S, t)$ is evaluated at $t = T$.*

From (25.2) we have

$$\frac{\partial V}{\partial T} = -rV + e^{-r(T-t^*)} \int_E^\infty (S - E) \frac{\partial p}{\partial T}\,dS.$$

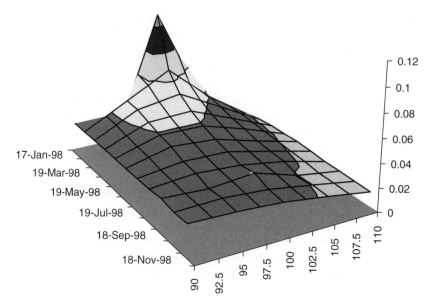

Figure 25.5 Risk-neutral transition probability density function calculated from European call prices.

This can be written as

$$\frac{\partial V}{\partial T} = -rV + e^{-r(T-t^*)} \int_E^\infty \left(\frac{1}{2} \frac{\partial^2(\sigma^2 S^2 p)}{\partial S^2} - \frac{\partial(rSp)}{\partial S} \right) (S - E) \, dS.$$

using the forward Equation (25.4). Integrating this by parts twice, assuming that p and its first S derivative tend to zero sufficiently fast as S goes to infinity we get

$$\frac{\partial V}{\partial T} = -rV + \frac{1}{2} e^{-r(T-t^*)} \sigma^2 E^2 p + r e^{-r(T-t^*)} \int_E^\infty Sp \, dS. \tag{25.5}$$

In this expression $\sigma(S, t)$ has $S = E$ and $t = T$. Writing

$$\int_E^\infty Sp \, dS = \int_E^\infty (S - E) p \, dS + E \int_E^\infty p \, dS$$

and collecting terms, we get

$$\frac{\partial V}{\partial T} = \frac{1}{2} \sigma^2 E^2 \frac{\partial^2 V}{\partial E^2} - rE \frac{\partial V}{\partial E}.$$

Rearranging this we find that

$$\sigma = \sqrt{\frac{\dfrac{\partial V}{\partial T} + rE \dfrac{\partial V}{\partial E}}{\dfrac{1}{2} E^2 \dfrac{\partial^2 V}{\partial E^2}}}.$$

This gives us $\sigma(E, T)$ and hence, by *relabelling the variables*, $\sigma(S, t)$.

This calculation of the volatility surface from option prices worked because of the particular form of the payoff, the call payoff, which allowed us to derive the very simple relationship between derivatives of the option price and the transition probability density function.

When there is a constant and continuous dividend yield on the underlying the relationship between call prices and the local volatility is

$$
\sigma = \sqrt{\frac{\frac{\partial V}{\partial T} + (r - D)E\frac{\partial V}{\partial E} + DV}{\frac{1}{2}E^2\frac{\partial^2 V}{\partial E^2}}}
\tag{25.6}
$$

There is no change in this expression when the interest rate and dividend yield are time-dependent, just use the relevant forward-rates.

One of the problems with this expression concerns data far into or far out of the money. Unless we are close to at the money both the numerator and denominator of (25.6) are small, leading to inaccuracies when we divide one small number by another. One way of avoiding this is to relate the local volatility surface to the implied volatility surface as I now show.

In the same way that we found a relationship between the local volatility and the implied volatility in the purely time-dependent case, Equation (25.1), we can find a relationship in the general case of asset- and time-dependent local volatility. This relationship is obviously quite complicated and I omit the details of the derivation. The result is

$$
\sigma(E, T) =
$$

$$
\sqrt{\frac{\sigma_{\text{imp}}^2 + 2(T - t^*)\sigma_{\text{imp}}\frac{\partial \sigma_{\text{imp}}}{\partial T} + 2(r - D)E(T - t^*)\sigma_{\text{imp}}\frac{\partial \sigma_{\text{imp}}}{\partial E}}{\left(1 + Ed_1\sqrt{T - t^*}\frac{\partial \sigma_{\text{imp}}}{\partial E}\right)^2 + E^2(T - t^*)\sigma_{\text{imp}}\left(\frac{\partial^2 \sigma_{\text{imp}}}{\partial E^2} - d_1\left(\frac{\partial \sigma_{\text{imp}}}{\partial E}\right)^2\sqrt{T - t^*}\right)}}
\tag{25.7}
$$

$$
d_1 = \frac{\log\left(\frac{S^*}{E}\right) + \left(r - D + \frac{1}{2}\sigma_{\text{imp}}^2\right)(T - t^*)}{\sigma_{\text{imp}}\sqrt{T - t^*}}
$$

In terms of the implied volatility the implied risk-neutral probability density function is

$$
p(S^*, t^*; E, T) = \frac{1}{E\sigma_{\text{imp}}\sqrt{2\pi(T - t^*)}}e^{-(1/2)d_2^2}
$$

$$
\times \left(\left(1 + Ed_1\sqrt{T - t^*}\frac{\partial \sigma_{\text{imp}}}{\partial E}\right)^2 + E^2(T - t^*)\sigma_{\text{imp}}\left(\frac{\partial^2 \sigma_{\text{imp}}}{\partial E^2} - d_1\left(\frac{\partial \sigma_{\text{imp}}}{\partial E}\right)^2\sqrt{T - t^*}\right)\right)
$$

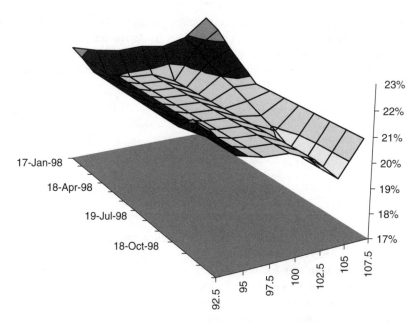

Figure 25.6 Local volatility surface calculated from European call prices.

One of the advantages of writing the local volatility and probability density function in terms of the implied volatility surface is that if you put in a flat implied volatility surface you get out a flat local surface and a lognormal distribution.

In practice there only exists a finite, discretely-spaced set of call prices. To deduce a local volatility surface from this data requires some interpolation and extrapolation. This can be done in a number of ways and there is no correct way. One of the problems with these approaches is that the final result depends sensitively on the form of the interpolation. The problem is actually 'ill-posed', meaning that a small change in the input can lead to a large change in the output. There are many ways to get around this ill-posedness, coming under the general heading of 'regularization'. Several suggestions for further reading in this area are given at the end of the chapter.

An example of a local volatility surface is plotted in Figure 25.6.

25.7 A SIMPLE VOLATILITY SURFACE PARAMETERIZATION

Backing out the deterministic local volatility surface from the market prices of traded instruments has become increasingly popular since the introduction of the idea in the early 90s. The jury is still out on whether the method is consistent with future market prices. What is clear is that the practical implementation of the method suffers from serious numerical instabilities and can be slow to compute. In this section we examine an approximation to the local volatility surface that is simple and fast to implement.

There is a major problem with using the results of the previous section, and that concerns the loss of accuracy due to the discreteness of the implied volatility data: we only know σ_{imp} at a finite number of points (E, T). The resulting local volatility surface often looks unrealistic, and is very sensitive to the input implied volatilities. There is also some evidence (Dumas,

Fleming & Whaley, 1998) that what information is contained in the market prices is lost if the local volatility surface is over-fitted. In other words, a parsimonious representation of volatility surfaces is best.

In this section I am going to assume that the implied volatilities are of the simple form

$$\sigma_{\text{imp}}(E, T) = a(T)(E - S^*) + b(T). \tag{25.8}$$

The motivation behind assumption (25.8) is that the most important information in market prices concerns the at-the-money volatility and the volatility skew, both of which may be time-dependent.

Later we will see how to calculate $a(T)$ and $b(T)$ from the market prices of an at-the-money straddle and a risk reversal. First, however, we will find the local volatility surface assuming (25.8).

25.8 AN APPROXIMATE SOLUTION

Substituting (25.8) into (25.7) we find that

$$\sigma(E, T) =$$

$$\sqrt{\frac{(a(E-S^*)+b)^2 + 2(T-t^*)(a(E-S^*)+b)(a'(E-S^*)+b') + 2(r-D)E(T-t^*)(a(E-S^*)+b)a}{(1+Ed_1\sqrt{T-t^*}a)^2 - E^2(T-t^*)^{3/2}(a(E-S^*)+b)d_1a^2}},$$

where

$$d_1 = \frac{\log\left(\dfrac{S^*}{E}\right) + \left(r - D + \frac{1}{2}(a(E - S^*) + b)^2\right)(T - t^*)}{(a(E - S^*) + b)\sqrt{T - t^*}}.$$

Here $'$ denotes d/dT. Rewriting σ in terms of its more natural arguments, S and t, we have

$$\sigma(S, t) =$$

$$\sqrt{\frac{(a(S-S^*)+b)^2 + 2(t-t^*)(a(S-S^*)+b)(a'(S-S^*)+b') + 2(r-D)S(t-t^*)(a(S-S^*)+b)a}{(1+Sd_1\sqrt{t-t^*}a)^2 - S^2(t-t^*)^{3/2}(a(S-S^*)+b)d_1a^2}},$$

where

$$d_1 = \frac{\log\left(\dfrac{S^*}{S}\right) + \left(r - D + \frac{1}{2}(a(S - S^*) + b)^2\right)(t - t^*)}{(a(S - S^*) + b)\sqrt{t - t^*}}.$$

If we can find a and b then our simple volatility surface parameterization is complete.

25.9 VOLATILITY INFORMATION CONTAINED IN AN AT-THE-MONEY STRADDLE

The straddle position is made up of a long call and a long put with the same strikes and expiries. Practitioners use the market prices of at-the-money straddles to deduce the at-the-money volatility. The Black–Scholes value of a straddle is given by

$$V_S = C + P = C + C - Se^{-D(T-t)} + Ee^{-r(T-t)} = 2C - Se^{-D(T-t)} + Ee^{-r(T-t)},$$

where C and P are the values of the call and the put respectively and we have used put-call parity. We can therefore deduce the price of a single call from

$$C = \tfrac{1}{2}(V_S + Se^{-D(T-t)} - Ee^{-r(T-t)})$$

and hence the implied volatility. Since the straddle is at-the-money we have $S = E = S^*$ and, of course, $t = t^*$.

From our assumed form (25.8) for the implied volatility we have

$$e^{-D(T-t^*)}N(d_1) - e^{-r(T-t^*)}N(d_2) = \tfrac{1}{2}\left(\frac{V_S}{S^*} + e^{-D(T-t^*)} - e^{-r(T-t^*)}\right),$$

where

$$d_1 = \frac{\left(r - D + \tfrac{1}{2}b(T)^2\right)\sqrt{T - t^*}}{b(T)} \quad \text{and} \quad d_2 = d_1 - b(T)\sqrt{T - t^*}.$$

Since V_S is known from the market we can calculate $b(T)$.

25.10 VOLATILITY INFORMATION CONTAINED IN A RISK-REVERSAL

The risk-reversal is a long call, with strike above the current spot, and a short put with a strike below the current spot. Both have the same expiry. Practitioners use the market price of the risk-reversal to deduce the volatility skew.

In the following we will assume that the strikes of the call and the put are a short distance ε away from the current spot: the strike of the call is thus $S^* + \varepsilon$ and the strike of the put is $S^* - \varepsilon$; this can easily be generalized. I will shortly be expanding various quantities in Taylor series in ε.

I will now need a slightly more informative notation. I will include the strike and implied volatility of each option as arguments e.g. $C(E, \sigma_{imp})$ means a call with strike E and implied volatility σ_{imp}, similarly for puts.

The Black–Scholes value of the risk-reversal is given by

$$V_{RR} = C(S^* + \varepsilon, \sigma_{imp}(S^* + \varepsilon, T)) - P(S^* - \varepsilon, \sigma_{imp}(S^* - \varepsilon, T))$$

$$= C(S^* + \varepsilon, \sigma_{imp}(S^* + \varepsilon, T)) - C(S^* - \varepsilon, \sigma_{imp}(S^* - \varepsilon, T))$$

$$+ S^*e^{-D(T-t^*)} - (S^* - \varepsilon)e^{-r(T-t^*)}.$$

I have directly written everything in terms of the current spot and time, and again used put-call parity.

If ε is small, that is if the two strikes are close together, which is usually the case, we can expand the above. We find that

$$V_{RR} - S^*(e^{-D(T-t^*)} - e^{-r(T-t^*)}) = \varepsilon\left(e^{-r(T-t^*)} + 2\frac{\partial C}{\partial E}(S^*, b) + 2\frac{\partial C}{\partial \sigma_{imp}}(S^*, b)a\right).$$

Using the formulae for a call we get

$$V_{RR} - S^*(e^{-D(T-t^*)} - e^{-r(T-t^*)}) = \varepsilon(e^{-r(T-t^*)} - 2e^{-r(T-t^*)}N(d_2)$$

$$+ 2aS^*\sqrt{T - t^*}N'(d_1)e^{-D(T-t^*)}),$$

where

$$d_1 = \frac{\left(r - D + \frac{1}{2}b(T)^2\right)\sqrt{T - t^*}}{b(T)} \quad \text{and} \quad d_2 = d_1 - b(T)\sqrt{T - t^*}.$$

If we have found b from the at-the-money straddle then finding a from the above is very easy:

$$a(T) = \frac{e^{D(T-t^*)}}{2\varepsilon S^* \sqrt{T - t^*}N'(d_1)}(V_{RR} - S^*(e^{-D(T-t^*)} - e^{-r(T-t^*)})) + \frac{e^{(D-r)(T-t^*)}N(d_2)}{S^* \sqrt{T - t^*}N'(d_1)}.$$

25.11 TIME DEPENDENCE

We have dealt with the discreteness of the data in the direction of the strike, but we must still deal with the discreteness in the expiry dates. In practice we only have a small number of expiries from which to deduce the term structure of the parameters a and b.

There are several approaches we can take depending on how much smoothness we want to insist upon. Fortunately, the time-dependence of the local volatility surfaces is less extreme than the typical asset price-dependence. The simplest interpolation between data points in the time direction is linear. This is adequate in most situations. If we know $a(T_i)$ and $b(T_i)$ we take the implied volatility parameters to be

$$\frac{a(T_i)(T_{i+1} - T) + a(T_{i+1})(T - T_i)}{T_{i+1} - T_i} \quad \text{and} \quad \frac{b(T_i)(T_{i+1} - T) + b(T_{i+1})(T - T_i)}{T_{i+1} - T_i}$$

for times T between the two expiries T_i and T_{i+1}.

25.12 A MARKET CONVENTION

PRACTITIONERS OFTEN THINK IN TERMS OF DELTA INSTEAD OF STRIKE

Instead of quoting an option price by specifying its expiry, maturity and volatility, it is common to quote its expiry, *delta* and volatility. For instance, one might be told that the six-month 25 delta call has a volatility of 13%. This means that the option with an expiry in six months and a theoretical Black–Scholes delta of 0.25 will cost the theoretical Black–Scholes value using a volatility of 0.13. What seems to be missing from this? The strike of the option. But we can work this out from the rest of the information. Let's continue with this example. Recalling the formula for the delta of a call, we know that

$$e^{-D(T-t)}N(d_1) = 0.25$$

where

$$d_1 = \frac{\log\left(\frac{S}{E}\right) + \left(r - D + \frac{1}{2}0.13^2\right)(T - t)}{0.13\sqrt{(T - t)}}.$$

Also in this example

$$T - t = 0.5,$$

and we presumably know the spot price S, the interest rate r and the dividend yield D. We now solve for E.

Another example might be that we are told the volatility of a specified maturity zero-delta straddle. Again, we can work backwards to find the location of the strike.[2]

The vol surface parameterization can straightforwardly be put into the form of this market convention.

25.13 **HOW DO I USE THE LOCAL VOLATILITY SURFACE?**

There are two ways to look at the local volatility surface. One is to say that it is the market's view of future volatility and that these predictions will come to pass. We then price other, more complex, products using this asset- and time-dependent volatility. This is a very naive belief. Not only do the predictions not come true but even if we come back a few days later to look at the 'prediction', i.e. to refit the surface, we see that it has changed.

The other way of using the surface is to acknowledge that it is only a snapshot of the market's view and that tomorrow it may change. But it can be used to price non-traded contracts in a way that is consistent across all instruments. As long as we price our exotic contract consistently with the vanillas, that is, with the same volatility structure, *and simultaneously hedge with these vanillas* then we are reducing our exposure to model error. This approach is readily justifiable, although it is a bit difficult to estimate by how much we have reduced our model exposure. However, if you price using the calibrated volatility surface but only delta hedge, then you are asking for trouble. Suppose you price, and sell, a volatility-sensitive instrument such as an up-and-out call with a fitted volatility surface which increases with stock level. If it turns out that when the volatility is realized it is a downward-sloping function of the asset then you are in big trouble.

As an example, calculate the local volatility surface using vanilla calls. Now price a barrier option using this volatility structure. This means solving the Black–Scholes partial differential equation with the asset- and time-dependent volatility and with the relevant boundary conditions. This must be done numerically, by the methods explained in Part Six, for example. Now statically hedge the barrier by buying and selling the vanilla contracts to mimic as closely as possible the payoff and boundary conditions of the barrier option. This is described more fully in Chapter 32.

25.14 **SUMMARY**

Whether you believe in them or not, local volatility surfaces have taken the practitioner world by storm. Now that they are commonly used for pricing and hedging exotic contracts there is no way back to the world of constant volatility.

Personally I am in two minds about this issue. But as long as you hedge as much of the cashflows as possible using traded vanilla options then you will be reducing your model exposure anyway. Once you have done this then you have done your best to reduce dependence on volatility. The danger is always going to be there if you never bother to statically hedge, only delta hedge. This is discussed in detail in Chapter 32.

---

[2] It will be close to, but not exactly at-the-money forward, because of the skew caused by lognormality.

FURTHER READING

- Merton (1973) was the first to find the explicit formulae for European options with time-dependent volatility.

- See Dupire (1993, 1994) and Derman & Kani (1994) for more details of fitting the local volatility surface.

- Rubinstein (1994) constructs an implied tree using an optimization approach, and this has been generalized by Jackwerth & Rubinstein (1996).

- To get around the problem of ill-posedness, Avellaneda, Friedman, Holmes & Samperi (1997) propose calibrating the local volatility surface by entropy minimization. Their article is also a very good source of references to the volatility surface literature.

- For trading strategies involving views on the direction of volatility see Connolly (1997).

- See Dewynne, Ehrlichman & Wilmott (1998) for more details of the simple volatility surface parameterization.

- I suspect that there is not much information about future volatility contained in the local volatility surface. This is demonstrated in Dumas, Fleming & Whaley (1998).

CHAPTER 26
stochastic volatility

In this Chapter...

- modeling volatility as a stochastic variable
- discontinuous volatility
- how to price contracts when volatility is stochastic

26.1 INTRODUCTION

Volatility does not behave how the Black–Scholes equation would like it to behave; it is not constant, it is not predictable, it is not even directly observable. This makes it a prime candidate for modeling as a random variable. There is plenty of evidence that returns on equities, currencies and commodities are not Normally distributed; they have higher peaks and fatter tails than predicted by a Normal distribution. We have seen this in several places in this book. This has also been cited as evidence for non-constant volatility.

26.2 RANDOM VOLATILITY

If we draw a number at random from a distribution with either a 10% volatility, 20% volatility or 30% volatility with equal probability, then the resulting distribution would have the higher peak-fatter tails properties we have seen in the data. However, as long as the standard deviation of this distribution is finite (and scales with the square root of the timestep) we can price options using the standard deviation in place of the volatility in Black–Scholes (see also Chapter 23). Pricing only becomes a problem with random volatility when the timescale of the evolution is of the same order as the evolution of the underlying. For example, if we have a stochastic differential equation model for the volatility then we must move beyond the Black–Scholes world.

26.3 A STOCHASTIC DIFFERENTIAL EQUATION FOR VOLATILITY

Figure 26.1 shows four estimates of volatility of the Dow Jones Index, using weighted averages of the daily changes in the index over the previous 10, 30, 50 and 100 days. First of all, this graph shows that volatility is not constant. Moreover, it's not at all clear how to measure it.

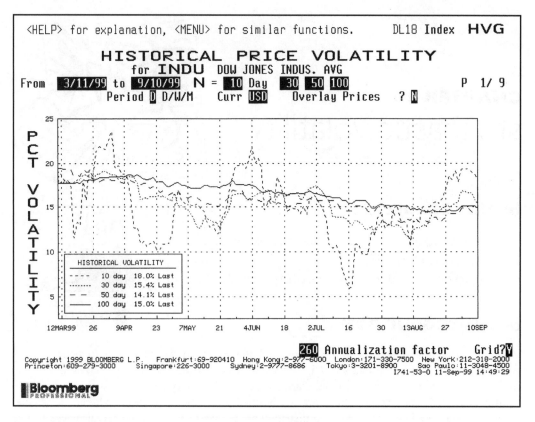

Figure 26.1 Volatility against time. Source: Bloomberg L.P.

Each estimate gives a different answer. Modeling volatility as a stochastic process requires there to be such a quantity as volatility... but maybe observability isn't that important.

We continue to assume that S satisfies

$$dS = \mu S\, dt + \sigma S\, dX_1,$$

but we further assume that volatility satisfies

$$d\sigma = p(S, \sigma, t)\, dt + q(S, \sigma, t)\, dX_2. \qquad (26.1)$$

The two increments dX_1 and dX_2 have a correlation of ρ. The choice of functions $p(S, \sigma, t)$ and $q(S, \sigma, t)$ is crucial to the evolution of the volatility, and thus to the pricing of derivatives. The choice of these functions is discussed later in this chapter and in Chapter 28. For the moment, I only comment that a mean-reverting process is a natural choice.

The value of an option with stochastic volatility is a function of three variables, $V(S, \sigma, t)$.

26.4 **THE PRICING EQUATION**

The new stochastic quantity that we are modeling, the volatility, is not a traded asset. Thus, when volatility is stochastic we are faced with the problem of having a source of randomness that cannot be easily hedged away. Because we have two sources of randomness we must hedge our option with two other contracts, one being the underlying asset as usual, but now we also need another option to hedge the volatility risk. We therefore must set up a portfolio containing one option, with value denoted by $V(S, \sigma, t)$, a quantity $-\Delta$ of the asset and a quantity $-\Delta_1$ of another option with value $V_1(S, \sigma, t)$. We have

$$\Pi = V - \Delta S - \Delta_1 V_1. \tag{26.2}$$

The change in this portfolio in a time dt is given by

$$d\Pi = \left(\frac{\partial V}{\partial t} + \tfrac{1}{2}\sigma^2 S^2 \frac{\partial^2 V}{\partial S^2} + \rho\sigma q S \frac{\partial^2 V}{\partial S \partial \sigma} + \tfrac{1}{2}q^2 \frac{\partial^2 V}{\partial \sigma^2} \right) dt$$

$$- \Delta_1 \left(\frac{\partial V_1}{\partial t} + \tfrac{1}{2}\sigma^2 S^2 \frac{\partial^2 V_1}{\partial S^2} + \rho\sigma q S \frac{\partial^2 V_1}{\partial S \partial \sigma} + \tfrac{1}{2}q^2 \frac{\partial^2 V_1}{\partial \sigma^2} \right) dt$$

$$+ \left(\frac{\partial V}{\partial S} - \Delta_1 \frac{\partial V_1}{\partial S} - \Delta \right) dS + \left(\frac{\partial V}{\partial \sigma} - \Delta_1 \frac{\partial V_1}{\partial \sigma} \right) d\sigma.$$

where I have used Itô's lemma on functions of S, σ and t.

To eliminate all randomness from the portfolio we must choose

$$\frac{\partial V}{\partial S} - \Delta - \Delta_1 \frac{\partial V_1}{\partial S} = 0,$$

to eliminate dS terms, and

$$\frac{\partial V}{\partial \sigma} - \Delta_1 \frac{\partial V_1}{\partial \sigma} = 0,$$

to eliminate $d\sigma$ terms. This leaves us with

$$d\Pi = \frac{\partial V}{\partial t} dt + \tfrac{1}{2}\sigma^2 S^2 \frac{\partial^2 V}{\partial S^2} dt + \rho\sigma S q \frac{\partial^2 V}{\partial S \partial \sigma} dt + \tfrac{1}{2}q^2 \frac{\partial^2 V}{\partial \sigma^2} dt$$

$$- \Delta_1 \left(\frac{\partial V_1}{\partial t} dt + \tfrac{1}{2}\sigma^2 S^2 \frac{\partial^2 V_1}{\partial S^2} dt + \rho\sigma S q \frac{\partial^2 V_1}{\partial S \partial \sigma} dt + \tfrac{1}{2}q^2 \frac{\partial^2 V_1}{\partial \sigma^2} dt \right)$$

$$= r\Pi \, dt = r(V - \Delta S - \Delta_1 V_1) \, dt,$$

where I have used arbitrage arguments to set the return on the portfolio equal to the risk-free rate.

As it stands, this is *one* equation in the *two* unknowns, V and V_1. This contrasts with the earlier Black–Scholes case with one equation in the one unknown.

Collecting all V terms on the left-hand side and all V_1 terms on the right-hand side we find that

$$\frac{\dfrac{\partial V}{\partial t} + \tfrac{1}{2}\sigma^2 S^2 \dfrac{\partial^2 V}{\partial S^2} + \rho\sigma S q \dfrac{\partial^2 V}{\partial S \partial \sigma} + \tfrac{1}{2}q^2 \dfrac{\partial^2 V}{\partial \sigma^2} + rS \dfrac{\partial V}{\partial S} - rV}{\dfrac{\partial V}{\partial \sigma}}$$

$$= \frac{\dfrac{\partial V_1}{\partial t} + \tfrac{1}{2}\sigma^2 S^2 \dfrac{\partial^2 V_1}{\partial S^2} + \rho\sigma S q \dfrac{\partial^2 V_1}{\partial S \partial \sigma} + \tfrac{1}{2}q^2 \dfrac{\partial^2 V_1}{\partial \sigma^2} + rS \dfrac{\partial V_1}{\partial S} - rV_1}{\dfrac{\partial V_1}{\partial \sigma}}.$$

We are lucky that the left-hand side is a function of V but not V_1 and the right-hand side is a function of V_1 but not V. Since the two options will typically have different payoffs, strikes or expiries, the only way for this to be possible is for both sides to be independent of the contract type. Both sides can only be functions of the *independent* variables, S, σ and t. Thus we have

$$\frac{\partial V}{\partial t} + \tfrac{1}{2}\sigma^2 S^2 \frac{\partial^2 V}{\partial S^2} + \rho\sigma Sq\frac{\partial^2 V}{\partial S \partial \sigma} + \tfrac{1}{2}q^2\frac{\partial^2 V}{\partial \sigma^2} + rS\frac{\partial V}{\partial S} - rV = -(p - \lambda q)\frac{\partial V}{\partial \sigma},$$

for some function $\lambda(S, \sigma, t)$.

Reordering this equation, we usually write

$$\frac{\partial V}{\partial t} + \tfrac{1}{2}\sigma^2 S^2 \frac{\partial^2 V}{\partial S^2} + \rho\sigma Sq\frac{\partial^2 V}{\partial S \partial \sigma} + \tfrac{1}{2}q^2\frac{\partial^2 V}{\partial \sigma^2} + rS\frac{\partial V}{\partial S} + (p - \lambda q)\frac{\partial V}{\partial \sigma} - rV = 0 \qquad (26.3)$$

The function $\lambda(S, \sigma, t)$ is called the **market price of (volatility) risk**.

I HAVE A PHOBIA ABOUT MARKET PRICES OF RISK

26.5 THE MARKET PRICE OF VOLATILITY RISK

If we can solve Equation (26.3) then we have found the value of the option, and the hedge ratios. Generally, this must be done numerically. But note that we find *two* hedge ratios, $\partial V/\partial S$ and $\partial V/\partial \sigma$. We have two hedge ratios because we have two sources of randomness that we must hedge away.

Because one of the modeled quantities, the volatility, is not traded we find that the pricing equation contains a market price of risk term. What does this term mean? Suppose we hold one of the options with value V, and satisfying the pricing Equation (26.3), delta hedged with the underlying asset only i.e. we have

$$\Pi = V - \Delta S.$$

The change in this portfolio value is

$$d\Pi = \left(\frac{\partial V}{\partial t} + \tfrac{1}{2}\sigma^2 S^2 \frac{\partial^2 V}{\partial S^2} + \rho\sigma qS\frac{\partial^2 V}{\partial S \partial \sigma} + \tfrac{1}{2}q^2\frac{\partial^2 V}{\partial \sigma^2}\right) dt$$

$$+ \left(\frac{\partial V}{\partial S} - \Delta\right) dS + \frac{\partial V}{\partial \sigma} d\sigma.$$

Because we are delta hedging the coefficient of dS is zero. We find that

$$d\Pi - r\Pi\, dt = \left(\frac{\partial V}{\partial t} + \tfrac{1}{2}\sigma^2 S^2 \frac{\partial^2 V}{\partial S^2} + \rho\sigma qS\frac{\partial^2 V}{\partial S \partial \sigma} + \tfrac{1}{2}q^2\frac{\partial^2 V}{\partial \sigma^2} + rS\frac{\partial V}{\partial S} - rV\right) dt + \frac{\partial V}{\partial \sigma} d\sigma$$

$$= q\frac{\partial V}{\partial \sigma}(\lambda\, dt + dX_2).$$

This has used both the pricing Equation (26.3) and the stochastic differential equation for σ, (26.1). Observe that for every unit of volatility risk, represented by dX_2, there are λ units of extra return, represented by dt. Hence the name 'market price of risk'.

The quantity $p - \lambda q$ is called the **risk-neutral drift rate** of the volatility. Recall that the risk-neutral drift of the underlying asset is r and not μ. When it comes to pricing derivatives, it is the risk-neutral drift that matters and not the real drift, whether it is the drift of the asset or of the volatility.

26.5.1 Aside: The Market Price of Risk for Traded Assets

Let us return briefly to the Black–Scholes world of constant volatility. In Chapter 5 we derived the Black–Scholes equation for equities by constructing a portfolio consisting of one option and a number $-\Delta$ of the underlying asset. We were able to do this because the underlying asset, the equity, was traded. Suppose that instead we were to follow the analysis above and construct a portfolio of two *options* with different maturity dates (or different exercise prices, for that matter) instead of an option and the underlying. We would have

$$\Pi = V - \Delta_1 V_1.$$

Note that there are none of the underlying asset in this portfolio. The same argument as used above leads us to

$$\frac{\partial V}{\partial t} + \tfrac{1}{2}\sigma^2 S^2 \frac{\partial^2 V}{\partial S^2} + (\mu - \lambda_S \sigma)S \frac{\partial V}{\partial S} - rV = 0. \tag{26.4}$$

What is special about the variable S? It is the value of a traded asset. This means that $V = S$ must itself be a solution of (26.4). Substituting $V = S$ into (26.4) we find that

$$(\mu - \lambda_S \sigma)S - rS = 0,$$

i.e.

$$\lambda_S = \frac{\mu - r}{\sigma};$$

this is the market price of risk for a traded asset. Now putting $\lambda_S = (\mu - r)/\sigma$ into (26.4) we arrive at

$$\frac{\partial V}{\partial t} + \tfrac{1}{2}\sigma^2 S^2 \frac{\partial^2 V}{\partial S^2} + rS \frac{\partial V}{\partial S} - rV = 0.$$

We are back at the Black–Scholes equation, which contains no mention of μ or λ_S.

26.6 **AN EXAMPLE**

Figure 26.2 shows the theoretical value of a call option assuming stochastic volatility. The details of the volatility model and the code that produced the data are given in Chapter 67. One of the inputs into the model is the correlation between the movements in the underlying and the movements in the volatility. In Figure 26.3 are plots of call values for three different values of that correlation, for the same volatility. In this example the correlation effect is quite large for slightly out-of-the-money options. For in-the-money options the effect on the option's time value is also significant.

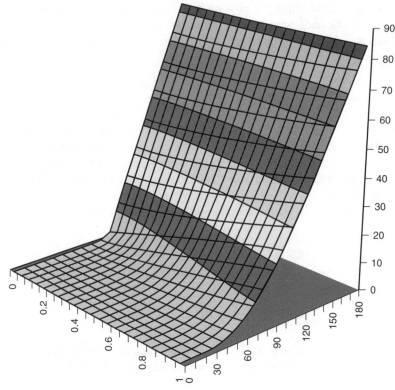

Figure 26.2 Call value when volatility is stochastic.

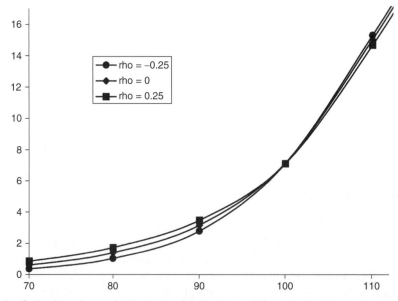

Figure 26.3 Call value when volatility is stochastic; three different correlations.

26.7 **NAMED MODELS**

Hull & White (1987)

Hull & White considered both general and specific volatility modeling. The most important result of their analysis is that when the stock and the volatility are uncorrelated and the risk-neutral dynamics of the volatility are unaffected by the stock (i.e. $p - \lambda q$ and q are independent of S) then the fair value of an option is the average of the Black–Scholes values for the option, with the average taken over the distribution of σ^2.

One of the (risk-neutral) stochastic volatility models considered by Hull & White was

$$d(\sigma^2) = a(b - \sigma^2)\,dt + c\sigma^2\,dX_2.$$

Usually, the value of an option must be found numerically, but there are some simple approximations using Taylor series.

Heston (1993)

In Heston's model

$$dv = (a - bv)\,dt + c\sqrt{v}\,dX_2,$$

where $v = \sigma^2$. This has arbitrary correlation between the underlying and its volatility.

26.8 **GARCH**

Generalized autoregressive conditional heteroskedasticity, or GARCH for short, is a model for an asset and its associated volatility. The simplest such model is GARCH(1,1) which takes the form

$$\sigma_{t+1}^2 = \omega + \beta\sigma_t^2 + \alpha\varepsilon_t^2,$$

where the ε_t are the asset price returns after removing the drift. That is,

$$S_{t+1} = S_t(1 + \mu + \sigma_t\varepsilon_t).$$

Note how the notation is different from, but related to, that which we are used to. This is because, historically, GARCH was developed in an econometrical and not a financial environment. There should not be any confusion as this is the only place in the book where this notation is used. It can be shown that this simplest GARCH model becomes the same as the stochastic volatility model

$$d(\sigma^2) = \phi(\theta - \sigma^2)\,dt + v\sigma^2\,dX_2,$$

as the timestep tends to zero. The parameters in the above are related to the parameters in the original GARCH specification and to the timestep.

There are many references to GARCH in the Further Reading section.

26.9 **STOCHASTIC IMPLIED VOLATILITY: THE MODEL OF SCHÖNBUCHER**

The problem with stochastic volatility models is that it is usually very hard to make the outputs consistent with the current prices of liquid instruments. In other words, calibrating or fitting is difficult. The reason that you might want to 'get right' the prices of liquid instruments is that you want to hedge your exotic contract with them. And it's no good hedging with something when you can't even get the price of that something right.[1]

One way around this problem is to model instead the *implied* volatilities of the liquid instruments.

I am not going to go into the details, just give an overview of the approach. The details can be found in one of the papers mentioned in the Further Reading section.

The implied volatility, for each strike, E, and expiry, T, that we want to model is to satisfy some stochastic differential equation. In that stochastic differential equation there will be a deterministic drift term, a Brownian motion correlated with the underlying asset process and a part uncorrelated with that process. Let's just look at the single traded option case, so that we only have to worry about modeling one implied volatility.

The stochastic differential equation for the implied volatility $\hat{\sigma}$ will be

$$d\hat{\sigma} = u\,dt + \gamma\,dX_1 + v\,dX_2.$$

The dX_1 is the same as in the asset price process

$$dS = \mu S\,dt + \sigma S\,dX_1,$$

but the dX_2 is uncorrelated with that. The relationship between the actual spot volatility σ and the other variables and functions will be made precise later on.

To be clear, $\hat{\sigma}$ is the implied volatility of a vanilla option with a strike E and an expiry T. Let's assume that it is a call option.

The value of this call option is given by

$$Se^{-D(T-t)}N(d_1) - Ee^{-r(T-t)}N(d_2)$$

in terms of the implied volatility etc., where

$$d_1 = \frac{\log\left(\dfrac{S}{E}\right) + \left(r - D + \frac{1}{2}\hat{\sigma}^2\right)(T-t)}{\hat{\sigma}\sqrt{T-t}}$$

and

$$d_2 = d_1 - \hat{\sigma}\sqrt{T-t}.$$

The process for the call price is given by

$$dC = \frac{\partial C}{\partial t}\,dt + \frac{\partial C}{\partial S}\,dS + \frac{\partial C}{\partial \hat{\sigma}}\,d\hat{\sigma} + \frac{1}{2}\sigma^2 S^2\frac{\partial^2 C}{\partial S^2}\,dt + \sigma S\gamma\frac{\partial^2 C}{\partial S\partial\hat{\sigma}}\,dt + \frac{1}{2}v^2\frac{\partial^2 C}{\partial\hat{\sigma}^2}\,dt.$$

We could go through the usual business of setting up a risk-free portfolio etc. and applying the no-arb principle. Equivalently, we can say that the expected risk-adjusted drift rate of the call must be the same as the risk-free rate i.e.

$$E[dC] = rC\,dt.$$

---

[1] I talk about fitting in an interest rate context in Chapter 41. There you'll see that I don't necessarily approve of the practice.

Putting this all together we get

$$rC = \frac{\partial C}{\partial t} + rS\frac{\partial C}{\partial S} + u\frac{\partial C}{\partial \hat{\sigma}} + \frac{1}{2}\sigma^2 S^2 \frac{\partial^2 C}{\partial S^2} + \sigma S\gamma\frac{\partial^2 C}{\partial S\partial\hat{\sigma}} + \frac{1}{2}v^2\frac{\partial^2 C}{\partial\hat{\sigma}^2}. \tag{26.5}$$

Since we know that C must also satisfy the Black–Scholes equation with a volatility of $\hat{\sigma}$ we can simplify (26.5) to get

$$\frac{1}{2}(\sigma^2 - \hat{\sigma}^2)S^2\frac{\partial^2 C}{\partial S^2} + u\frac{\partial C}{\partial \hat{\sigma}} + \sigma S\gamma\frac{\partial^2 C}{\partial S\partial\hat{\sigma}} = 0.$$

We know the formula for C so we can write this as

$$\hat{\sigma}u = \frac{\sigma^2 - \hat{\sigma}^2}{2(T-t)} - \frac{1}{2}d_1 d_2 v^2 + \frac{d_2\sigma\gamma}{\sqrt{T-t}}. \tag{26.6}$$

Equation (26.6) is a relationship between the risk-adjusted drift of the implied volatility u, the implied volatility, the actual volatility and the two components of the volatility of the implied volatility. These cannot all be independent of each other. You have to make a decision what you are going to model, and which will be deduced from the no-arbitrage condition (26.6).

It's natural to specify the two functions v and γ, which could be found from an examination of a time series of the volatility of the implied volatility. You can't get the drift u from the data because it is the *risk-neutral* drift of the implied volatility. So you are left with either specifying the risk-neutral drift, and from that deducing the process for the actual volatility σ, or vice versa.

26.10 **SUMMARY**

Because of the profound importance of volatility in the pricing of options, and because volatility is hard to estimate, observe or predict, it is natural to model it as a random variable. For some contracts, most notably barriers, a constant volatility model is just too inaccurate. In Chapter 29 we'll return to modeling volatility, but there we will examine jump volatility models.

FURTHER READING

- See Hull & White (1987) and Heston (1993) for more discussion of pricing derivatives when volatility is stochastic.

- GARCH is explained in Bollerslev (1986).

- Nelson (1990) shows the relationship between GARCH models and diffusion processes.

- There are many articles in *Risk* magazine, starting in 1992, covering GARCH and its extensions in detail. Recent developments are described in Engle & Mezrich (1996).

- For further work on GARCH see the papers by Engle (1982), Engle & Bollerslev (1987), Alexander (1995, 1996b, 1997b), Alexander & Riyait (1992), Alexander & Chibuma (1997), Alexander & Williams (1997), and the collection of papers edited by Engle (1995).

- Derman & Kani (1997) model the stochastic evolution of the whole local volatility surface.

- Ahn, Arkell, Choe, Holstad & Wilmott (1998) examine the risk involved in delta and static hedging under stochastic volatility.

- For details of the market model for stochastic *implied* volatility see Schönbucher (1999).

CHAPTER 27
uncertain parameters

In this Chapter...

- why the parameters in the Black–Scholes model are not reliable
- the difference between 'random' and 'uncertain'
- how to price contracts when volatility, interest rate and dividend are uncertain

27.1 INTRODUCTION

The Black–Scholes equation

$$\frac{\partial V}{\partial t} + \frac{1}{2}\sigma^2 S^2 \frac{\partial^2 V}{\partial S^2} + (r - D)S\frac{\partial V}{\partial S} - rV = 0$$

is a parabolic partial differential equation in two variables, S and t, with three parameters, σ, r and D, not to mention other parameters such as strike price, barrier levels etc. specific to the contract. Out of these variables and parameters, which ones are easily measurable?

- *Asset price*: The asset price is quoted and therefore easy to measure in theory. In practice, two prices are quoted, the bid and the ask prices; and even these prices will differ between market makers. This issue of transaction costs and their effect on option prices was discussed in Chapter 24.

- *Time to expiry*: Today's date and the expiry date are the easiest quantities to measure. (There is some question about how to treat weekends, but this is more a question of modeling asset price movements than of parameter estimation.)

- *Volatility*: There are two traditional ways of measuring volatility: implied and historical. Whichever way is used, the result cannot be the future value of volatility; either it is the market's estimate of the future or an estimate of values in the past. The correct value of volatility to be used in an option calculation cannot be known until the option has expired. A time series plot of historical volatility, say, might look something like Figure 27.1, and it is certainly not constant as assumed in the simple Black–Scholes formulae. We can see in

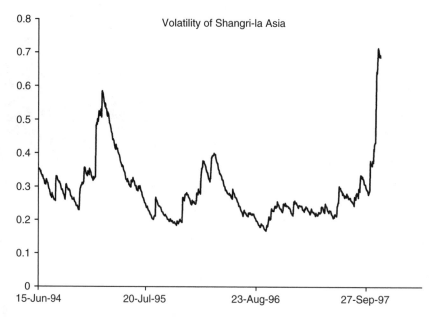

Figure 27.1 A typical time series for historical volatility; an implied volatility time series would look similar.

this figure that volatility for this stock typically ranges between 20 and 60%. The exception to this was during the October/November 1997 crash, for which jump/crash models are perhaps more relevant.

- *Risk-free interest rate*: Suppose we are valuing an option with a lifespan of six months. We can easily find the yield to maturity of a six-month bond and this could be our value for the risk-free rate. However, because the hedged portfolio earns the instantaneous spot rate, the Black–Scholes theory requires knowledge of the future behavior of the *spot* interest rate, and this is not the same as the six-month rate. We can, of course, couple an asset price model *and* an interest rate model, as in Chapter 43, but then we have even more problems with the accuracy of our interest rate model and estimating its parameters.

- *Dividends*: Dividends are declared a few months before they are paid. However, before then what value do we use in our option value calculation? Again, we have to make a guess at the dividend value, obviously using the past as a guide. See Chapter 8 for a detailed discussion of modeling dividend structure.

In this chapter we address the problem of how to value options when parameter values are *uncertain*. Rather than assuming that we know the precise value for a parameter, we assume that all we know about the parameters is that *they lie within specified ranges*. With this assumption, we do not find a *single* value for an option, instead we find that the option's value can also lie within a range: there is no such thing as *the* value. In fact, the correct value cannot be determined until the expiry of the option when we know the path taken by the parameters. Until then, there are many possible values, any of which *might* turn out to be correct.

We will see that this problem is nonlinear, and thus an option valued in isolation has a different range of values from an option valued as part of a portfolio: we find that the range of possible option values depends on what we use to hedge the contract. If we put other options into the portfolio this will change the value of the original portfolio. This leads to the idea of incorporating traded options into an OTC portfolio in such a way as to maximize its value. This is called optimal static hedging, discussed in depth in Chapter 32. In that chapter, I also show how to apply the idea to path-dependent contracts such as barrier options.

Two of the advantages of the approach we adopt in this chapter are obvious. First, we can be more certain about the correctness of a range of values than a single value: we will be happier to say that the volatility of a stock lies within the range 20–30% over the next nine months than to say that the average volatility over this period will be 24%. Another advantage concerns the crucial matter of whether to believe market prices for a contract or the value given by some model. We have seen in Chapter 25 (and will see in Chapter 41) that it is common practice to 'adjust' a model so that it gives theoretical values for liquid contracts that exactly match the market values. Since the uncertain parameter model gives ranges for option values this means that we no longer have to choose between the correctness of a prediction and of a market price. Now they can differ; all we can say is that, according to the model, arbitrage is only certain if the market value lies outside our predicted range.

27.2 **BEST AND WORST CASES**

The first step in valuing options with uncertain parameters is to acknowledge that we can do no better than give ranges for the future values of the parameters. For volatility, for example, this range may be the range of past historical volatility, or implied volatilities, or encompass both of these. Then again, it may just be an educated guess. The range we choose represents our estimate of the upper and lower bounds for the parameter value

for the life of the option or portfolio in question. These ranges for parameters lead to ranges for the option's value. Thus it is natural to think in terms of a lowest and highest possible option value; if you are long the option, then we can also call the lowest value the *worst* value and the highest the *best*. Work in this area was started by Avellaneda, Levy, Parás and Lyons.

We begin by considering uncertain volatility and shortly address the problem of uncertain interest rate (very similar) and then uncertain discretely-paid dividends (slightly different).

27.2.1 Uncertain Volatility: The Model of Avellaneda, Levy & Parás and Lyons (1995)

Let us suppose that the volatility lies within the band

$$\sigma^- < \sigma < \sigma^+.$$

We will follow the Black–Scholes hedging and no-arbitrage argument as far as we can and see where it leads us.

Construct a portfolio of one option, with value $V(S, t)$, and hedge it with $-\Delta$ of the underlying asset. The value of this portfolio is thus

$$\Pi = V - \Delta S.$$

We still have

$$dS = \mu S \, dt + \sigma S \, dX,$$

even though σ is unknown, and so the change in the value of this portfolio is

$$d\Pi = \left(\frac{\partial V}{\partial t} + \tfrac{1}{2}\sigma^2 S^2 \frac{\partial^2 V}{\partial S^2} \right) dt + \left(\frac{\partial V}{\partial S} - \Delta \right) dS.$$

Even with the volatility unknown, the choice of $\Delta = \partial V / \partial S$ eliminates the risk:

$$d\Pi = \left(\frac{\partial V}{\partial t} + \tfrac{1}{2}\sigma^2 S^2 \frac{\partial^2 V}{\partial S^2} \right) dt.$$

At this stage we would normally say that if we know V then we know $d\Pi$. This is no longer the case since we do not know σ. The argument now deviates subtly from the vanilla Black–Scholes argument. What we will now say is that we will be pessimistic: we will assume that the volatility over the next timestep is such that our portfolio increases by the least amount. If we have a long position in a call option, for example, we assume that the volatility is at the lower bound σ^-; for a short call we assume that the volatility is high. This amounts to considering the *minimum* return on the portfolio, where the minimum is taken over all possible values of the volatility within the given range. The return on this worst-case portfolio is then set equal to the risk-free rate:

$$\min_{\sigma^- < \sigma < \sigma^+} (d\Pi) = r\Pi \, dt.$$

Thus we set

$$\min_{\sigma^- < \sigma < \sigma^+} \left(\frac{\partial V}{\partial t} + \tfrac{1}{2}\sigma^2 S^2 \frac{\partial^2 V}{\partial S^2} \right) dt = r \left(V - S \frac{\partial V}{\partial S} \right) dt.$$

Now observe that the volatility term in the above is multiplied by the option's gamma. Therefore the value of σ that will give this its minimum value depends on the sign of the gamma. When the gamma is positive we choose σ to be the lowest value σ^- and when it is negative we choose σ to be its highest value σ^+. We find that the worst-case value V^- satisfies

$$\frac{\partial V^-}{\partial t} + \tfrac{1}{2}\sigma(\Gamma)^2 S^2 \frac{\partial^2 V^-}{\partial S^2} + rS \frac{\partial V^-}{\partial S} - rV^- = 0 \qquad (27.1)$$

where

$$\Gamma = \frac{\partial^2 V^-}{\partial S^2}$$

and

$$\sigma(\Gamma) = \begin{cases} \sigma^+ & \text{if } \Gamma < 0 \\ \sigma^- & \text{if } \Gamma > 0. \end{cases}$$

We can find the best option value V^+, and hence the range of possible values, by solving

$$\frac{\partial V^+}{\partial t} + \tfrac{1}{2}\sigma(\Gamma)^2 S^2 \frac{\partial^2 V^+}{\partial S^2} + rS \frac{\partial V^+}{\partial S} - rV^+ = 0$$

where

$$\Gamma = \frac{\partial^2 V^+}{\partial S^2}$$

but this time

$$\sigma(\Gamma) = \begin{cases} \sigma^+ & \text{if } \Gamma > 0 \\ \sigma^- & \text{if } \Gamma < 0. \end{cases}$$

We won't find much use for the problem for the best case in practice since it would be financially suicidal to assume the best outcome. (Note that just by changing the signs in Equation (27.1) we go from the worst-case equation to the best. In other words, the problem for the worst price for long and short positions in a particular contract is mathematically equivalent to valuing a long position only, but in worst and best cases. This distinction between long and short positions is an important consequence of the nonlinearity of the equation and we discuss this in depth shortly.)

Equation (27.1), derived by Avellaneda, Levy & Parás and Lyons, is exactly the same as the Hoggard–Whalley–Wilmott transaction cost model that we saw in Chapter 24. The partial differential equation may be the same, but the reason for it is completely different.

27.2.2 Example: An Up-and-out Call

Consider the following up-and-out call option. The strike is 100, the spot interest rate is 5%, the barrier is at 120 and there is one year to expiry. Let's suppose we are not sure what the volatility is, but it should be in the range 17–23%.

If we priced the option using first a 17% volatility and then a 23% volatility we would get two curves looking like those in Figure 27.2. If the asset value were at around 80 you might think that the option was insensitive to the volatility. But actually it is around that asset value where the option is very sensitive to the volatility ranging over 17–23%. Let's see why this is by solving the worst-case scenario problem with the range 17–23%; after all, the volatility could do anything within this range, it doesn't have to remain constant.

Equation (27.1) must in general be solved numerically, because it is nonlinear. In Figure 27.3 are shown the best and worst prices for an up-and-out call option. As I said above, the best and worst prices could be interpreted as worst-case prices for short and long positions respectively. In the figure is a Black–Scholes value, the middle line, using a volatility of 20%. The other two bold lines give the worst-case long and short values assuming a volatility ranging from 17% to 23%.

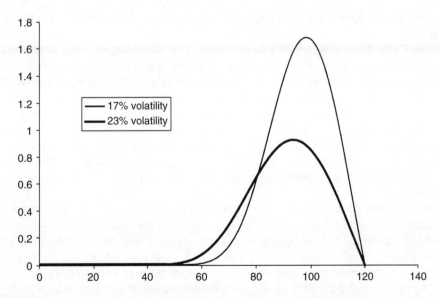

Figure 27.2 Up-and-out call values in a Black–Scholes world with volatilities of 17% and 23%.

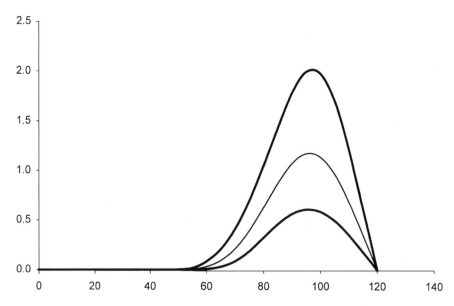

Figure 27.3 Up-and-out call value assuming a range for volatility, and the Black–Scholes price.

We must solve the nonlinear equation numerically because the gamma for this contract is not single-signed. The problem is genuinely nonlinear, and we cannot just substitute each of 17% and 23% into a Black–Scholes formula.

Observe that the best/worst of the two curves (the 'envelope') in Figure 27.2 is not the same as the best/worst of Figure 27.3. If you were to price the option using the worst out of the two standard Black–Scholes prices, with constant volatility, you would significantly overestimate the value of the option in the worst-case scenario. For this reason it can be extremely dangerous to calculate a contract's vega when the contract has a gamma that changes sign. Having said that, practitioners do have ways of fudging their vega calculations to make some allowance for this effect.

Continuing with this barrier option example, let us look at implied volatilities. In Figure 27.4 is shown the Black–Scholes value of an up-and-out call option as a function of the volatility. The strike is 100, the stock is at 80, the spot interest rate is 5%, the barrier is at 120 and there is one year to expiry. This contract has a gamma that changes sign, and a price that is not monotonic in the volatility. This figure shows that there is a maximum option value of 0.68 when the volatility is about 20%. Now turn the problem around. Suppose that the market is pricing this contract at 0.55. From the figure we can see that there are *two* volatilities that correspond to this market price. One has a positive vega and the other negative. Which, if either, is correct? The question is probably meaningless because of the non-single-signed gamma of this contract.

Take this example further. What if the market price is 0.72? This value cannot be reached by any single volatility. Does this mean that there are arbitrage opportunities? Not necessarily. This could be due to the market pricing with a non-constant volatility, either with a volatility surface, stochastic volatility or a volatility range. As we have seen from the best/worst prices for this contract, the uncertainty in the option value may be large enough to cover the market price of the option, and there may be no guaranteed arbitrage at all. The best value is greater than 1 for the volatility range 17–23%, as can be seen in Figure 27.3.

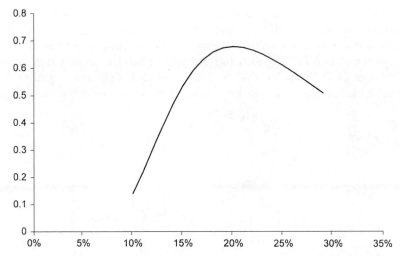

Figure 27.4   Black–Scholes values for an up-and-out call option against volatility.

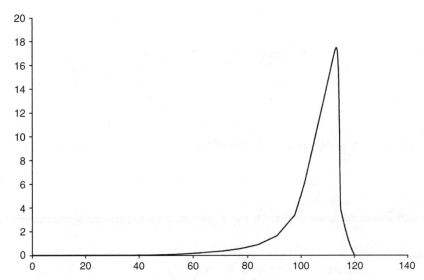

Figure 27.5   The envelope of all Black–Scholes values for an up-and-out call option.

In Figure 27.5 is shown the envelope of all possible Black–Scholes up-and-out call option values assuming a *constant* volatility ranging from *zero to infinity*. Option values outside this envelope cannot be attained with constant volatility. With uncertain volatility, prices outside this envelope *can* easily be attained.

27.2.3   Uncertain Interest Rate

Exactly the same idea can be applied to the case of uncertain interest rate. Let us suppose that the risk-free interest rate lies within the band

$$r^- < r < r^+.$$

Continuing with the worst case, we assume that if our portfolio has a net positive value ($\Pi > 0$) then interest rates will be high, and if it has a negative value then rates will be low. The reason for this is that if we have a positive amount invested in options then in the worst case we are failing to benefit from high interest rates. Thus, the interest rate we choose will depend on the sign of Π. As before, we set the return on the portfolio equal to the risk-free rate. With our special choice for this rate we arrive at the equation for the worst-case scenario:

$$\frac{\partial V^-}{\partial t} + \frac{1}{2}\sigma^2 S^2 \frac{\partial^2 V^-}{\partial S^2} + r(\Pi)\left(S\frac{\partial V^-}{\partial S} - V^-\right) = 0 \tag{27.2}$$

where

$$\Pi = V^- - S\frac{\partial V^-}{\partial S}$$

and

$$r(\Pi) = \begin{cases} r^+ & \text{if } \Pi > 0 \\ r^- & \text{if } \Pi < 0. \end{cases}$$

The equation for the best case is obvious.

27.2.4 Uncertain Dividends

Consider a dividend yield that is independent of the asset price. There are two cases to consider here: continuously paid and discretely paid. In the former case (and with FX options where the dividend yield is replaced by the foreign interest rate) we assume that the dividend yield lies between two values:

$$D^- \leq D \leq D^+.$$

Now, for the option's worst value we simply solve

$$\frac{\partial V^-}{\partial t} + \frac{1}{2}\sigma^2 S^2 \frac{\partial^2 V^-}{\partial S^2} + r\left(S\frac{\partial V^-}{\partial S} - V^-\right) - D(\Delta)S\frac{\partial V^-}{\partial S} = 0 \tag{27.3}$$

where

$$D(\Delta) = \begin{cases} D^+ & \text{if } \Delta > 0 \\ D^- & \text{if } \Delta < 0. \end{cases}$$

The case of discretely paid dividends is slightly different. Consider what happens across a dividend date, t_i. Let us suppose that the discretely-paid dividend yield lies within the band

$$D^- < D < D^+.$$

The jump condition across a discretely-paid dividend date is

$$V(S, t_i^-) = V((1 - D)S, t_i^+)$$

as shown in Chapter 8. When this dividend yield is uncertain we simply minimize over the possible values of D and this gives the following jump condition

$$V^-(S, t_i^-) = \min_{D^- < D < D^+} V^-((1 - D)S, t_i^+). \tag{27.4}$$

There is a corresponding jump condition for the best case.

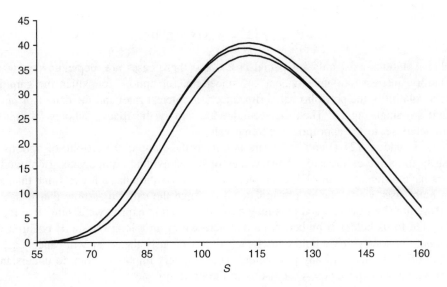

Figure 27.6 Portfolio value under uncertain dividends, best and worst cases.

Pricing with an uncertain dividend, instead of dividend yield, is just the same. Figure 27.6 gives an example of a portfolio consisting of long two calls struck at 80 and three short struck at 110, all expire in six months. The risk-free rate is 7% and volatility 20%. A dividend is paid after four months, and this dividend lies in the range zero to five.

Of course, it is a simple matter to put together the results of the above sections to model uncertainty in all of volatility, interest rate and dividend.

I emphasize that from now on we only consider the pessimistic case. We assume the worst outcome and price contracts accordingly. By so doing we guarantee that we never lose money provided that our uncertain parameter ranges are not violated.

27.3 **UNCERTAIN CORRELATION**

Exactly the same idea can be applied to multi-asset instruments which have a dependence on the correlation between the underlyings. Correlation is something that is particularly difficult to calculate or predict so there is obviously a role to be played by uncertainty. The downside is that the spread on correlation is likely to be so large that the spread on the option price will be unrealistic. And since there are so few correlation instruments with which to hedge it may not be possible to reduce this spread to anything tradeable.

27.4 **NONLINEARITY**

All of the new partial differential equations that we have derived are nonlinear. In particular, the uncertain volatility model results in the same nonlinear equation, Equation (27.1), as the Hoggard–Whalley–Wilmott model for pricing options in the presence of transaction costs, discussed in Chapter 24. Because of this nonlinearity, we must therefore distinguish between long and short positions; for example, for a long call we have

$$V^-(S, T) = \max(S - E, 0)$$

and for a short call

$$V^-(S, T) = -\max(S - E, 0).$$

Explicit solutions exist only in special cases, and these cases are, depending on which of the volatility, interest rate and dividend are uncertain, for options for which the gamma (for uncertain volatility), the portfolio value (for uncertain interest rate) and the delta (for uncertain dividend) are single-signed. Then the formulae are simply the Black–Scholes formulae with the parameters set to the appropriate extreme values.

Because Equations (27.1) and (27.2) are nonlinear the value of a portfolio of options is not necessarily the same as the sum of the values of the individual components. Long and short positions have different values. For example, a long call position has a lower value than a short call. In both cases we are being pessimistic: if we own the call we assume that it has a low value, if we are short the call and thus may have to pay out at expiry, we assume that the value of the option to its holder is higher. Note that here we mean a long (or short) position valued *in isolation*. Obviously, if we hold one of each simultaneously then they will cancel each other regardless of the behavior of any parameters. This is a very important point to understand: the value of a contract depends on what else is in the portfolio.

Unfortunately, the model as it stands predicts very wide spreads on options. For example, suppose that we have a European call, strike price $100, today's asset price is $100, there are six months to expiry, no dividends but a spot interest rate that we expect to lie between 5% and 6% and a volatility between 20% and 30%. We can calculate the values for long and short calls assuming these ranges for the parameters directly from the Black–Scholes formulae. This is because the gamma and the portfolio value are single-signed for a call. A long call position is worth $6.85 (the Black–Scholes value using a volatility of 20% and an interest rate of 5%) and a short call is worth $9.85 (the Black–Scholes value using a volatility of 30% and an interest rate of 6%). This spread is much larger than that in the market and, in this example, is mostly due to the uncertain volatility. The market prices may, for example, be based on an interest rate of 5.5% with a volatility between 24% and 26%. Unless the model can produce narrower spreads the model will be useless in practice.

The spreads *can* be tightened by 'static hedging', which means the purchase and sale of traded option contracts so as to improve the marginal value of our original position. This only works because we have a *nonlinear* governing equation: the price of a contract depends on what else is in the portfolio. This static hedge can be optimized so as to give the original contract its best value; we can squeeze even more value out of our contract with the best hedge.

The issues of static hedging and optimal static hedging are covered in detail in Chapter 32.

27.5 **SUMMARY**

Greeks such as vega and rho can be completely misleading if used carelessly and without understanding. The uncertain parameter model gives a consistent way to eliminate dependence of a price on a parameter, and to some extent reduce model dependence. Out of the three volatility models (deterministic calibrated smile, stochastic and uncertain), uncertain volatility is easily my favorite. As presented in this chapter the spreads on the best/worst prices are too large for the model to be of any practical use. Fortunately this situation will be remedied in Chapter 32.

FURTHER READING

- See Avellaneda, Levy & Parás (1995), Avellaneda & Parás (1996) and Lyons (1995) for more details about the modeling of uncertain volatility.

- See Bergman (1995) for the derivation of Equation (27.2) for a world in which there are different rates for borrowing and lending.

- See Oztukel (1996) for the uncertain parameter technique applied to correlation.

- A 'Virtual Option Pricer' for uncertain volatility can be found at home.cs.nyu.edu:8000/cgi-bin/vop.

- Ahn, Muni & Swindle (1996) explain how to modify the final payoff to reduce the effect of volatility errors in the worst-case scenario.

- See Ahn, Muni & Swindle (1998) for a stochastic control approach to utility maximization in the worst-case model.

CHAPTER 28
empirical analysis of volatility

In this Chapter...

- how to analyze volatility data to determine the most suitable time-independent stochastic model
- how to determine the probability that the volatility will stay within any specified range
- how to assign a degree of confidence to your uncertain volatility model price

28.1 INTRODUCTION

In this chapter we focus closely on real data for the behavior of volatility. Our principal aim in this is to derive a good stochastic volatility model. This volatility model can then be used in a number of ways: in a two-factor option-pricing model; to examine the time evolution of volatility from an initial known value today to, in the long run, a steady-state distribution; to estimate the probability of our uncertain volatility model price being correct. The approach we adopt must be contrasted with the traditional approach to stochastic volatility which seems to be to write down something nice and tractable and then fit the parameters.

28.2 STOCHASTIC VOLATILITY AND UNCERTAIN PARAMETERS REVISITED

The classical way of dealing with random variables is to model them stochastically. This we do here for volatility, deriving a stochastic model such that drift and variance, and the steady-state mean and dispersion of volatility, are *compatible with historical data*. We can then, among other things, determine the evolution of volatility from the known value today to the steady-state distribution in the long term.

Recall also, the approach of Avellaneda, Levy & Parás and Lyons for modeling volatility. In their models they allow volatility to do just about anything as long as it doesn't move outside some given range. With this model it is natural to calculate not a single option value but a worst price and a best price for the option. This results in a nonlinear partial differential equation. This procedure yields a 'certainty interval' for the price of the option, driven by the input volatility band. Hence we know that, for example, if we could access the option in the market below our worst price then, within the limits of our certainty band assumptions, we are guaranteed a profit. But what if we are not one hundred percent sure about the volatility range? Can we use our stochastic volatility model to see how likely volatility is to stay in the range?

28.3 DERIVING AN EMPIRICAL STOCHASTIC VOLATILITY MODEL

In this chapter we are going to work with daily Dow Jones Industrial Average spot data from January 1975 until August 1995. The method that I am going to demonstrate for determining the volatility model can be used for many financial time series; indeed later we will later use the same method for modeling US short-term interest rates.

In Figure 28.1 is shown the calculated annualized 30-day volatility of daily returns for the DJIA over the period of interest. (Whether or not the volatility that has been calculated here is representative of the 'actual' volatility is debatable. Nevertheless, the techniques I describe below are applicable to whatever volatility series we have.)

Previous attempts at modeling volatility stochastically are numerous. However, since adding another stochastic variable complicates the problem of pricing options, the emphasis to date has been on deriving a tractable model. We are going to try to determine a stochastic model for volatility by fitting the drift and variance functions to empirical data, letting tractability take a secondary role. We model the volatility in isolation, assuming that it does not depend on the level of the index. Thus we assume that the stochastic process for volatility is given by

$$d\sigma = \alpha(\sigma)\,dt + \beta(\sigma)\,dX. \tag{28.1}$$

We use twenty years of daily Dow Jones closing prices (in total over 5000 observations) to calculate daily returns and 30-day volatility of these daily returns.

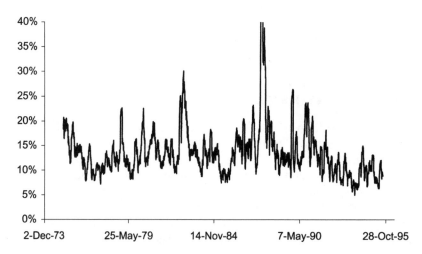

Figure 28.1 Thirty-day volatility of daily returns for the Dow Jones Industrial Average from 1975 to 1995.

28.4   **ESTIMATING THE VOLATILITY OF VOLATILITY**

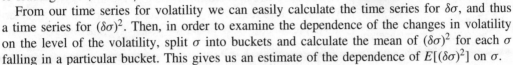

From (28.1), the square of the day-to-day change in volatility is given by

$$(\delta\sigma)^2 = \beta(\sigma)^2 \phi^2 \, \delta t \qquad (28.2)$$

to leading order, where ϕ is a standardized Normal variable.

From our time series for volatility we can easily calculate the time series for $\delta\sigma$, and thus a time series for $(\delta\sigma)^2$. Then, in order to examine the dependence of the changes in volatility on the level of the volatility, split σ into buckets and calculate the mean of $(\delta\sigma)^2$ for each σ falling in a particular bucket. This gives us an estimate of the dependence of $E[(\delta\sigma)^2]$ on σ.

In Figure 28.2 is plotted $\log(E[(\delta\sigma)^2])$ versus $\log(\sigma)$. Superimposed on this figure is a straight line, fitted to the dots by ordinary least squares.

The straight line suggests that

$$\log(E[(\delta\sigma)^2]) = 2\log(\beta(\sigma)) + \log \delta t = a + b\log(\sigma)$$

is a good approximation. From the data we can estimate the parameters a and b and so deduce that a good model is

$$\beta(\sigma) = \nu\sigma^\gamma. \qquad (28.3)$$

From the data we find that

$$\nu = 50.88 \quad \text{and} \quad \gamma = 1.05.$$

In our model we take

$$d\sigma = \alpha(\sigma)\,dt + \nu\sigma^\gamma \, dX.$$

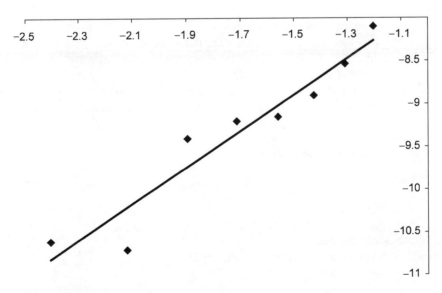

Figure 28.2   Plot of $\log(E[(\delta\sigma)^2])$ versus $\log(\sigma)$ and the fitted line.

This gives a model with volatility of volatility that is consistent with empirical data. To complete the modeling we still need to find the drift of the volatility $\alpha(\sigma)$.

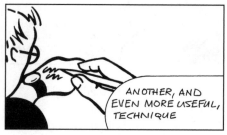

ANOTHER, AND EVEN MORE USEFUL, TECHNIQUE

28.5 ESTIMATING THE DRIFT OF VOLATILITY

To calculate $\alpha(\sigma)$ we need to examine the time series for σ in a slightly different way. In particular we are going to determine the steady-state probability density function for σ and from that deduce the drift function. The equation governing the probability density function for σ is of the form derived in Chapter 10, the Fokker–Planck equation (or forward Kolmogorov equation)

$$\frac{\partial p}{\partial t} = \tfrac{1}{2} \frac{\partial^2}{\partial \sigma^2}(\beta^2 p) - \frac{\partial}{\partial \sigma}(\alpha p) \tag{28.4}$$

where $p(\sigma, t)$ is the probability density function for σ and I am using σ and t to denote the forward variables. Suppose that somehow we know the steady-state distribution, $p_\infty(\sigma)$, for σ. This will satisfy

$$0 = \tfrac{1}{2} \frac{d^2}{d\sigma^2}(\beta^2 p_\infty) - \frac{d}{d\sigma}(\alpha p_\infty).$$

This is simply a steady-state version of (28.4). Integrating this once we get

$$\alpha(\sigma) = \frac{1}{2 p_\infty} \frac{d}{d\sigma}(\beta^2 p_\infty). \tag{28.5}$$

The constant of integration is zero, as can be shown by examining the behavior of α for large and small σ.

From (28.5), we see that if we know p_∞ then we can find the drift term. But can we find p_∞? Yes, we can. If we assume that all parameters are independent of time, then we can determine the steady-state probability density function by using the 'ergodic property' of random walks; that is, the equivalence of ensemble and time averages.

We find p_∞ from the data by plotting the frequency distribution of σ versus buckets of σ; that is, how many observations fall into each bucket. The empirical distribution, shown in Figure 28.3, closely resembles a lognormal curve. Since we need to differentiate this function to find the drift according to (28.5) we fit a curve to the distribution. We shall assume p_∞ to be of the form of a lognormal distribution:

$$p_\infty = \frac{1}{\sqrt{2\pi}a\sigma} e^{-(1/2a^2)(\log(\sigma/\bar{\sigma}))^2}$$

where $\log \bar{\sigma}$ represents the mean of the distribution of $\log \sigma$ and a describes the dispersion of the distribution about the mean.

This graph and fitted curve are shown in Figure 28.3.

We find from the data that,

$$a = 0.33 \quad \text{and} \quad \bar{\sigma} = 13.4\%.$$

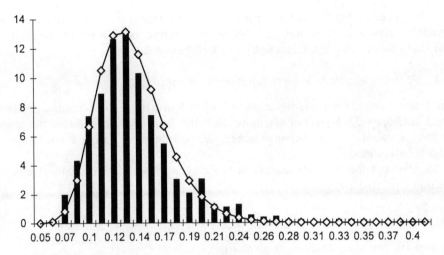

Figure 28.3 The steady-state distribution of σ and the fitted lognormal curve.

We now have that

$$\alpha(\sigma) = v^2\sigma^{2\gamma-1}\left(\gamma - \tfrac{1}{2} - \frac{1}{2a^2}\log\left(\frac{\sigma}{\bar{\sigma}}\right)\right).$$

28.6 **HOW TO USE THE MODEL**

We can use our model in a number of ways, here I just make a few suggestions. The three possibilities that I will describe are straightforward option pricing, determining the future distribution for volatility and estimating our confidence in volatility ranges.

28.6.1 Option Pricing with Stochastic Volatility

The first way we can use the model is in a two-factor model for pricing options. The two factors would be the underlying asset and its volatility. The only problem with this is that we have estimated from data the *real* drift rate and not the *risk-neutral* rate that we need for option pricing. Whether this matters or not is debatable.

28.6.2 The Time Evolution of Stochastic Volatility

From the stochastic volatility model we can derive the probabilistic evolution of volatility from an initial value. We turn once again to the Fokker–Planck equation given by (28.4) which describes the evolution over time of the probability density function of a random variable defined by a stochastic differential equation. We can use the functions α and β derived above to find the probability density function of volatility over a specified time horizon. This way we can observe the evolution of the distribution of volatility.

As the initial condition for Equation (28.4) we would apply a delta function — meaning that we know the value of today's volatility with certainty — and solve for the resulting probability density function as time evolves.

What we would see would be the density function starting out with a delta function spike, and gradually smoothing out until it stabilized at the input steady-state probability density function.

Because the volatility has a sensible distribution in the steady state we would not see any unreasonable behavior from the model. Other models, having nice but otherwise arbitrary drift rate and volatility of volatility, might not be so well behaved.

28.6.3 Stochastic Volatility, Certainty Bands and Confidence Limits

The third way we could use the model is in conjunction with the uncertain volatility model described in Chapter 27. From our stochastic volatility model we can estimate the probability of our chosen volatility range being breached. We can therefore assign a probability to our uncertain volatility model price.

The likelihood of the volatility staying in any range can be mathematically represented by the function $C(\sigma, t)$ satisfying

$$\frac{\partial C}{\partial t} + \frac{1}{2}\beta(\sigma)^2 \frac{\partial^2 C}{\partial \sigma^2} + \alpha(\sigma)\frac{\partial C}{\partial \sigma} = 0,$$

see Chapter 10. The final condition is given by

$$C(\sigma, T) = 0 \quad \text{for all } \sigma$$

since at $t = T$ the likelihood of σ breaching the barriers is zero. The boundary conditions are given by

$$C(\sigma^+, t) = C(\sigma^-, t) = 1 \quad \text{for all } t.$$

Here σ^- and σ^+ are the lower and upper barriers respectively. These conditions are obvious: if $\sigma = \sigma^-$ or $\sigma = \sigma^+$ then the barrier is certain to be breached. The final and boundary conditions are shown schematically below in Figure 28.4.

This third use of the model is quite a nice compromise. One reason I like it is that it allows the volatility to be random, satisfying a given model, yet gives an option price that doesn't depend on the market price of risk.

28.7 SUMMARY

Generally speaking I don't like to accept a model because it looks nice. If at all possible I prefer to look at the data to try and find the best model. In this chapter we examined volatility

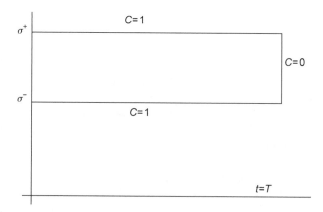

Figure 28.4 The final and boundary conditions for the exit probability.

data to find the best model having time-independent functions describing the drift and volatility of volatility. Crucially we modeled the volatility of volatility by examining the daily changes of volatility. After all, the randomness in the level of the volatility is seen over the shortest timescale. However, to find the best drift we looked over a longer timescale, and we estimated the steady-state distribution for the volatility. The drift only really comes into its own over longer timescales. To examine the short timescale to estimate the drift would at best not make much sense and at worst result in a model with poor long-term properties.

The resulting stochastic differential equation model for the volatility can be used in several ways. One is to just accept it as the model for volatility in a two-factor option model. Of course, then you have to estimate the market price of risk; not nice. Another use for the model, which gets around the market price of risk problem, is in conjunction with an uncertain volatility model, such as described in Chapter 27. Prescribe a range for volatility and then use the stochastic differential equation to estimate the probability of the volatility staying within this range.

FURTHER READING

- The subject of bounds for option values when volatility is stochastic was first addressed by El Karoui, Jeanblanc-Picque & Viswanathan (1991).
- See Oztukel (1996) and Oztukel & Wilmott (1998) for further details of the analysis of volatility.

CHAPTER 29
jump diffusion

In this Chapter...

- the Poisson process for modeling jumps
- hedging in the presence of jumps
- how to price derivatives when the path of the underlying can be discontinuous
- jump volatility

29.1 INTRODUCTION

There is plenty of evidence that financial quantities, be they equities, currencies or interest rates, for example, do not follow the lognormal random walk that has been the foundation of almost everything in this book, and almost everything in the financial literature. We look at some of this evidence in a moment. One of the striking features of real financial markets is that every now and then there is a sudden unexpected fall or crash. These sudden movements occur far more frequently than would be expected from a Normally-distributed return with a reasonable volatility. On all but the shortest timescales the move looks discontinuous; the prices of assets have jumped. This is important for the theory and practice of derivatives because *it is usually not possible to hedge through the crash*. One certainly cannot delta hedge as the stock market tumbles around one's ankles, and to offload all one's positions will lead to real instead of paper losses, and may even make the fall worse.

In this chapter I explain classical ways of pricing and hedging when the underlying follows a jump-diffusion process.

29.2 EVIDENCE FOR JUMPS

Let's look at some data to see just how far from Normal the returns really are. There are several ways to visualize the difference between two distributions, in our case the difference between the empirical distribution and the Normal distribution. One way is to overlay the two probability distributions. In Figure 29.1 we see the distribution of Xerox returns, from 1986 until 1997, normalized to unit standard deviation. The

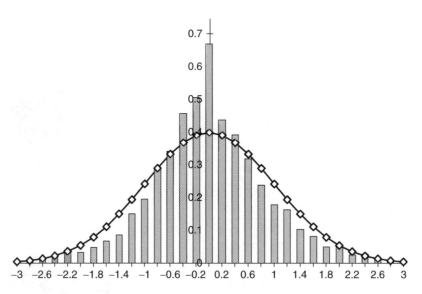

Figure 29.1 The probability density functions for Xerox daily returns, scaled to zero mean and unit standard deviation, and the standardized Normal distribution.

peak of the real distribution is clearly higher than the Normal distribution. Because both of these distributions have the same standard deviation then the higher peak must be balanced by fatter tails, it's just that they would be too small to see on this figure. They may be too small to see here, but they are still very important.

This difference is typical of all markets, even typical of the changes in interest rates. The empirical distribution diverges from the Normal distribution quite markedly. The peak being much higher means that there is a greater likelihood of a small move than we would expect from the lognormal random walk. More importantly, and concerning the subject of this chapter, the tails are much fatter. There is a greater chance of a large rise or fall than the Normal distribution predicts.

In Figure 29.2 is shown the difference between the cumulative distribution functions for the standardized returns of Xerox, and the Normal distribution. If you look at the figure you will see that there is more weight in the empirical distribution than the Normal distribution from about two standards deviations away from the mean. If you couple this likelihood of an extreme movement with the *importance* of an extreme movement, assuming perhaps that people are hedged against *small* movements, you begin to get a very worrying scenario.

Another useful plot is the logarithm of the pdf for the actual distribution and the Normal distribution. The latter is simply a parabola, but what is the former? Such a plot is shown in Figure 29.3. The log of the pdf of the actual distribution looks more linear than quadratic in the tails.

The final picture that I plot, in Figure 29.4, is called a **Quantile-Quantile** or **Q-Q plot**. This is a common way of visualizing the difference between two distributions when you are particularly interested in the tails of the distribution. This plot is made up as follows. Rank the empirical returns in order from smallest to largest, call these y_i with an index i going from 1 to n. For the Normal distribution find the returns x_i such that the cumulative distribution function at x_i has value i/n. Now plot each point (x_i, y_i). The better the fit between the two

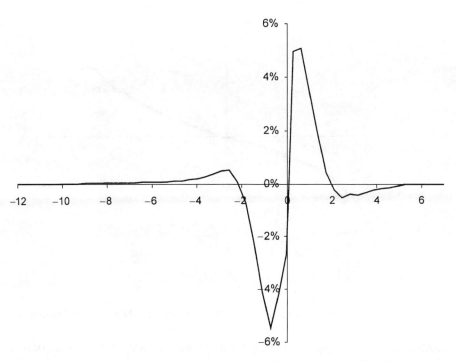

Figure 29.2 The difference between the cumulative distribution functions for Xerox daily returns, scaled to zero mean and unit standard deviation, and the standardized Normal distribution.

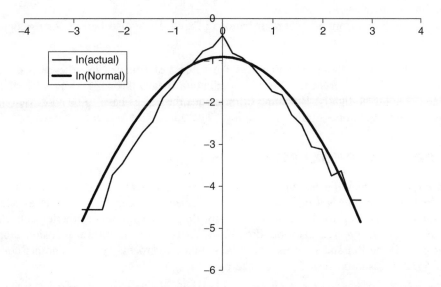

Figure 29.3 Logarithms of the pdfs for Xerox daily returns and the standardized Normal distribution.

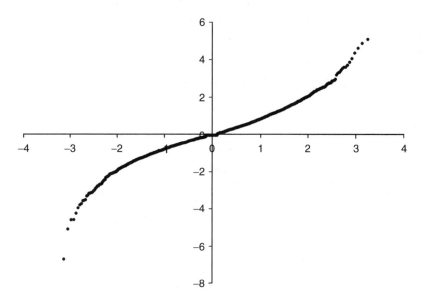

Figure 29.4 Q-Q plot for Xerox daily returns and the standardized Normal distribution.

distributions, the closer the line is to straight. In the present case the line is far from straight, due to the extra weight in the tails.

Several theories have been put forward for the non-Normality of the empirical distribution. Three of these are

- Volatility is stochastic
- Returns are drawn from another distribution, a Pareto-Levy distribution, for example
- Assets can jump in value

There is truth in all of these. The first was the subject of Chapter 26. The second is a can of worms; moving away from the Normal distribution means throwing away 99% of current theory and is not done lightly. But some of the issues this raises have to be addressed in jump diffusion (such as the impossibility of hedging). The third is the present subject.

29.3 **POISSON PROCESSES**

The basic building block for the random walks we have considered so far is continuous Brownian motion based on the Normally-distributed increment. We can think of this as adding to the return from one day to the next a Normally distributed random variable with variance proportional to timestep. The extra building block we need for the **jump-diffusion model** for an asset price is the **Poisson process**. We'll be seeing this process again in another context in Chapter 55. A Poisson process dq is defined by

$$dq = \begin{cases} 0 & \text{with probability } 1 - \lambda\,dt \\ 1 & \text{with probability } \lambda\,dt. \end{cases}$$

There is therefore a probability $\lambda\,dt$ of a jump in q in the timestep dt. The parameter λ is called the **intensity** of the Poisson process. The scaling of the probability of a jump with the size of the timestep is important in making the resulting process 'sensible', i.e. there being a finite chance of a jump occurring in a finite time, with q not becoming infinite.

This Poisson process can be incorporated into a model for an asset in the following way:

$$dS = \mu S \, dt + \sigma S \, dX + (J - 1)S \, dq. \tag{29.1}$$

We assume that there is no correlation between the Brownian motion and the Poisson process. If there is a jump ($dq = 1$) then S immediately goes to the value JS. We can model a sudden 10% fall in the asset price by $J = 0.9$.

We can generalize further by allowing J to also be a random quantity. We assume that it is drawn from a distribution with probability density function $P(J)$, again independent of the Brownian motion and Poisson process.

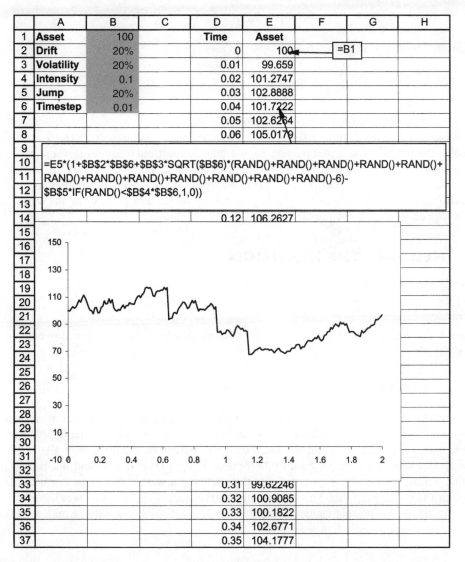

Figure 29.5 Spreadsheet simulation of a jump-diffusion process.

The random walk in $\log S$ follows from (29.1):

$$d(\log S) = \left(\mu - \tfrac{1}{2}\sigma^2\right) dt + \sigma \, dX + (\log J) \, dq.$$

This is just a jump-diffusion version of Itô.

Figure 29.5 is a spreadsheet showing how to simulate the random walk for S. In this simple example the stock jumps by 20% at random times given by a Poisson process.

29.4 HEDGING WHEN THERE ARE JUMPS

Now let us build up a theory of derivatives in the presence of jumps. Begin by holding a portfolio of the option and $-\Delta$ of the underlying:

$$\Pi = V(S, t) - \Delta S.$$

The change in the value of this portfolio is

$$d\Pi = \left(\frac{\partial V}{\partial t} + \tfrac{1}{2}\sigma^2 S^2 \frac{\partial^2 V}{\partial S^2}\right) dt + \left(\frac{\partial V}{\partial S} - \Delta\right) dS + (V(JS, t) - V(S, t) - \Delta(J - 1)S) \, dq.$$

$$(29.2)$$

Again, this is a jump-diffusion version of Itô.

If there is no jump at time t so that $dq = 0$, then we could have chosen $\Delta = \partial V/\partial S$ to eliminate the risk. If there is a jump and $dq = 1$ then the portfolio changes in value by an $O(1)$ which amount *cannot be hedged away*. In that case perhaps we should choose Δ to minimize the variance of $d\Pi$. This presents us with a dilemma. We don't know whether to hedge the small(ish) diffusive changes in the underlying which are always present, or the large moves which happen rarely. Let us pursue both of these possibilities.

29.5 HEDGING THE DIFFUSION

If we choose

$$\Delta = \frac{\partial V}{\partial S}$$

we are following a Black–Scholes type of strategy, hedging the diffusive movements. The change in the portfolio value is then

$$d\Pi = \left(\frac{\partial V}{\partial t} + \tfrac{1}{2}\sigma^2 S^2 \frac{\partial^2 V}{\partial S^2}\right) dt + \left(V(JS, t) - V(S, t) - (J - 1)S \frac{\partial V}{\partial S}\right) dq.$$

The portfolio now evolves in a deterministic fashion, except that every so often there is a non-deterministic jump in its value. It can be argued (see Merton, 1976) that if the jump component of the asset price process is uncorrelated with the market as a whole, then the risk in the discontinuity should not be priced into the option. Diversifiable risk should not be rewarded. In other words, we can take expectations of this expression and set that value equal to the risk-free return from the portfolio. This is not completely satisfactory, but is a common assumption whenever there is a risk that cannot be fully hedged; default risk is another example of this. If we take such an approach then we arrive at

$$\frac{\partial V}{\partial t} + \tfrac{1}{2}\sigma^2 S^2 \frac{\partial^2 V}{\partial S^2} + rS \frac{\partial V}{\partial S} - rV + \lambda E[V(JS, t) - V(S, t)] - \lambda \frac{\partial V}{\partial S} SE[(J - 1)] = 0. \quad (29.3)$$

$E[\cdot]$ is the expectation taken over the jump size J, which can also be written

$$E[x] = \int xP(J)\,dJ,$$

where $P(J)$ is the probability density function for the jump size.

This is a pricing equation for an option when there are jumps in the underlying. The important point to note about this equation that makes it different from others we have derived is its non-local nature. That is, the equation links together option values at distant S values, instead of just containing local derivatives. Naturally, the value of an option here and now depends on the prices to which it can instantaneously jump. When $\lambda = 0$ the equation reduces to the Black–Scholes equation.

There is a simple solution of this equation in the special case that the logarithm of J is Normally distributed. If the logarithm of J is Normally distributed with standard deviation σ' and if we write

$$k = E[J - 1]$$

then the price of a European non-path-dependent option can be written as

$$\sum_{n=0}^{\infty} \frac{1}{n!} e^{-\lambda'(T-t)} (\lambda'(T-t))^n V_{BS}(S, t; \sigma_n, r_n).$$

In the above

$$\lambda' = \lambda(1 + k), \quad \sigma_n^2 = \sigma^2 + \frac{n\sigma'^2}{T-t} \quad \text{and} \quad r_n = r - \lambda k + \frac{n \log(1 + k)}{T - t},$$

and V_{BS} is the Black–Scholes formula for the option value in the absence of jumps. This formula can be interpreted as the sum of individual Black–Scholes values, each of which assumes that there have been n jumps, and they are weighted according to the probability that there will have been n jumps before expiry.

If one does not make the assumption that jumps should not be priced in, then one has to play around with concepts such as the market price of risk.

29.6 HEDGING THE JUMPS

In the above we hedged the diffusive element of the random walk for the underlying. Another possibility is to hedge both the diffusion and jumps as much as we can. For example, we could choose Δ to minimize the variance of the hedged portfolio; after all, this is ultimately what hedging is about.

The change in the value of the portfolio with an arbitrary Δ is, to leading order,

$$d\Pi = \left(\frac{\partial V}{\partial S} - \Delta\right) dS + (-\Delta(J - 1)S + V(JS, t) - V(S, t))\,dq + \cdots.$$

The variance in this change, which is a measure of the risk in the portfolio, is

$$\text{var}[d\Pi] = \left(\frac{\partial V}{\partial S} - \Delta\right)^2 \sigma^2 S^2\,dt + \lambda E[(-\Delta(J - 1)S + V(JS, t) - V(S, t))^2]\,dt + \cdots.$$

$$(29.4)$$

This is minimized by the choice

$$\Delta = \frac{\lambda E[(J - 1)(V(JS, t) - V(S, t))] + \sigma^2 S \dfrac{\partial V}{\partial S}}{\lambda S E[(J - 1)^2] + \sigma^2 S}.$$

(To see this, differentiate (29.4) with respect to Δ and set the resulting expression equal to zero.)

If we value the options as a pure discounted real expectation under this best-hedge strategy then we find that

$$\frac{\partial V}{\partial t} + \tfrac{1}{2}\sigma^2 S^2 \frac{\partial^2 V}{\partial s^2} + S \frac{\partial V}{\partial S}\left(\mu - \frac{\sigma^2}{d}(\mu + \lambda k - r)\right) - rV$$

$$+ \lambda E\left[(V(JS, t) - V(S, t))\left(1 - \frac{J - 1}{d}(\mu + \lambda k - r)\right)\right] = 0,$$

where

$$d = \lambda E[(J - 1)^2] + \sigma^2.$$

When $\lambda = 0$ this collapses to the Black–Scholes equation. At the other extreme, when there is no diffusion, so that $\sigma = 0$, we have

$$\Delta = \frac{E[(J - 1)(V(JS, t) - V(S, t))]}{SE[(J - 1)^2]}$$

and

$$\frac{\partial V}{\partial t} + \mu S \frac{\partial V}{\partial S} - rV + \lambda E\left[(V(JS, t) - V(S, t))\left(1 - \frac{J - 1}{d}(\mu + \lambda k - r)\right)\right] = 0.$$

All of the pricing equations we have seen in this chapter are integro-differential equations. (The integral nature is due to the expectation taken over the jump size.) Because of the convoluted nature of these equations they are candidates for solution by Fourier transform methods.

29.7 HEDGING THE JUMPS AND RISK NEUTRALITY

The above talks about taking expectations, rather than using the usual hedging arguments, to get to a pricing equation. It is kinda possible to set up a perfectly hedged portfolio, as I'll explain briefly now.

Suppose we know the size of the jump, but not its timing. We can construct a portfolio of *two* options and the underlying that is risk-free. We can hedge the jump risk. But, as always, because we will have one equation (no arbitrage) in two unknowns (the values of the two options) we will inevitably end up with a pricing model that contains a market price of jump risk, an unknown function.

If the jump takes one of two values then we set up a hedged portfolio with three options and the underlying. If there are n jump states then we need $n + 1$ options and the underlying for the perfect hedge. If we have a distribution of jump sizes we need a distribution of options.

I won't go into the details, not being a great fan of anything resulting in market prices of risk.

29.8 THE DOWNSIDE OF JUMP-DIFFUSION MODELS

Jump diffusion as described above is unsatisfying. Why bother to delta hedge at all when the portfolio will anyway be exposed to extreme movements? Hedging 'on average' is fine; after all, that is being done whenever hedging is discrete, but after a crash the portfolio change is so dramatic that it makes hedging appear pointless. The other possibility is to examine the worst-case scenario. What is the worst that could happen, crashwise? Assume that this does happen and price it into the contract. This is discussed in depth in Chapter 30.

29.9 JUMP VOLATILITY

In this section I'm going to return to volatility modeling, now that we know what a Poisson process is. In Chapter 26 we saw Brownian motion models for volatility, but now let's model volatility as a jump process.

Perhaps volatility is constant for a while, then randomly jumps to another value. A bit later it jumps back. Let us model volatility as being in one of two states σ^- or $\sigma^+ > \sigma^-$. The jump from lower to higher value will be modeled by a Poisson process with intensity λ^+ and intensity λ^- going the other way. A realization of this process for σ is shown in Figure 29.6.

If we hedge the random movement in S with the underlying, then take *real* expectations, and set the return on the portfolio equal to the risk-free rate, we arrive at

$$\frac{\partial V^+}{\partial t} + \frac{1}{2}\sigma^{+2}S^2\frac{\partial^2 V^+}{\partial S^2} + r\frac{\partial V^+}{\partial S} - rV^+ + \lambda^-(V^- - V^+) = 0,$$

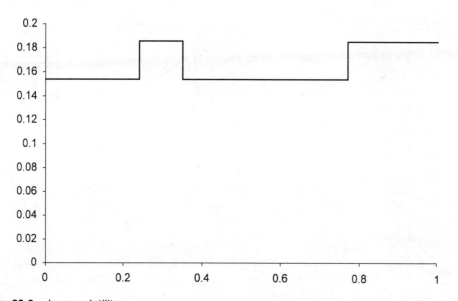

Figure 29.6 Jump volatility.

for the value V^+ of the option when the volatility is σ^+. Similarly, we find that

$$\frac{\partial V^-}{\partial t} + \tfrac{1}{2}\sigma^{-2}S^2\frac{\partial^2 V^-}{\partial S^2} + r\frac{\partial V^-}{\partial S} - rV^- + \lambda^+(V^+ - V^-) = 0$$

for the value V^- when the volatility is σ^-.

29.10 JUMP VOLATILITY WITH DETERMINISTIC DECAY

A more sophisticated jump process for the volatility, which resembles the real behavior of volatility, also contains exponential decay of the volatility after the jump. We can write

$$\sigma(\tau) = \sigma^- + (\sigma^+ - \sigma^-)e^{-\nu\tau},$$

where τ is the time since the last sudden jump in the volatility and ν is a decay parameter. It doesn't actually matter what form this function takes as long as it depends only on τ (and S and t if you want). At any time, governed by a Poisson process with intensity λ, the volatility can jump from its present level to σ^+. A realization of this process for σ is shown in Figure 29.7.

The value of an option is given by $V(S, t, \tau)$, the solution of

$$\frac{\partial V}{\partial t} + \frac{\partial V}{\partial \tau} + \tfrac{1}{2}\sigma(\tau)^2 S^2\frac{\partial^2 V}{\partial S^2} + rS\frac{\partial V}{\partial S} - rV + \lambda(V(S, t, 0) - V(S, t, \tau)) = 0.$$

Note that the time since last jump τ is incremented at the same rate as real time t. (In this, 'value' means that we have delta hedged with the underlying to eliminate risk due to the movement of the asset but we have taken real expectations with respect to the volatility jump.)

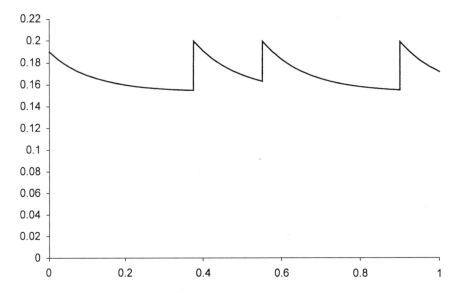

Figure 29.7 Jump volatility with exponential decay.

29.11 **SUMMARY**

Jump diffusion models undoubtedly capture a real phenomenon that is missing from the Black–Scholes model. Yet they are rarely used in practice. There are three main reasons for this: difficulty in parameter estimation, solution, and impossibility of perfect hedging.

In order to use any pricing model one needs to be able to estimate parameters. In the lognormal model there is just the one parameter to estimate. This is just the right number. More than one parameter is too much work, but no parameters to estimate and trading could be done by machine. The jump diffusion model in its simplest form needs an estimate of probability of a jump, measured by λ, and its size, J. This can be made more complicated by having a distribution for J.

The governing equation is no longer a diffusion equation (about the easiest problem to solve numerically), but a difference-diffusion equation. The equation does not just contain local derivatives of the value with respect to its variables, but now links option values at different asset values. The numerical solution of such an equation is certainly not impossible, but is harder than the solution of the basic Black–Scholes equation.

Finally, perfect risk-free hedging is impossible when there are jumps in the underlying. This is because of the non-local nature of the option pricing equation. We have seen two approaches to this hedging, neither of which matches the elegance of the Black–Scholes hedge. The use of a jump-diffusion model acknowledges that one's hedge is less than perfect, which bothers some people. If one sticks to the Black–Scholes model then one can pretend to be hedging perfectly. Of course, the reality of the situation is that there are many reasons why risk-free hedging is impossible, nevertheless the 'eyes wide shut' approach has become market standard.

FURTHER READING

- The original jump-diffusion model in finance was due to Merton (1976).
- An impressive tome on the modeling of extremal events is by Embrechts, Klüppelberg & Mikosch (1997).
- Jump volatility is described by Naik (1993).

CHAPTER 30
crash modeling

In this Chapter...

- how to price contracts in a worst-case scenario when there are crashes in the prices of underlyings
- how to reduce the effect of these crashes on your portfolio; the Platinum Hedge

30.1 INTRODUCTION

Jump diffusion models have two weaknesses: they don't allow you to hedge and the parameters are very hard to measure. Nobody likes a model that tells you that hedging is impossible (even though that may correspond to common sense) and in the classical jump-diffusion model of Merton the best that you can do is a kind of average hedging. It may be quite easy to estimate the impact of a rare event such as a crash, but estimating the probability of that rare event is another matter. In this chapter we discuss a model for pricing and hedging a portfolio of derivatives that takes into account the effect of an extreme movement in the underlying but we will make *no assumptions about the timing of this 'crash' or the probability distribution of its size*, except that we put an upper bound on the latter. This effectively gets around the difficulty of estimating the likelihood of the rare event. The pricing follows from the assumption that the worst scenario actually happens i.e. the size and time of the crash are such as to give the option its worst value. And hedging, delta and static hedging, will continue to play a key role. The optimal static hedge follows from the desire to make the best of this worst value. This, latter, static hedging follows from the desire to optimize a portfolio's value. I also show how to use the model to evaluate the value at risk for a portfolio of options.

30.2 VALUE AT RISK

The true business of a financial institution is to manage risk.

The trader manages 'normal event' risk, where the world operates close to a Black–Scholes one of random walks and dynamic hedging. The institution, however, views its portfolio on a 'big picture' scale and focuses on 'tail events' where liquidity and large jumps are important (Figure 30.1).

Value at Risk (VaR) is a measure of the potential losses due to a movement in underlying markets. It usually has associated with it a timeframe and an estimate of the maximum sudden

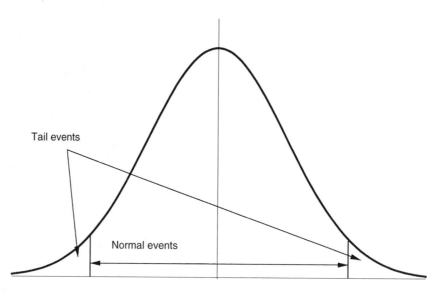

Figure 30.1 'Normal events' and 'tail events'.

change thought likely in the markets. There is also a 'confidence interval'; for example, the daily VaR is $15 million with a degree of confidence of 99%. The details of VaR and its measurement are discussed in Chapter 53.

A more general and more encompassing definition of VaR will give a useful tool to both book runners and senior management. A true measure of the risk in a portfolio will answer the question 'What is the value of any realistic market movement to my portfolio?'

The approach taken here in finding the value at risk for a portfolio is to model the cost to a portfolio of a crash in the underlying. I show how to value the cost of a crash in a worst-case scenario, and also how to find an optimal static hedge to minimize this cost and so reduce the value at risk.

30.3 A SIMPLE EXAMPLE: THE HEDGED CALL

To motivate the problem and model, consider this simple example. You hold a long call position, delta hedged in the Black–Scholes fashion. What is the worst that can happen, in terms of crashes, for the value of your portfolio? One might naively say that a crash is bad for the portfolio, after all, look at the Black–Scholes value for a call as a function of the underlying, the lower the underlying the lower the call value. Wrong. Remember you hold a *hedged* position; look at Figure 30.2 to see the value of the option, the short asset position and the whole portfolio. The last is the bold line in the figure. Observe that the position is currently delta neutral. Also observe that the portfolio's value is currently at its minimum; a sudden fall (or, indeed, a rize) will result in a higher portfolio value; a crash is beneficial. If we are assuming a worst-case scenario, then the worst that could happen is that there is no crash. Changing all the signs to consider a short call position we find that a crash is bad, but how do we find the worst case? If there is going to be one crash of 10% when is the worst time for this to happen? This is the motivation for the model below. Note first that, generally speaking, a positive gamma position benefits from a crash, while a negative gamma position loses.

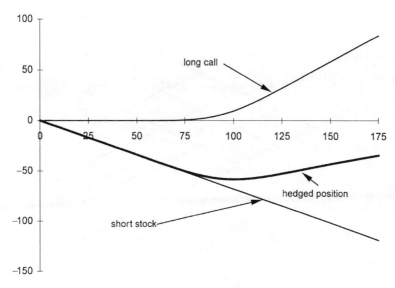

Figure 30.2 The delta-hedged long call portfolio according to Black–Scholes.

30.4 A MATHEMATICAL MODEL FOR A CRASH

The main idea in the following model is simple. We assume that the worst will happen. We value all contracts assuming this and then, unless we are very unlucky and the worst does happen, we will be pleasantly surprized. In this context, 'pleasantly surprized' means that we make more money than we expected. We can draw an important distinction between this model and the models of Chapter 29, the jump diffusion models. In the latter we make bold statements about the frequency and distribution of jumps and finally take expectations to arrive at a value for a derivative. Here *we make no statements about the distribution of either the jump size or when it will happen*. At most the number of jumps is limited. Finally, we examine the worst-case scenario so that no expectations are taken.

I will model the underlying asset price behavior as the classical binomial tree, but with the addition of a third state, corresponding to a large movement in the asset. So, really, we have a trinomial walk but with the lowest branch being a significantly more distant asset value. The up and down diffusive branches are modeled in the usual binomial fashion. For simplicity, assume that the crash, when it happens, is from S to $(1 - k)S$ with k given; this assumption will later be dropped to allow k to cover a range of values, or even to allow a dramatic rise in the value of the underlying. Introduce the subscript 1 to denote values of the option before the crash i.e. with one crash allowed, and 0 to denote values after. Thus V_0 is the value of the option position after the crash. This is a function of S and t and since, I am only permitting one crash, V_0 must be exactly the Black–Scholes option value.

As shown in Figure 30.3, if the underlying asset starts at value S (point O) it can go to one of three values: uS, if the asset rises; vS, if the asset falls; $(1 - k)S$, if there is a crash. These three points are denoted by A, B and C respectively. The values for uS and vS are chosen in the usual manner for the traditional binomial model; see Chapter 12.

Before the asset price moves, we set up a 'hedged' portfolio, consisting of our option position and $-\Delta$ of the underlying asset. At this time our option has value V_1. We must find both an optimal Δ and then V_1. The hedged portfolio is shown in Figure 30.4.

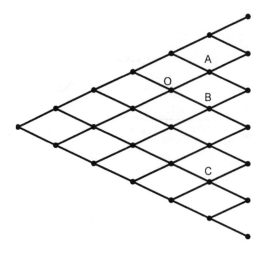

Figure 30.3 The tree structure.

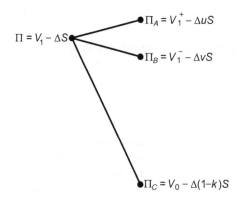

Figure 30.4 The tree and portfolio values.

A time δt later the asset value has moved to one of the three states, A, B or C, and at the same time the option value becomes either V_1^+ (for state A), V_1^- (for state B) or the Black–Scholes value V_0 (for state C).

The change in the value of the portfolio, between times t and $t + \delta t$ (denoted by $\delta \Pi$) is given by the following expressions for the three possible states:

$$\delta \Pi_A = V_1^+ - \Delta uS + \Delta S - V_1 \quad \text{(diffusive rise)}$$
$$\delta \Pi_B = V_1^- - \Delta vS + \Delta S - V_1 \quad \text{(diffusive fall)}$$
$$\delta \Pi_C = V_0 + \Delta kS - V_1 \quad \text{(crash)}.$$

These three functions are plotted against Δ in Figures 30.5 and 30.6. I will explain the difference between the two figures very shortly. My aim in what follows is to choose the hedge ratio Δ so as to minimize the pessimistic, worst outcome among the three possible.

There are two cases to consider, shown in Figures 30.5 and 30.6. The former, Case I, is when the worst-case scenario is not the crash but the simple diffusive movement of S. In this

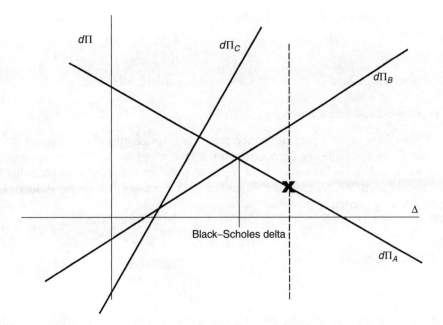

Figure 30.5 Case I: worst case is diffusive motion.

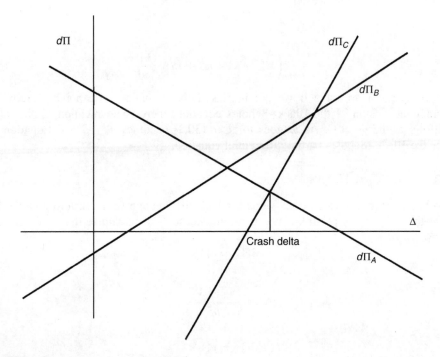

Figure 30.6 Case II: worst case is a crash.

case V_0 is sufficiently large for a crash to be beneficial:

$$V_0 \geq V_1^+ + (S - uS - kS)\frac{V_1^+ - V_1^-}{uS - vS}. \tag{30.1}$$

If V_0 is smaller than this, then the worst scenario is a crash; this is Case II.

30.4.1 Case I: Black–Scholes Hedging

Refer to Figure 30.5. In this figure we see the three lines representing $\delta\Pi$ for each of the moves to A, B and C. Pick a value for the hedge ratio Δ (for example, see the dashed vertical line in Figure 30.5), and determine on which of the three lines lies the worst possible value for $\delta\Pi$ (in the example in the figure, the point is marked by a cross and lies on the A line). Change your value of Δ to maximize this worst value.

In this case the maximal-lowest value for $\delta\Pi$ occurs at the point where

$$\delta\Pi_A = \delta\Pi_B,$$

that is

$$\Delta = \frac{V_1^+ - V_1^-}{uS - vS}. \tag{30.2}$$

This will be recognized as the expression for the hedge ratio in a Black–Scholes world.

Having chosen Δ, we now determine V_1 by setting the return on the portfolio equal to the risk-free interest rate. Thus we set

$$\delta\Pi_A = r\Pi\,\delta t$$

to get

$$V_1 = \frac{1}{1 + r\,\delta t}\left(V_1^+ + (S - uS + rS\,\delta t)\frac{V_1^+ - V_1^-}{uS - vS}\right). \tag{30.3}$$

This is the equation to solve if we are in Case I. Note that it corresponds exactly to the usual binomial version of the Black–Scholes equation, there is no mention of the value of the portfolio at the point C. As δt goes to zero, (30.2) becomes $\partial V/\partial S$ and Equation (30.3) becomes the Black–Scholes partial differential equation.

30.4.2 Case II: Crash Hedging

Refer to Figure 30.6. In this case the value for V_0 is low enough for a crash to give the lowest value for the jump in the portfolio. We therefore choose Δ to maximize this worst case. Thus we choose

$$\delta\Pi_A = \delta\Pi_C,$$

that is,

$$\Delta = \frac{V_0 - V_1^+}{S - uS - kS}. \tag{30.4}$$

Now set

$$\delta\Pi_A = r\Pi\,\delta t$$

to get

$$V_1 = \frac{1}{1 + r\,\delta t}\left(V_0 + S(k + r\,\delta t)\frac{V_0 - V_1^+}{S - uS - kS}\right). \tag{30.5}$$

This is the equation to solve when we are in Case II. Note that this is different from the usual binomial equation, and does not give the Black–Scholes partial differential equation as δt goes to zero (see later in the chapter). Also (30.4) is not the Black–Scholes delta. To appreciate that delta hedging is not necessarily optimal, consider the simple example of the butterfly spread. If the butterfly spread is delta hedged on the right 'wing' of the butterfly, where the delta is negative, a large fall in the underlying will result in a large loss from the hedge, whereas the loss in the butterfly spread will be relatively small. This could result in a negative value for a contract, even though its payoff is everywhere positive.

30.5   AN EXAMPLE

All that remains to be done is to solve Equations (30.3) and (30.5) (which one is valid at any asset value and at any point in time depends on whether or not (30.1) is satisfied). This is easily done by working backwards down the tree from expiry in the usual binomial fashion.

As an example, examine the cost of a 15% crash on a portfolio consisting of the call options in Table 30.1

At the moment the portfolio only contains the first two options. Later I will add some of the third option for static hedging, that is when the bid-ask prices will concern us. The volatility of the underlying is 17.5% and the risk-free interest rate is 6%.

The solution to the problem is shown in Figure 30.7. Observe how the value of the portfolio assuming the worst (21.2 when the spot is 100) is lower than the Black–Scholes value (30.5).

Table 30.1   Available contracts.

Strike	Expiry	Bid	Ask	Quantity
100	75 days			−3
80	75 days			2
90	75 days	11.2	12	0

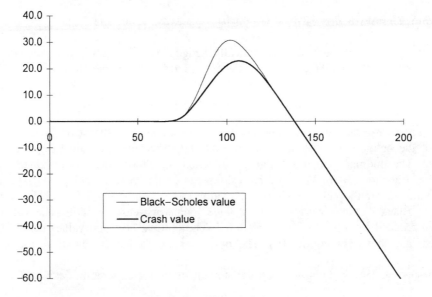

Figure 30.7   Example showing crash value and Black–Scholes value.

This is especially clear where the portfolio's gamma is highly negative. This is because when the gamma is positive, a crash is beneficial to the portfolio's value. When the gamma is close to zero, the delta hedge is very accurate and the option is insensitive to a crash. If the asset price is currently 100, the difference between the before and after portfolio values is $30.5 - 21.2 = 9.3$. This is the 'Value at Risk' under the worst-case scenario.

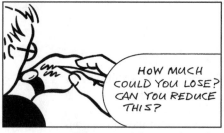

30.6 **OPTIMAL STATIC HEDGING: VAR REDUCTION**

The 9.3 value at risk is due to the negative gamma around the asset price of 100. An obvious hedging strategy that will offset some of this risk is to buy some positive gamma as a 'static' hedge. In other words, we should buy an option or options having a counterbalancing effect on the value at risk. We are willing to pay a premium for such an option. We may even pay more than the Black–Scholes fair value for such a static hedge because of the extra benefit that it gives us in reducing our exposure to a crash. Moreover, if we have a choice of contracts with which to statically hedge we should buy the most 'efficient' one. To see what this means consider the above example in more detail.

Recall that the value of the initial portfolio under the worst-case scenario is 21.2. How many of the 90 calls should we buy (for 12) or sell (for 11.2) to make the best of this scenario? Suppose that we buy λ of these calls. We will now find the optimal value for λ.

The cost of this hedge is

$$\lambda C(\lambda)$$

where $C(\lambda)$ is 12 if λ is positive and 11.2 otherwise. Now solve Equations (30.3) and (30.5) with the final total payoffs

$$V_0(S, T) = V_1(S, T) = 2\max(S - 80, 0) - 3\max(S - 100, 0) + \lambda \max(S - 90, 0).$$

This is the payoff at time T for the statically hedged portfolio. The *marginal* value of the original portfolio (that is, the portfolio of the 80 and 100 calls) is therefore

$$V_1(100, 0) - \lambda C(\lambda) \tag{30.6}$$

i.e. the worst-case value for the new portfolio less the cost of the static hedge. The arguments of the before-crash option value are 100 and 0 because they are today's asset value and date. The optimality in this hedge arises when we choose the quantity λ to maximize the value; Equation (30.6). With the bid-ask spread in the 90 calls being 11.2–12, we find that buying 3.5 of the calls maximizes Equation (30.6). The value of the new portfolio is 70.7 in a Black–Scholes world and 65.0 under our worst-case scenario. The value at risk has been reduced from 9.3 to $70.7 - 65 = 5.7$. The optimal portfolio values before and after the crash are shown in Figure 30.8. The optimal static hedge is known as the **Platinum Hedge**.

The issues of static hedging and optimal static hedging are covered in detail in Chapter 32.

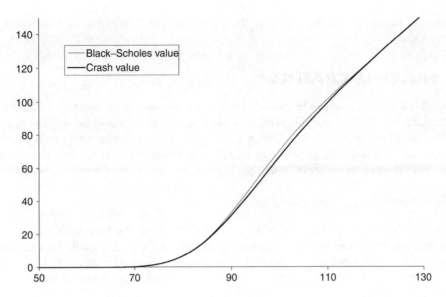

Figure 30.8 Optimally-hedged portfolio, before and after crash.

30.7 **CONTINUOUS-TIME LIMIT**

If we let $\delta t \to 0$ in Equations (30.1), (30.2), (30.3), (30.4) and (30.5) we find that the Black–Scholes equation is still satisfied by $V_1(S, t)$ but we also have the constraint

$$V_1(S, t) - kS\frac{\partial V_1}{\partial S}(S, t) \le V_0(S(1 - k), t)$$

Such a problem is similar in principle to the American option valuation problem, where we also saw a constraint on the derivative's value. Here the constraint is more complicated. To this we must add the condition that the first derivative of V_1 must be continuous for $t < T$.

30.8 **A RANGE FOR THE CRASH**

In the above model, the crash has been specified as taking a certain value. Only the timing was left to be determined for the worst-case scenario. It is simple to allow the crash to cover a range of values, so that S goes to $(1 - k)S$ where

$$k^- \le k \le k^+.$$

A negative k^- corresponds to a rise in the asset.

In the discrete setting the worst-case option value is given by

$$V_1 = \min_{k^- \le k \le k^+}\left(\frac{1}{1 + r\,\delta t}\left(V_0 + S(k + r\,\delta t)\frac{V_0 - V_1^+}{S - uS - kS}\right)\right).$$

This contains the min($\cdot$) because we want the worst-case crash. When a crash is beneficial we still have (30.3).

30.9 MULTIPLE CRASHES

The model described above can be extended in many ways, one of the most important of which is to consider the effect of multiple crashes. I describe two possibilities below. The first puts a constraint on the total number of crashes in a time period; there can be three crashes within the horizon of one year, say. The second puts a limit on the time between crashes; there cannot be another crash if there was a crash in the last six months, say.

30.9.1 Limiting the Total Number of Crashes

We will allow up to N crashes. We make no statement about the time these occur. We will assume that the crash size is given, allowing a fall of $k\%$. This can easily be extended to a range of sizes, as described above. Introduce the functions $V_i(S, t)$ with $i = 0, 1, \ldots, N$, such that V_i is the value of the option with i more crashes still allowed. Thus, as before, V_0 is the Black–Scholes value.

We must now solve N coupled equations of the following form. If

$$V_{i-1} \geq V_i^+ + (S - uS - kS)\frac{V_i^+ - V_i^-}{uS - vS}$$

then we are in Case I, a crash is beneficial and is assumed not to happen. In this case we have

$$V_i = \frac{1}{1 + r\,\delta t}\left(V_i^+ + (S - uS + rS\,\delta t)\frac{V_i^+ - V_i^-}{uS - vS}\right).$$

Otherwise a crash is bad for the hedged option; this is Case II. We then have

$$V_i = \frac{1}{1 + r\,\delta t}\left(V_{i-1} + S(k + r\,\delta t)\frac{V_{i-1} - V_i^+}{S - uS - kS}\right).$$

In continuous time the equations become

$$\frac{\partial V_i}{\partial t} + \tfrac{1}{2}\sigma^2 S^2 \frac{\partial^2 V_i}{\partial S^2} + rS\frac{\partial V_i}{\partial S} - rV_i = 0$$

for $i = 0, \ldots, N$, subject to

$$V_i(S, t) - kS\frac{\partial V_i}{\partial S}(S, t) \leq V_{i-1}(S(1 - k), t)$$

for $i = 1, \ldots, N$. Each of the V_i has the same final condition, representing the payoff at expiry.

30.9.2 Limiting the Frequency of Crashes

Finally, we model a situation where the time between crashes is limited; if there was a crash less than a time ω ago another is not allowed.

This is slightly harder than the N-crash model and we have to introduce a new variable τ measuring the time since the last crash. We now have two functions to consider, $V_c(S, t)$ and

$V_n(S, t, \tau)$. The former is the worst-case option value when a crash is allowed (and therefore we don't need to know how long it has been since the last crash) and the latter is the worst-case option value when a crash is not yet allowed.

The governing equations, which are derived in the same way as the original crash model, are, in continuous time, simply

$$\frac{\partial V_c}{\partial t} + \tfrac{1}{2}\sigma^2 S^2 \frac{\partial^2 V_c}{\partial S^2} + rS\frac{\partial V_c}{\partial S} - rV_c = 0$$

subject to

$$V_c(S, t) - kS\frac{\partial V_c}{\partial S}(S, t) \leq V_n(S(1-k), t, 0),$$

and for $V_n(S, t, \tau)$,

$$\frac{\partial V_n}{\partial t} + \frac{\partial V_n}{\partial \tau} + \tfrac{1}{2}\sigma^2 S^2 \frac{\partial^2 V_n}{\partial S^2} + rS\frac{\partial V_n}{\partial S} - rV_n = 0$$

with the condition

$$V_n(S, t, \omega) = V_c(S, t).$$

Observe how the time τ and real time t increase together in the equation when a crash is not allowed.

30.10 CRASHES IN A MULTI-ASSET WORLD

When we have a portfolio of options with many underlyings we can still examine the worst-case scenario, but we have two choices. Either (a) we allow a crash to happen in any underlyings completely independently of all other underlyings or (b) we assume some relationship between the assets during a crash. Clearly the latter is not as bad a worst case as the former. It is also easier to write down, so we will look at that model only. Assuming that all assets fall simultaneously by the same percentage k we have

$$V_1(S_1, \ldots, S_n, t) - k\sum_{i=1}^{n} S_i \frac{\partial V_1}{\partial S_i}(S_1, \ldots, S_n, t) \leq V_0((1-k)S_1, \ldots, (1-k)S_n, t, t).$$

We will pursue stock market crashes in CrashMetrics, see Chapter 57.

30.11 FIXED AND FLOATING EXCHANGE RATES

Many currencies are linked directly to the currency in another country. Some countries have their currency linked to the US dollar; the Argentine Peso is tied at a rate of one to the dollar. The European Monetary Union is another example of linked exchange rates.

Once an exchange rate is fixed in this way the issue of fluctuating rates becomes a credit risk issue. All being well, the exchange rate will stay constant with all the advantages of stability that this brings with it. If economic conditions in the two countries start to diverge then the exchange rate will come under pressure. In Figure 30.9 is a plot of the possible exchange rate, showing a fixed rate for a while, followed by a sudden discontinuous drop and then a random

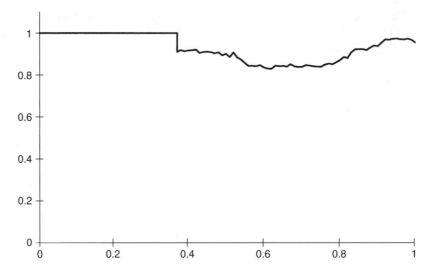

Figure 30.9 Decoupling of an exchange rate.

fluctuation. How can we model derivatives of the exchange rate? The models of this chapter are ideally suited to this situation.

I'm going to ignore interest rates in the following. This is because of the complex issues this would otherwise raise. For example, the pressure on the exchange rate and the decoupling of the currencies would be accompanied by changing interest rates. This can be modeled but would distract us from the application of the crash model.

While the exchange rate is fixed, before the 'crash', the price of an option, $V_1(S, t)$, satisfies

$$\frac{\partial V_1}{\partial t} = 0,$$

since I am assuming zero interest rates. Here S is the exchange rate. After the 'crash' we have

$$\frac{\partial V_0}{\partial t} + \tfrac{1}{2}\sigma^2 S^2 \frac{\partial^2 V_0}{\partial S^2} = 0.$$

(But with what volatility?) This is just the Black–Scholes model, with the relevant Black–Scholes value for the particular option payoff.

The worst-case crash model is now almost directly applicable. I will leave the details to the reader.

30.12 **SUMMARY**

I have presented a model for the effect of an extreme market movement on the value of portfolios of derivative products. This is an alternative way of looking at value at risk. I have shown how to employ static hedging to minimize this VaR. In conclusion, note that the above is not a jump-diffusion model since I have deliberately not specified any probability distribution for the size or the timing of the jump: we model the worst-case scenario.

We have examined several possible models of crashes, of increasing complexity. One further thought is that we have not allowed for the rise in volatility that accompanies crashes. This can be done with ease. There is no reason why the after-crash model (V_0 in the simplest case above) cannot have a different volatility from the before-crash model. Of particular interest is the final model where there is a minimum time between crashes; we could easily have the volatility post-crash being a decaying function of the time since the crash occured, τ. This would involve no extra computational effort.

FURTHER READING

- For further details about crash modeling see Hua & Wilmott (1997) and Hua (1997).
- Derman & Zou (1997) describe and model the behavior of implied volatility after a large move in an index.

CHAPTER 31
speculating with options

In this Chapter...

- how to find the present value of the *real* expected payoff (and why you should want to know this)
- several ways to model asset price drift
- how to optimally close your option position
- when and when not to hedge

31.1 INTRODUCTION

Almost everything I have shown you so far is about finding pricing equations for derivative contracts using hedging arguments. There is a very powerful reason why such arguments should give option prices in the market regardless of whether or not investors actually hedge the option. The reason is simple: if the option is not priced at the Black–Scholes fair value and hedging is possible then either

- the option price is too low, in which case someone will buy it, hedge away the risk and make a riskless profit or
- the option price is too high, in which case someone will artificially replicate the payoff and charge less for this contract.

Both of these are examples of market inefficiencies which would disappear quickly in practice (to a greater or lesser extent).

However, a few minutes spent on a typical trading floor will convince you that option contracts are often bought for speculation and that the value of an option to a speculator and the choice of expiry date and strike price depend strongly on his **market view**, something that is irrelevant in the Black–Scholes world. Moreover, many OTC and other contracts are used to offset risks outside the market, and if every derivative is hedged by both writer and purchaser then the whole thing is pointless, or at best a series of hedging and modeling competitions. In this chapter I show possible ways in which a speculator can choose an option contract so as to profit from his view if it turns out to be correct.

Throughout this chapter I use the word 'value' to mean the *worth of an option to a speculator* and this will not necessarily be the Black–Scholes 'fair' value. To get around the obvious criticism that what follows is nonsense in a complete market, in which delta hedging is possible, I'm going to take the point of view that the investor can value a contract at other than the Black–Scholes value because of market incompleteness, for whatever reason you like. The cause of incompleteness may be transaction costs, restrictions on sales or purchases, uncertain parameters, general model errors etc. Thus any 'arbitrage' that you may see because values are different from Black–Scholes cannot be exploited by our investor. He is going to buy a contract as an investment, a risky investment.

One of the motivations for this chapter is the observation that in the classical Black–Scholes theory the drift of the asset plays no part. Two people will agree on the price of an option even if they differ wildly in their estimates for μ — indeed, they may not even have estimates for μ — as long as they agree on the other parameters such as the volatility.

31.2 A SIMPLE MODEL FOR THE VALUE OF AN OPTION TO A SPECULATOR

Imagine that you hold very strong views about the future behavior of a particular stock: you have an estimate of the volatility and you believe that for the next twelve months the stock will have an upwards drift of 15%, say, and that this is much greater than the risk-free rate. How can you benefit from this view if it is correct? One simple way is to follow the principles of Modern Portfolio Theory, Chapter 51, and buy a call, just out of the money, maybe, perhaps with an expiry of about one year because this may have an appealing risk/return profile (also see later in this chapter). You might choose an out of-the-money option because it is cheap, but not so far out of the money that it is likely to expire worthless. You might choose a twelve-month contract to benefit from as much of the asset price rise as possible. Now imagine it is a couple of months later, and the asset has fallen 5% instead of rising. What do you do? Maybe, the stock *did* have a drift of 15% but, since we only see one realization of the random walk, the volatility caused the drop. Alternatively, maybe you were wrong about the 15%.

This example illustrates several points:

- How do you determine the future parameters for the stock: the drift and volatility?
- How do you subsequently know whether you were right?
- And if not, what should you do about it?
- If you have a good model for the future random behavior of a stock, how can you use it to measure an option's value *to you*?
- Which option do you buy?

A thorough answer to the first of these questions is outside the scope of this book; but I have made suggestions in that direction. The other four questions fall more into the area of modeling and we discuss them here.

31.2.1 The Present Value of Expected Payoff

Speculation is the opposite side of the coin to hedging. The speculator is taking risks and hopes to profit from this risk by an amount greater than the risk-free rate. When gambling like this, it is natural to ask what is your expected profit. In option terms, this means we would ask what is the present value of the expected payoff. I say 'present value' because the payoff, if it comes, will be at expiry (assuming that we do not exercise early if the option is American, or close the position).

If the asset price random walk is

$$dS = \mu S \, dt + \sigma S \, dX$$

and the option has a payoff of

$$\max(S - E, 0)$$

at time T then the *present value of the expected payoff*[1] satisfies

$$\frac{\partial V}{\partial t} + \tfrac{1}{2}\sigma^2 S^2 \frac{\partial^2 V}{\partial S^2} + \mu S \frac{\partial V}{\partial S} - rV = 0$$

with

$$V(S, T) = \max(S - E, 0).$$

This is related to the ideas in Chapter 10 for transition density functions.

This equation differs from the Black–Scholes equation in having a μ instead of an r in front of the delta term. Seemingly trivial, this difference is of fundamental importance. The replacement of μ by r is the basis of 'risk-neutral valuation'. The absence of μ from the Black–Scholes option pricing equations means that the only asset price parameters needed are the volatility and the dividend structure. For the above equation, however, we need to know the drift rate μ in order to calculate the expected payoff.

Let us return to the earlier example, and fill in some details. Suppose we buy a call, struck at 100, with an expiry of one year, the underlying has a volatility of 20% and, we believe, a drift of 15%. The interest rate is 5%. In Figure 31.1 is shown the Black–Scholes value of the option (the lower curve) together with the present value of the *real* expected payoff (the upper curve) both plotted against S. These two curves would be identical if μ and r were equal. The lower curve is a risk-neutral expectation and the upper curve a real expectation.

It is easy to show, and financially obvious, that if $\mu > r$ then the present value of the expected payoff for a call is greater than the Black–Scholes fair value. In other words, if we expect the stock to drift upwards with a rate higher than the interest rate then we expect to make more money than if we had invested in bonds. Here the word 'expect' is of paramount importance: there is nothing certain about this investment. The greater expected return is compensation for the greater risk of speculation, absent in a hedged portfolio. Nevertheless, for the speculator the expected payoff is an important factor in deciding which option to buy. The other important factor is a measure of the risk; when we know the risk and the return for an option we are in a position to apply ideas from Modern Portfolio Theory, Chapter 51, for example.

31.2.2 Standard Deviation

As I have said, the return from an unhedged option position is uncertain. That uncertainty can be measured by the standard deviation of the return about its mean level.[2] This standard

---

[1] This is not dissimilar to how options used to be priced in the 1960s before Black and Scholes.

[2] This is not entirely satisfactory for many products, especially those with an non-unimodal payoff distribution.

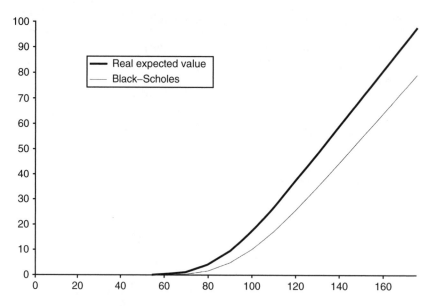

Figure 31.1 The Black–Scholes value of a call option (the lower curve) and the present value of the expected payoff (the upper curve). See text for data.

deviation can be determined as follows. Suppose that today's date is $t = 0$, and introduce the function $G(S, t)$ as the expected value of the square of the present value of the payoff. This function, being an expectation, satisfies the backward Kolmogorov equation (see Chapter 10):

$$\frac{\partial G}{\partial t} + \tfrac{1}{2}\sigma^2 S^2 \frac{\partial^2 G}{\partial S^2} + \mu S \frac{\partial G}{\partial S} = 0 \tag{31.1}$$

(which, of course, is simply the Black–Scholes equation with no discounting and with μ instead of r).[3] The final condition is

$$G(S, T) = (e^{-rT} \max(S - E, 0))^2.$$

Then from its definition, the standard deviation today is given by

$$\sqrt{G(S, 0) - (V(S, 0))^2}.$$

In Figure 31.2 is shown this standard deviation plotted against S, using the same data as in Figure 31.1.

A natural way to view these results is in a risk/reward plot. In this model the reward is the logarithm of the ratio of the present value of the expected payoff to the cost of the option, divided by the time to expiry of the option. But how much will the option cost? A not unreasonable assumption is that the rest of the market is using Black–Scholes to value the option so we shall assume that the cost of the option is simply the Black–Scholes value V_{BS}. The risk is the standard deviation, again scaled with the Black–Scholes value of the option and with the

---

[3] The astute reader will notice that there is no $-rG$ in this 'present value' calculation; we have absorbed this into the final condition. Read on.

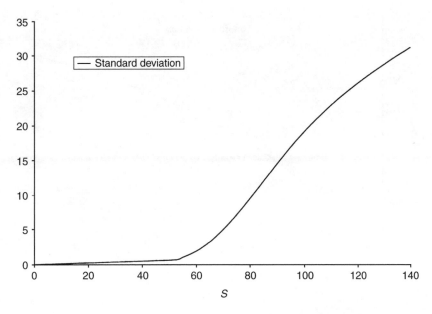

Figure 31.2 The standard deviation of the present value of the expected payoff for a call option. See text for data.

square root of the time to expiry. Thus we define

$$\text{Reward} = \frac{1}{T} \log \left(\frac{V(S, 0)}{V_{BS}(S, 0)} \right)$$

and

$$\text{Risk} = \frac{\sqrt{G(S, 0) - (V(S, 0))^2}}{V_{BS}(S, 0) \sqrt{T}}.$$

These definitions, in particular the scalings with time, have been chosen to tie in with the traditional measures of reward (or return) and risk.

In Figure 31.3 we show a plot of this reward versus the risk for a call option and the same data as the previous two figures. In Figure 31.4 we show the same plot but for a put option and with the same data. Observe how, with our data assumptions, the put option has decreasing return for increasing risk. Obviously, in this case, call options are more attractive.

(In the above we have considered only the standard deviation as a measure of risk. This measure attaches as much weight to a better than average outcome as it does to a worse outcome. A better choice as a measure of risk may be a one-sided estimate of only the downside risk; I leave the formulation of this problem as an exercise for the reader.)

31.3 MORE SOPHISTICATED MODELS FOR THE RETURN ON AN ASSET

Despite modeling the 'value' of an option to a speculator, we know from experience that as time progresses we are likely to change our view of the market and have to make a decision

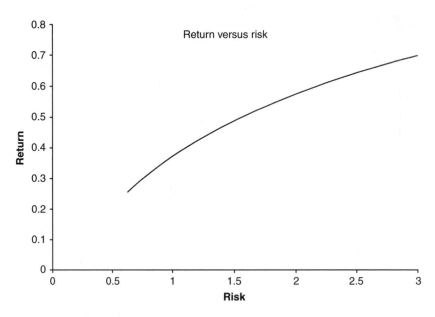

Figure 31.3 The scaled present value of expected payoff versus the scaled standard deviation for a call option. See text for details and data.

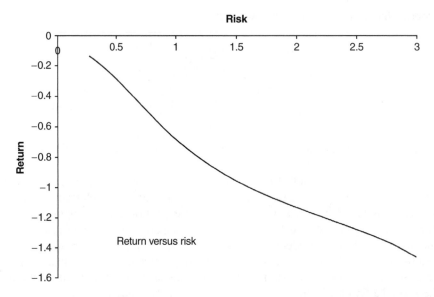

Figure 31.4 The scaled present value of expected payoff versus the scaled standard deviation for a put option. See text for details and data.

regarding the success, or otherwise, of our position: perhaps six months into the life of our call option we decide that we were wrong about the direction of the market and we may therefore close our position. Is there any way in which we can build into our model of the market, *a priori*, our experience that market conditions change?

I am now going to present two models for a randomly varying drift rate. The first assumes that the drift rate follows some stochastic differential equation. The second model assumes that the drift can be in one of two states: high drift or low drift. The drift jumps randomly between these two states.

31.3.1 Diffusive Drift

The asset price still satisfies

$$dS = \mu S \, dt + \sigma S \, dX_1.$$

Now assuming that the drift rate μ satisfies

$$d\mu = \eta(S, \mu, t) \, dt + v(S, \mu, t) \, dX_2$$

then the present value of the expected return $V(S, \mu, t)$ satisfies the two-factor partial differential equation

$$\frac{\partial V}{\partial t} + \tfrac{1}{2}\sigma^2 S^2 \frac{\partial^2 V}{\partial S^2} + \rho\sigma v S \frac{\partial^2 V}{\partial S \partial \mu} + \tfrac{1}{2}v^2 \frac{\partial^2 V}{\partial \mu^2} + \mu S \frac{\partial V}{\partial S} + \eta \frac{\partial V}{\partial \mu} - rV = 0$$

where ρ is the correlation coefficient between the two random walks.[4]

In Figure 31.5 I show the value of a call option for the model

$$d\mu = (a - b\mu) \, dt + \beta \, dX_2.$$

In this figure I have used $a = 0.3$, $b = 3$ and $\beta = 0.1$ with $\rho = 0$. The option is a call struck at 100 with an expiry of one year and a volatility of 20%. The interest rate is 5%. When the asset is 100 and the drift is zero the option value is 11.64.

The standard deviation of these payoffs may be found in a similar manner to that above for constant drift. The relevant equation is simply the two-factor version of (31.1).

31.3.2 Jump Drift

The above model for diffusive drift is perhaps unnecessarily complicated for describing the market view of a typical trader. A simpler model, which I describe now, allows the drift to be in one of two states, either high or low. The asset random walk is allowed to jump from one state to the other. For example, we believe that in the short term the asset in question will have a drift of 15%. However, this may only last for six months. If the drift does change then it will drop to a level of 0%. In this example, call options look appealing in the short term, but we may not want to buy longer-dated calls since we will probably suffer when the downturn arrives. We assume that the two states have constant drift and that the jump from one state to another is governed by a Poisson process. In other words, the probability of changing from a

---

[4] At this point had we been using a Black–Scholes hedging argument we would have arrived at this equation but with r as the coefficient of the $\partial V/\partial S$ term. With final conditions independent of μ we would then find that $\partial V/\partial \mu = 0$ everywhere and thus that V satisfies the Black–Scholes equation. Thus even with stochastic drift the Black–Scholes option value is independent of μ. There are some technical details, concerning the 'support' of the real and risk-neutral random walks, that must be satisfied. Simply put, the range of values that can be attained by the two random walks must be the same.

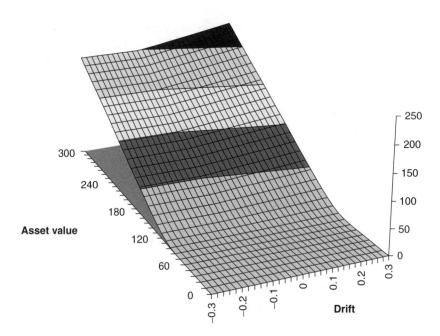

Figure 31.5 The value of a call option when the drift is diffusive. See text for details and data.

high drift to a low drift in a period dt is $\lambda^- dt$, and the probability of changing from a low drift to a high drift is $\lambda^+ dt$.

We can value options to a speculator by again looking at the present value of the expected payoff. First, we must introduce the notation. Let $V^+(S, t)$ be the option 'value' when the asset has a drift of μ^+. Similarly, the option has value $V^-(S, t)$ when the asset has a drift of $\mu^- < \mu^+$. We can easily examine expected payoffs by considering how the option value changes at each timestep.

At time t the asset value is S and we assume without loss of generality that it is in the higher drift state, with a drift of μ^+. At the later time $t + dt$ the asset price changes to $S + dS$ and at the same time the drift rate may jump to μ^- but only with a probability of $\lambda^- dt$.

With the real expected return equal to the risk-free rate we arrive at the following equation:

$$\frac{\partial V^+}{\partial t} + \frac{1}{2}\sigma^2 S^2 \frac{\partial^2 V^+}{\partial S^2} + \mu^+ S \frac{\partial V^+}{\partial S} - rV^+ + \lambda^-(V^- - V^+) = 0.$$

Similarly, we find that

$$\frac{\partial V^-}{\partial t} + \frac{1}{2}\sigma^2 S^2 \frac{\partial^2 V^-}{\partial S^2} + \mu^- S \frac{\partial V^-}{\partial S} - rV^- + \lambda^+(V^+ - V^-) = 0$$

when the drift starts off in the lower state. The final conditions for these equations are, for example,

$$V^+(S, T) = V^-(S, T) = \max(S - E, 0)$$

for a call option.

The standard deviation also takes two forms, depending on which state the drift is in. We must solve

$$\frac{\partial G^+}{\partial t} + \frac{1}{2}\sigma^2 S^2 \frac{\partial^2 G^+}{\partial S^2} + \mu^+ S \frac{\partial G^+}{\partial S} + \lambda^- (G^- - G^+) = 0$$

and

$$\frac{\partial G^-}{\partial t} + \frac{1}{2}\sigma^2 S^2 \frac{\partial^2 G^-}{\partial S^2} + \mu^- S \frac{\partial G^-}{\partial S} + \lambda^+ (G^+ - G^-) = 0$$

with

$$G^+(S, T) = G^-(S, T) = (e^{-rT} \max(S - E, 0))^2.$$

Then the standard deviation at time $t = 0$ is given by

$$\sqrt{G^\pm(S, 0) - (V^\pm(S, 0))^2},$$

depending on whether the drift is in the high or low drift state at that time.[5]

In Figure 31.6 we show the value of a call option against asset price. The top curve shows the option value when the asset is in the high drift state, and the bottom curve is when the asset is in the low drift state. Between the two, the bold curve, is the Black–Scholes fair value. The option is a call struck at 100 with one year to expiry. The underlying has a volatility of 20% and the interest rate is 5%. The high drift is 15% and the low drift is zero. The intensity of the Poisson process taking the asset from high to low is 1 and 0.5 going from low to high. In this example it is easier for the asset to sink to the low drift state than to recover, and this is why the expected value in the low drift case is below Black–Scholes.

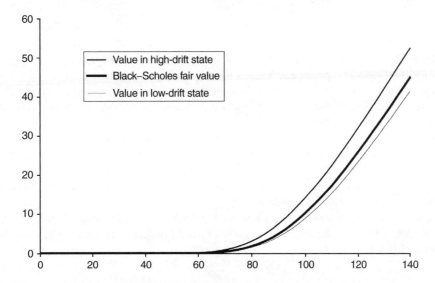

Figure 31.6 The value of a call option when the drift jumps between two levels. See text for details and data.

---

[5] How do we know what state we are in? That's a statistical question. Alternatively, we could say that if the underlying has fallen by a certain amount this means a change of state; this is also easy to model.

31.4 **EARLY CLOSING**

GET OUT WHILE YOU CAN

So far we have modeled the value to a speculator of an option on which he has a strong view on the drift rate and which he will hold until expiry. The diffusive drift and the jump drift models allow for the possibility that the drift leaves its advantageous state and changes to a disadvantageous state. In our example of Section 31.2, the asset drift was initially high, but has fallen to below the risk-free rate, and our call option no longer seems like such a good bet. What can we do? Although our models price this into the option's value, they do not allow for the obvious trader reaction: sell the option and close the position.

Within the present models it is natural for the trader to close his position if the value to him falls below the market value. Why hold onto a position in which you expect to make a loss? He may want to sell earlier than this but we use this very simple constraint as an example. We also assume that the market value of the option is simply the Black–Scholes value, with the same risk-free rate of interest and volatility; again, this could be generalized.

Having seen the mathematical analysis for American options in Chapter 9, it is obvious that our early-closing problem is identical in spirit: take either the diffusion model or the jump model for the drift and impose the constraint that the option value must always be greater than the Black–Scholes value and that the first derivatives must be continuous. In the diffusion model this amounts to

$$V(S, \mu, t) \geq V_{BS}(S, t) \quad \text{and} \quad \frac{\partial V}{\partial S} \text{ continuous.}$$

For the jump model we have

$$V^+(S, t) \geq V_{BS}(S, t), \quad V^-(S, t) \geq V_{BS}(S, t) \quad \text{and} \quad \frac{\partial V^\pm}{\partial S} \text{ continuous.}$$

In Figure 31.7 is shown the value of the option against the underlying and the drift for the diffusive drift model. The parameter values are as above. The option value at an asset value of 100 and a drift of zero is now 11.84. The extra value comes from the ability to close the position. In Figure 31.8 is plotted the difference between the option values with and without the ability to close the position.

In Figure 31.9 is shown the value of the option against the underlying in the two-state drift model. This two-drift state model is particularly interesting because the option value in the low-drift state is the same as Black–Scholes. The interpretation of this is that you should sell the option as soon as you believe that the drift is in the low state. The extra value in being able to sell the position is plotted in Figure 31.10.

The problem for the standard deviation is slightly more complicated when there is early closing because of the appearance of the early-closing boundary. For example, with the two-state drift model we have

$$\frac{\partial G^+}{\partial t} + \tfrac{1}{2}\sigma^2 S^2 \frac{\partial^2 G^+}{\partial S^2} + \mu^+ S \frac{\partial G^+}{\partial S} + \lambda^-(G^- - G^+) = 0$$

$$G^+(S, T) = (e^{-rT} \max(S - E, 0))^2$$

and

$$G^+(S_e(t), t) = (e^{-rt} V^+(S_e(t), t))^2.$$

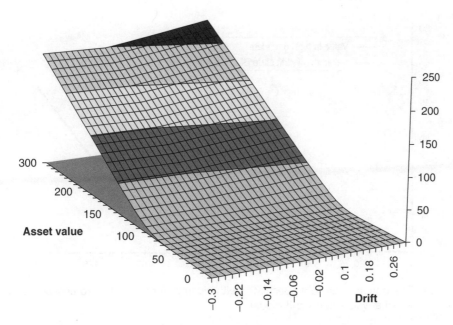

Figure 31.7   The option value to a speculator with closure of the position to ensure that it never falls below the market value: diffusive drift model.

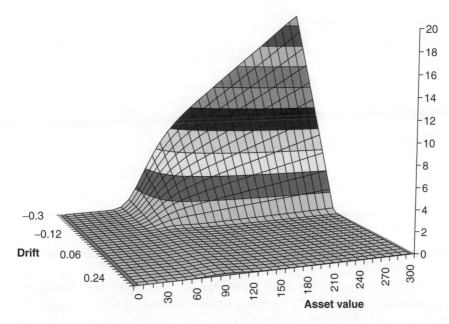

Figure 31.8   The added value of being able to close the position: diffusive drift model.

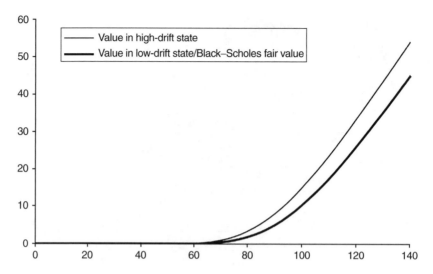

Figure 31.9 The option value to a speculator with closure of the position to ensure that it never falls below the market value: jump drift model.

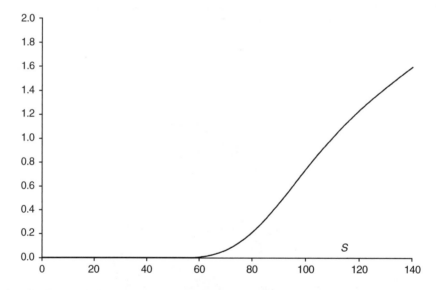

Figure 31.10 The added value of being able to close the position: two-state model.

Here $S_e(t)$ is the position of the early-closing boundary given by the solution of the free boundary problem for V^+ and V^-. There is a similar equation for G^-.

Again, the standard deviations are given by

$$\sqrt{G^\pm(S, 0) - V^\pm(S, 0)^2}.$$

The above illustrates the importance of not only having a model for the underlying but also, when relevant, having a model for the pricing mechanism *of the market*. This is relevant if one expects or hopes to sell the contract before its expiry, since one must sell it for its market value and not one's theoretical value. One of the benefits inherent in this is that having a model for the market value enables you to make even more money than simply having a model for the underlying. This gives you the option of either holding the contract to expiry or, if the market value rises higher than your theoretical value, closing out the position. The value of the contract to you, under this scenario, cannot be less than your theoretical value without the option to close the position. This principle can easily be extended. For example, suppose you have a model for the market's perception of volatility; you can use this to give a lower bound to the contract in question, regardless of the validity of the market's view.

In the above framework for valuing contracts it is obvious that some contracts are more appealing than others. For an asset with a drift above the risk-free rate calls would be preferred over puts for example. There are timescales associated with changes in the drift rate and these may have some bearing on the choice of expiry dates.

In practice you don't have to liquidate the entire portfolio, you may decide to sell off only part of the portfolio. Having this extra freedom means that you can squeeze even more value from your position. But how can this be done optimally? You should be able to do this, at least theoretically, by the end of this part of the book.

31.5 **TO HEDGE OR NOT TO HEDGE?**

So far in this chapter we have looked at 'valuing' options in the complete absence of hedging. As I have said, speculation is the opposite side of the coin to hedging. Perhaps it is possible to bring these two sides closer together.

In this section we look at the expected values of options, assuming a quite sophisticated hedging strategy, one that is intuitively highly appealing. This strategy is to speculate when we expect to make money and hedge when we would otherwise expect to lose it.

THERE'S A WHOLE SPECTRUM BETWEEN HEDGING AND SPECULATING

We will assume that the drift rate of the asset, μ, is constant and greater than the risk-free interest rate. This is the simplest problem that can be used to introduce the ideas. Extensions to other cases, $\mu < r$, non-constant μ etc. are obvious. With this assumption, our strategy can be modeled mathematically by a portfolio of value Π where

$$\Pi = V(S, t) - \overline{\Delta}S, \tag{31.2}$$

and V is the 'value' of our option. We choose the quantity $\overline{\Delta}$ so as to maximize the expected growth in the value of our portfolio. We also impose the constraints on the value of $\overline{\Delta}$

$$\alpha\frac{\partial V}{\partial S} \leq \overline{\Delta} \leq \beta\frac{\partial V}{\partial S}.$$

Such a constraint could be used to bound the risk in the portfolio.

Since the expected drift of the portfolio in excess of the risk-free rate is

$$\left(\mu S\frac{\partial V}{\partial S} - rV - (\mu - r)S\overline{\Delta}\right) dt,$$

it is simple to show that the choice

$$
\overline{\Delta} = \begin{cases} \alpha \dfrac{\partial V}{\partial S} & \text{if } \dfrac{\partial V}{\partial S} \geq 0 \\[2ex] \beta \dfrac{\partial V}{\partial S} & \text{if } \dfrac{\partial V}{\partial S} \leq 0 \end{cases}
$$

maximizes the expected drift of the portfolio. (These inequalities would swap for the case $\mu < r$.) The reasoning behind this choice is simple: if our option has a positive $\Delta = \partial V / \partial S$ then we *expect* to make more money by speculating than hedging since we expect the asset price to rise faster than the risk-free rate, and hence we expect the portfolio value to rise faster. When Δ is negative we expect to lose out on a speculating position in comparison to a hedged position. If we now calculate the present value of the expected payoff we find that it satisfies

$$
\frac{\partial V}{\partial t} + \tfrac{1}{2}\sigma^2 S^2 \frac{\partial^2 V}{\partial S^2} + S \frac{\partial V}{\partial S}(\mu - (\mu - r)F(\Delta)) - rV = 0,
$$

with

$$
F(\Delta) = \begin{cases} \alpha\Delta & \text{if } \Delta \geq 0 \\ \beta\Delta & \text{if } \Delta \leq 0 \end{cases}
$$

This is another nonlinear diffusion equation.

In Figure 31.11 we see the solution of this equation for a position of two long calls at 90 and three short 100 calls. A model such as this is an obvious improvement over pure speculation.

The hedger-speculator model can be combined with the early closing constraint.

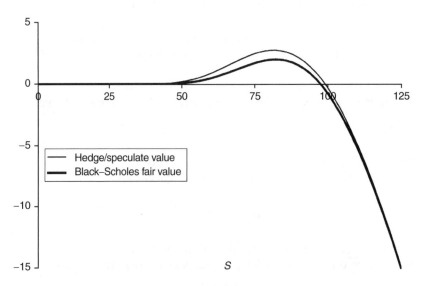

Figure 31.11 The option value to a hedger-speculator and the Black–Scholes value; see text for details.

With the parameters $\alpha = \beta = 1$ we have Black–Scholes and with $\alpha = \beta = 0$ we return to the hedging-speculating model (31.5).[6]

31.6 OTHER ISSUES

An accurate model of the drift of the underlying is especially important when speculating with barrier options or binary/digital options. Since binary options have a discontinuous payoff it is not possible in practice to follow the Black–Scholes hedging strategy no matter how frequently one hedges. Thus there is always some exposure to the drift; to some extent, one is always speculating with binary options.

We have seen how the drift rate of the underlying comes into the equation for the real expected payoff. Another simple but important calculation in a similar vein is the calculation of the real expected value of an option part-way into its life. For example, we hold a five year warrant and want to know how much this will be worth in two years' time. We may want to know this so as to have some idea of the value of our portfolio at that time. We calculate the *real expected* value of the warrant in two years as follows.

First, note that the value of the option in the Black–Scholes world satisfies the Black–Scholes equation and does not depend on μ. Let us assume that this is how the market will value the contract throughout its life. However, we want to know the real expected value in two years. We can do this by simply solving the equation

$$\frac{\partial V}{\partial t} + \tfrac{1}{2}\sigma^2 S^2 \frac{\partial^2 V}{\partial S^2} + \nu S \frac{\partial V}{\partial S} - rV = 0,$$

where

$$\nu = \begin{cases} \mu & \text{for } 0 \leq t \leq 2 \\ r & \text{for } 2 < t \leq 5. \end{cases}$$

31.7 SUMMARY

One mustn't get too hung up on delta hedging, no arbitrage and risk neutrality. The concepts are extremely important but there are times when their relevance is questionable, and one of those times is when you are speculating. Sure, the price that you buy a contract for has something to do with these concepts via Black–Scholes but, from then on, if you are not hedging, there are more important things to worry about.

FURTHER READING

- See Korn & Wilmott (1996a, b) for details about option pricing with a view.
- The best choice to make from a selection of risky investments is just asset allocation, see for example Markowitz (1959), but this is discussed in Chapter 51.

---

[6] There are many other obvious ways to bound risk. For example, by choosing α and β suitably we can overhedge if we think that this is advantageous. We could also allow α and β to depend on V or Π; we may want to take less risk with a larger portfolio.

CHAPTER 32
static hedging

In this Chapter...

- matching a contract's value at a set of points using standard contracts
- how nonlinear equations make a mockery of 'parameter fitting'
- statically hedging with traded options to improve your prices
- how to optimally statically hedge to make the most out of your contract

IF YOU'VE GOT A NONLINEAR EQN YOU'RE REALLY ROCKIN'

32.1 INTRODUCTION

Delta hedging is a wonderful concept. It leads to preference-free pricing (risk neutrality) and a risk-elimination strategy that can be used in practice. There are quite a few problems, though, on both the practical and the theoretical side. In practice, hedging must be done at discrete times and is costly. These issues were discussed in Chapters 23 and 24. Sometimes one has to buy or sell a prohibitively large number of the underlying in order to follow the theory. This is a problem with barrier options and options with discontinuous payoff.

On the theoretical side, we have to accept that the model for the underlying is not perfect, since at the very least we do not know parameter values accurately. Delta hedging alone leaves us very exposed to the model; this is model risk.

Many of these problems can be reduced or eliminated if we follow a strategy of static hedging as well as delta hedging: buy or sell more liquid contracts to reduce the cashflows in the original contract. The static hedge is put into place now, and left until expiry.[1]

32.2 STATIC REPLICATING PORTFOLIO

The value of a complex derivative is usually very model-dependent. Often these contracts are difficult to delta hedge, either because of transaction costs in the underlying or because of discontinuities in the payoff or the delta. Problems with delta hedging can be minimized by **static hedging** or **static replication**, a procedure which can reduce transaction costs, benefit from economies of scale and, importantly, reduce model risk.

---

[1] In practice, if conditions become favorable, one can reassess the static hedge. It is not set in stone.

There are two forms of static hedging that I want to discuss here. The first form of static hedging is about constructing a portfolio of traded options whose value matches the 'target' contract's value as closely as possible at a set of dates and asset values. The second form depends crucially on the governing equation being nonlinear.

32.3 **MATCHING A 'TARGET' CONTRACT**

The first idea is simple, and the implementation not much harder. Referring to Figure 32.1, the aim is to construct a portfolio of vanilla puts and calls, say, to have the same value as the target contract at expiry and at the points marked in the figure.

As a concrete example, suppose that the target contract has a call payoff at expiry T, but expires worthless on the upper boundary marked in the figure. This is a up-and-out call option. Thus on the upper boundary, our target contract has value zero. We have sold this contract and we want to statically hedge it by buying vanilla calls and puts. That is, we want to match the call payoff at T and have zero value on the boundary. This is easy. To match the payoff at expiry we buy a single call option with the same strike as the knock-out call. Ideally, we would like to have zero value on the remainder of the boundary, but this would involve us buying an infinite number of vanilla options. It is natural, therefore, since there are only a finite number of traded contracts, to match the portfolio value at the expiry dates of the traded options. In the figure, this means matching values at the four points on the upper boundary. Typically this will require us to buy four more vanilla options, in quantities to be decided.

We have matched the payoff at expiry with a single call option. Now, working back from expiry, value until time t_4 the contract consisting of the knock-out call held short and the single long vanilla call. In this 'valuation' we use whatever model we like. We use the model to give us the theoretical value of the two-contract portfolio at time t_4 at the point marked in

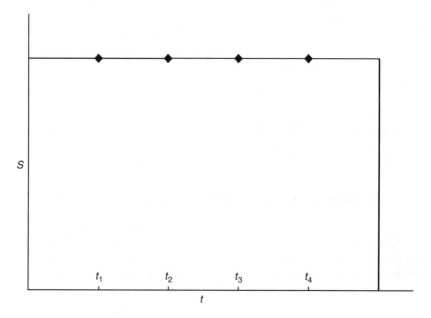

Figure 32.1 The target contract is to be replicated at the marked points.

Figure 32.1. We want to buy or sell further vanilla contracts to make the net portfolio value zero at this point. The obvious contracts to choose for this hedging are calls with expiry at T and strikes such that they are out-of-the money for all values of S on the target-option boundary at time T. We need only one contract to achieve this, an out-of-the money call, say.

Now value this new portfolio (it now contains our original target contract and two vanillas) back until time t_3. We can find its value at the marked point at this time. To hedge this value buy or sell vanilla contracts *that expire at time* t_4, that are out-of-the money for all values of S within the target-option boundary at time t_4, and such that between them they make the portfolio value zero on the upper boundary at time t_3. We now hold three options in our static hedge (plus the target contract) and are 'hedged' from t_3 until T. In this way we work backwards until the present.

We can obviously include as many expiry dates as exist in traded contracts, and the finer their resolution the better will be our static hedge. As long as we only use a finite number of options in our static hedge, then there will be some residual risk that is not statically hedged. We can either delta-hedge this or, if it is small and with bounded downside, we could leave it unhedged (assuming we have allowed for this when we sold the target contract).

I have illustrated the idea with a barrier option hedged with calls. Obviously the idea can be extended in any number of ways. One possibility is to use one-touch options, which are no longer thought of as being exotic and are actively traded.

Static hedging is obviously a useful technique. It is not perfect since there is still some exposure to the model. We had to value the portfolio using some 'model', and the accuracy of the hedge is reflected to some degree in the accuracy of the model. However, the resulting portfolio is not as model-dependent as a contract that is only delta hedged.

In the rest of this chapter we discuss static hedging when the governing equation is nonlinear. Some of these nonlinear models get a lot closer to model-independent pricing; at least they are 'parameter-insensitive' models. Note that the idea of 'optimal' static hedging that we will be seeing can also be applied to the above replicating portfolio: it may be cheaper to statically hedge with one contract than with another.

32.4 STATIC HEDGING: NONLINEAR GOVERNING EQUATION

Many of the models described in this part of the book are nonlinear. To summarize, these models are the following.

- **Transaction costs:** Purchase and sale of the underlying for delta hedging when there are costs leads to nonlinear equations for the option value. There are many models, all of them nonlinear. See Chapter 24.

- **Uncertain parameters:** When parameters, such as volatility, dividend rate or interest rate, are permitted to lie in a range, options can be valued in a worst-case scenario, for example the model of Avellaneda, Levy & Parás and Lyons. The governing equation for uncertain volatility is mathematically identical to the Hoggard–Whalley–Wilmott transaction cost model. See Chapter 27.

- **Crash modeling:** Allowing the underlying to jump and valuing contracts in a worst-case scenario, with no probabilistic assumptions about the crash, leads to a nonlinear equation. This contrasts with the linear classical jump diffusion model. See Chapter 30.

- **Speculating with options:** Some of the strategies for the 'valuation' of contracts when speculating lead to nonlinear equations. For example, optimal closure of an option position has a linear governing equation but the problem is nonlinear because it has a free boundary. Choosing to hedge or speculate also leads to a nonlinear governing equation. See Chapter 31.

Later, in Chapter 48, we will see a non-probabilistic interest rate model that is nonlinear: all of the ideas of this chapter can be applied to that model.

32.5 NONLINEAR EQUATIONS

Many of the new partial differential equations that we have derived are nonlinear. This nonlinearity has many important consequences. Most obviously, we must be careful when we refer to an option position, making it clear whether it is held long or short. For example, for a long call we have the final condition

$$V(S, T) = \max(S - E, 0)$$

and for a short call

$$V(S, T) = -\max(S - E, 0).$$

Because of the nonlinearity, the value of a portfolio of options is not necessarily the same as the sum of the values of the individual components. This is a very important point to understand: the value of a contract depends on what else is in the portfolio.

These two points are key to the importance of nonlinear pricing equations: they give us a bid-offer spread on option prices, and they allow *optimal* static hedging.

For the rest of this chapter we discuss the pricing and hedging of options when the governing equation is nonlinear. The ideas are applicable to any of the nonlinear models mentioned above. We use the notation $V_{NL}(S, t)$ to mean the solution of the model in question, whichever model it may be. So that the explanation of the issues does not get too confusing I will always refer to the concrete example of the uncertain volatility/transaction cost model, but remember *the ideas apply equally well to the other models.*

32.6 PRICING WITH A NONLINEAR EQUATION

One of the interesting points about nonlinear models is the prediction of a spread between long and short prices. If the model gives different values for long and short then this is in effect a spread on option prices. This can be seen as either a good or a bad point. It is good because it is realistic; spreads exist in practice. It only becomes bad when this spread is too large to make the model useful.

Let us consider a realistic, uncertain volatility/transaction cost model example. The reader is reminded that this model is

$$\frac{\partial V_{NL}}{\partial t} + \tfrac{1}{2}\sigma(\Gamma)^2 S^2 \frac{\partial^2 V_{NL}}{\partial S^2} + rS \frac{\partial V_{NL}}{\partial S} - rV_{NL} = 0 \tag{32.1}$$

where

$$\Gamma = \frac{\partial^2 V_{NL}}{\partial S^2}$$

and

$$\sigma(\Gamma) = \begin{cases} \sigma^+ & \text{if } \Gamma < 0 \\ \sigma^- & \text{if } \Gamma > 0. \end{cases}$$

(But we are using it as a proxy for any of the nonlinear equations.)

This model can predict very wide spreads on options. For example, suppose that we have a European call, strike price $100, today's asset price is $100, there are six months to expiry, no dividends on the underlying and a spot interest rate of 5%. We will assume that the volatility lies between 20% and 30%. We are lucky with this example. Because the gamma is single-signed for a vanilla call, we can calculate the values for long and short calls, assuming this range for volatility, directly from the Black–Scholes formulae. A long call position is worth $6.89 (the Black–Scholes value using a volatility of 20%) and a short call is worth $9.63 (the Black–Scholes value using a volatility of 30%). This spread is much larger than that in the market, and is due to the uncertain volatility. The market prices may, for example, be based on a volatility between 24% and 26%. The uncertain parameter model is useless unless it can produce tighter spreads. In the next section I show how the simple idea of static hedging can be used to significantly reduce these spreads. The idea was originally due to Avellaneda & Parás.

32.7 STATIC HEDGING

Suppose that we want to sell an option with some payoff that does not exist as a traded contract. (For the moment think in terms of a European contract with no path-dependent features; we will generalize later.) We want to determine how low a price we can sell it for (or how high a price we can buy it for), with the constraint that we guarantee that we will not lose money as long as our range for volatility is not breached. There are two

HEDGE WITH SOMETHING SIMILAR TO WHAT YOU ARE SELLING

related reasons for wanting to solve this problem. If we can sell it for more than this minimum then we are guaranteed to make money, and if our spread is tight then we will have more customers.

I motivate the idea of static hedging with a simple example. Suppose that options on a particular stock are traded with strikes of $90 and $110 and with six months to expiry. The stock price is currently $100. However, we have been asked to quote prices for long and short positions in $100 calls, again with six months before expiry. This call is *not* traded.

As above, assume that volatility lies between 20% and 30%. Remember that our pessimistic prices for the 100 call were $9.63 to sell and $6.85 to buy, calculated using the Black–Scholes formulae with volatilities 30% and 20% respectively. This spread is so large that we will not get the business. However, we are missing one vital point: the 90 and 110 calls are trading, can't we take advantage of them in some way?

Suppose that the market prices the 90 and 110 calls with an implied volatility of 25% (forget bid-offer spreads for the moment). The market prices, i.e. the Black–Scholes prices, are therefore 14.42 and 4.22 respectively. These numbers are shown in Table 32.1. The question marks are to emphasize that we can buy or sell as many of these contracts as we want, but in a fashion which will be made clear shortly. Shouldn't our quoted prices for the 100 call reflect the availability of contracts with which we can hedge?

Consider first the price at which we would sell the 100 call. If we sell the 100, and 'statically hedge' it by buying 0.5 of the 90 and 0.5 of the 110, then we have a residual payoff as shown

Table 32.1 Available contracts.

Strike	Expiry	Bid	Ask	Quantity
90	180 days	14.42	14.42	?
110	180 days	4.22	4.22	?

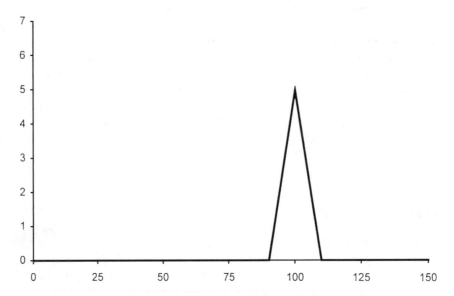

Figure 32.2 The residual risk after hedging a 100 call with 0.5 each of the 90 and 110.

in Figure 32.2. We call this a static hedge because we put it in place now and do not expect to change it. This contrasts with the delta hedge, for which we expect to hedge frequently (and would like to hedge continuously).

The statically-hedged portfolio has a much smaller payoff than the original unhedged call, as shown in Figure 32.2. It is this *new portfolio* that we value by solving the nonlinear Equation (32.1), and that we must delta hedge; I emphasize this last point, *the residual payoff must be delta-hedged*. To value the residual payoff in our uncertain parameter framework we solve Equation (32.1) with final condition.

$$V_{NL}(S, T) = -\max(S - 100, 0) + \tfrac{1}{2}(\max(S - 90, 0) + \max(S - 110, 0)).$$

Let us see what effect this has on the price at which we would *sell* the 100 call.

First of all, observe that we have paid

$$0.5 \times \$14.42 + 0.5 \times \$4.22 = \$9.32$$

for the static hedge made up of 0.5 of each of the 90 and 110 calls.

Now, solve Equation (32.1) using the payoff shown in Figure 32.2 as the final condition. The solution gives a value for the residual contract today of $0.58.

The net value of the call is therefore

$$\$9.32 - \$0.58 = \$8.74.$$

To determine how much we should pay to *buy* the 100 call we take as our starting point the sale of 0.5 of each of the 90 and 110 calls. This nets us \$9.32. But now we solve Equation (32.1) using the *negative* of the payoff shown in Figure 32.2 as the final condition; the effect of the nonlinearity is different from the previous case because $\sigma(\Gamma)$ takes different values in different places in S, t-space. We get a value of \$1.96 for the hedged position. Thus we find that we would pay

$$\$9.32 - \$1.96 = \$7.36.$$

Note how the use of a very simple static hedge has reduced the spread from \$6.85–9.63 to \$7.36–8.74. This is a substantial improvement, as it represents a volatility range of 22–27%; while our initial estimate for the volatility range was 20–30%. The reason for the smaller spread is that the residual portfolio has a smaller absolute spread, only \$0.58–1.96, and this is because it has a much smaller payoff than the unhedged 100 call.

In the trivial case where the option on which we are quoting is also traded then we would find that our quoted price was the same as the market price. This is because we would hedge one for one and the residual payoff, which we would normally delta hedge, would be identically zero. This means that we always match market prices; 'fitting' as described in Chapter 25 is redundant.

In the above example we decided to hedge the 100 call using 0.5 of each of the 90 and 110 calls. What prompted this choice? In this case we chose the numbers 0.5 for each option purely on the grounds of symmetry: it made the residual payoff look simple (a typical mathematician's choice). There was no reason why we should not choose other numbers. Suppose, for example, we decided to hedge with 0.4 of the 90 call and 0.7 of the 110. Would this have an effect on the value of our 100 call? Since our problem is nonlinear *the value of our OTC option depends on the combination of options with which we hedge*. So, generally speaking, we expect a different OTC option value if we choose a different static hedge portfolio. Of course, we now ask 'If we get different values for an option depending on what other contracts we hedge it with, then is there a *best* static hedge?'

32.8 OPTIMAL STATIC HEDGING

Suppose that we want to find the lowest price for which we can sell a particular OTC or 'exotic' option with payoff $\Lambda(S)$. Suppose that we can hedge our exotic with a variety of traded options, of which there are n. These options will have payoffs (at the same date as our exotic to keep things simple for the moment) which we call

OPTIMIZE TO MAKE THE MOST FROM THE AVAILABLE CONTRACTS

$$\Lambda_i(S).$$

At this point we can introduce bid and offer prices for the traded options: C_i^+ is the ask price of the ith option and C_i^- the bid, with $C_i^- < C_i^+$.

Now we set up our statically hedged portfolio: we will have λ_i of each option in our hedged portfolio. The cost of setting up this static hedge is

$$\sum_i \lambda_i C_i(\lambda_i),$$

where

$$C_i(\lambda_i) = \begin{cases} C_i^+ & \text{if } \lambda_i > 0 \\ C_i^- & \text{if } \lambda_i < 0. \end{cases}$$

If $\lambda_i > 0$ then we have a positive quantity of option i at the offer price in the market, C_i^+; if $\lambda_i < 0$ then we have a negative quantity of option i at the bid price in the market, C_i^-.

We let $V^-(S, t)$ be the pessimistic value of our *hedged* position. The residual payoff for our statically hedged option is

$$V^-(S, T) = \Lambda(S) + \sum_i \lambda_i \Lambda_i(S).$$

Now we solve (32.1) with this as final data, to find the *net* value of our position (today, at time $t = 0$, say) as

$$V^-(S(0), 0) - \sum_i \lambda_i C_i(\lambda_i) = F(\lambda_1, \ldots, \lambda_n).$$

This is a mathematical representation of the type of problem we solved in our first hedging example; $V^-(S, T)$ was the residual payoff shown in Figure 32.2.

Our goal now is to choose the λ_i to minimize $F(\cdots)$ if we are selling the exotic, and maximize if buying. (Thus the best hedge in the two cases will usually be different.) This is what we mean by 'optimal' static hedging.

Continuing with our earlier example, what are the optimal static hedges for long and short positions? Buying 0.5 each of the two calls to hedge the short 100 call we find a marginal value of $8.74 for the 100 call. This is actually a very good hedge. Slightly better, and optimal, is to buy 0.51 of each call giving a marginal value of $8.73. The optimal hedge for a long position is vastly different. We should sell 1.07 of the 90 call and sell 0.34 of the 110 call. The marginal value of the 100 call is then $7.47.

Sometimes the results are not always immediately obvious. This is because traditionally, when we think of hedging, we tend to choose hedges so as to minimize the risky payoff. In the above algorithm the cost of the hedge plays as important a role as its effect on the residual payoff. In an extreme case, if we had an option with an implied volatility outside our best-guess range then the algorithm would tell us to buy or sell an infinite amount of that option: the optimization algorithm finds arbitrage opportunities.

There are two other aspects to this static hedging. The first is that we are not necessarily restricted to hedging with options of the same maturity as the contract in question; they can have shorter or longer maturities. The optimization procedure will still find the best choice. The second point is that the static hedge need not be fixed. We could return at any time (or even continuously in theory) to examine the choice of optimal hedge portfolio. If the optimal hedge changes then, of course, we can buy and/or sell options to improve the worst-case value. This may not happen very often if the bid-offer spread on the traded options is large, but if it is negligible then the static hedge may often be rebalanced. However, this depends on the evolution of the implied volatilities, something which we are deliberately not modeling or predicting. If we do decide to change the static hedge then this is because we can improve the worst-case option value and thus we 'lock in' more guaranteed profit.

In this section we have concentrated on hedging unusual payoffs, or non-exchange traded contracts, with vanillas having the same expiry dates. The ideas are easily extended to hedging with options having other expiry dates, either before or after the expiry of the main contract.

This is simply done via jump conditions across each expiry date. As the nonlinear diffusion equation is solved backwards in time we must make the portfolio of options jump by the relevant payoff.

As a final thought in this section, observe how I never said that we 'believed' the price of market traded contracts, only that they *may possibly turn out to be correct*. This is in contrast to the less satisfying philosophy of Chapter 25 where I showed how to 'fit' the volatility structure to match traded prices.

It is a more complicated matter to extend the ideas to path-dependent options. In the following section I show how this is done for barrier options.

32.9 **HEDGING PATH-DEPENDENT OPTIONS WITH VANILLA OPTIONS**

Valuation with a nonlinear model and optimal static hedging are more complicated for barrier options (and other path-dependent contracts) because at the onset of the contract we do not know whether it will still exist by the final expiry date: we may start out with an excellent static hedge for a call payoff but after a few months we may find ourselves using the same hedge for a nonexistent option.

32.9.1 Barrier Options

The trick to valuing barrier options is to realize that there are two possible states for the option: untriggered and triggered. We can go from the former to the latter but not vice versa. Whatever static hedge we choose must take this into account. I introduce the terms 'active' and 'retired' to describe options that still exist and have been triggered, respectively.

Let $V_{NL}(S, t, 0)$ and $V_{NL}(S, t, 1)$ be the values of the portfolio of options before and after the barrier has been triggered. Both of these functions satisfy Equation (32.1), with final conditions at time T depending on the details of the option contracts. At the barrier the values of the two portfolios will be the same.

As a first example suppose that we hold a down-and-out barrier call only, i.e. there is no static hedge, and after the option is retired we hold an empty portfolio. The problem for $V_{NL}(S, t, 1)$ is simply Equation (32.1) with zero final conditions so that

$$V_{NL}(S, t, 1) = 0.$$

Of course, the solution for $V_{NL}(S, t, 1)$ is zero, since it corresponds to an empty portfolio. The problem for $V_{NL}(S, t, 0)$ is Equation (32.1) with

$$V_{NL}(S, T, 0) = \max(S - E_1, 0).$$

Finally, on the barrier

$$V_{NL}(X, t, 0) = V_{NL}(X, t, 1) = 0.$$

The problem for $V_{NL}(S, t, 1)$ only holds for $S > X$.

Now consider the slightly more complicated problem of hedging the barrier call with a vanilla call; we will hold short λ of a vanilla call, both contracts having the same expiry (although this can easily be generalized). The strike price is E_2. Thus we have hedged a barrier call with a vanilla call. The problem for $V_{NL}(S, t, 1)$ is simply Equation (32.1) with

$$V_{NL}(S, T, 1) = -\lambda \max(S - E_2, 0).$$

The problem for $V_{NL}(S, t, 0)$ is Equation (32.1) with

$$V_{NL}(S, T, 0) = \max(S - E_1, 0) - \lambda \max(S - E_2, 0).$$

Finally, on the barrier

$$V_{NL}(X, t, 0) = V_{NL}(X, t, 1).$$

The problem for $V_{NL}(S, t, 0)$ is to be solved for all S, but the problem for $V_{NL}(S, t, 1)$ only holds for $S > X$.

The ideas of optimal static hedging carry over; for example, what is the optimal choice for λ in the above?

Example

The stock price is $100, volatility is assumed to lie between 20% and 30%, the spot interest rate is 5%. What is the optimal static hedge, using vanilla calls with strike $110 and costing $4.22, for a long position in a down-and-out barrier call with strike $90? Both options expire in six months. In other words, what is the optimal number of vanilla calls to buy or sell to get as much value as possible out of the barrier option?

With no static hedge in place we simply solve the above problem with $V^-(S, t, 1) = 0$, since if the option knocks out that is the end of the story. In this case we find that the barrier option is worth $6.32.

With a hedge of -0.47 of the vanillas the barrier is now worth $6.6. This is the optimal static hedge.

32.9.2 Pricing and Optimally Hedging a Portfolio of Barrier Options

The pricing of a *portfolio* of barriers is a particularly interesting problem in the nonlinear equation framework. Again, this is best illustrated by an example. What is the worst value for a portfolio of one down-and-out call struck at $100 with a barrier at $90 and an up-and-in put struck at $110 with a barrier at $120? The volatility lies between 20 and 30%, the spot interest rate is 5%, there are no dividends, there are six months to expiry and the underlying asset has value $102.

The first step is to realize that instead of there being two states for the portfolio (as in the above where the single barrier is either active or retired), now there are *four* states: each of the two barrier options can be either active or retired. In general, for n barrier options we have 2^n states; the vanilla component of the portfolio is, of course, always active. As long as the barriers are not intermittent then there is a hierarchy of barrier options that can be exploited to dramatically reduce the computational time. This hierarchy exists because the triggering of one barrier means that barriers closer in are also triggered. This issue is discussed by Avellaneda & Buff (1997).

Returning to the two-option example, the four states are:

1. Both options are active; this is the initial state.
2. If the asset rises to 120 without first falling to 90 then the up barrier option is retired.
3. If the asset falls to 90 before rising to 120 then the down barrier option is retired.
4. If both barriers are triggered then both options are retired. The only active options will be any vanillas in the static hedge.

To solve this problem we must introduce the function $V_{NL}(S, t; i, j)$. The portfolio value with both options active is $V_{NL}(S, t; 0, 0)$, with the down-and-out option retired but the up-and-in still active $V_{NL}(S, t; 1, 0)$, with the down option active but the up retired $V_{NL}(S, t; 0, 1)$ and with both options retired $V_{NL}(S, t; 1, 1)$. In the absence of any static hedge, the problem to solve is Equation (32.1) for each V_{NL} with final conditions

- $V_{NL}(S, T; 0, 0) = \max(S - 90, 0)$ (since at expiry, if both options are active only the down-and-out call pays off),

- $V_{NL}(S, T; 1, 0) = 0$ (if the down-and-out has been triggered it has no payoff),

- $V_{NL}(S, T; 0, 1) = \max(S - 90, 0) + \max(120 - S, 0)$ (if the up knocks in it has the put payoff) and finally

- $V_{NL}(S, T; 1, 1) = \max(120 - S, 0)$ (if both options are triggered the payoff is that for the in option).

The retirement of an option is expressed through the boundary conditions

- $V_{NL}(90, t; 0, 0) = V_{NL}(90, t; 1, 0)$ and $V_{NL}(90, t; 0, 1) = V_{NL}(90, t; 1, 1)$ (the option value just before knock-out is equal to that just after, regardless of the status of the in option),

- $V_{NL}(120, t; 0, 0) = V_{NL}(120, t; 0, 1)$ and $V_{NL}(120, t; 1, 0) = V_{NL}(120, t; 1, 1)$ (the option value just before knock-in is equal to that just after regardless of the status of the out option).

32.10 **THE MATHEMATICS OF OPTIMIZATION**

We have encountered several nonlinear problems, and will meet some more later on. I have explained how the nonlinearity of these models allows static hedging to be achieved in an optimal fashion. Mathematically this amounts to finding the values of parameters such that some function is maximized. This is an example of an **optimization problem**. In this section I will briefly describe some of the issues involved in optimization problems and suggest a couple of methods for the solution.

Typically we are concerned with a maximization or a minimization. Since the minimization of a function is the same as the maximization of its negative, I will only talk about maximization problems. We will have some function to maximize, over a set of variables. Let's call the variables $\lambda_1, \ldots, \lambda_N$ and the function $f(\lambda_1, \ldots, \lambda_N)$. Note that there are N variables so that we are trying to find the maximum over an N-dimensional space. In our financial problems we have, for example, a set of λs representing the quantities of hedging contracts and the function f is the value of our target contract.

There are two kinds of maximum, **local** and **global**. There may be many local maxima, the highest point in its neighborhood, but it may not be the highest point over all of the N-dimensional space. What we really want to find is the global maximum, the point having the highest function value over the whole of the space of the variables. In Figure 32.3 is shown a one-dimensional function together with a few local maxima but only one global maximum. Whatever technique we use for finding the maximum must be able to find the global maximum.

32.10.1 Downhill Simplex Method

The first method I want to describe is the **downhill simplex method**. This method is not particularly fast, but it is very robust and easy to program. If there is more than one local

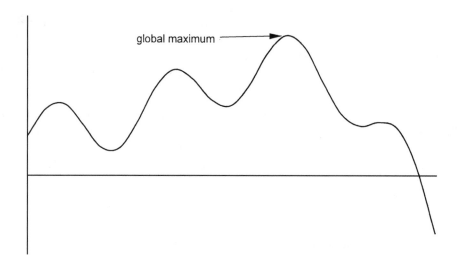

Figure 32.3 Local and global maxima.

maxima then it may not find the global maximum; instead it might get stuck at a local high spot. I will describe a way around this weakness later. Traditionally, one talks about function minimization, and so the word 'downhill' makes sense. Because our problems are typically ones of maximization perhaps we should use the word 'uphill'.

A **simplex** in N dimensions is made up of $N + 1$ vertices. In two dimensions this gives a triangle, in three a tetrahedron. The downhill simplex method starts with a initial simplex: think of it as a ballpark guess for the position of the maximum. Suppose that the $N + 1$ vertices of this simplex are at the points given by the vectors λ_i for $i = 1, \ldots, N + 1$. Calculate the values of the function at each of the vertices; these are $f(\lambda_i)$. One of these values will be lower than the other N values. If this value is particularly bad, i.e., low, perhaps we should look far away from this vertex for another guess at the maximum. One systematic way of doing this is to reflect the worst point in the 'plane' made up by the remaining N points. This is easy to see for the tetrahedral simplex shown in Figure 32.4. In this figure (a) is the initial state of the simplex. The worst/lowest point is reflected in the 'base' of the tetrahedron resulting in (b). If the function evaluated at the new vertex is even better/higher than all of the values of the function at the other (old) vertices, then expand the tetrahedron a bit to exploit this. This gives (c). On the other hand, if the new vertex has a function value that is worse than the second lowest then the tetrahedron should be contracted as shown in (d), otherwise the new simplex would be worse than before. Finally, if the one-dimensional contraction doesn't make things better, then the simplex is contracted about the best vertex in all dimensions simultaneously, giving (e). This algorithm sees the simplex tumbling over itself, rising to the peak of the function where it begins to contract. The process is repeated until the simplex has shrunk enough to give a sufficiently precise estimate of the maximum.

The following code finds a local maximum of the function func in N dimensions. The code is a translation of the C code in Press *et al.* into VB. The number of dimensions, the maximum number of iterations MaxIts and the tolerance tol must be input, as must the initial simplex. The initial guess goes in the $N + 1$ by N array FirstGuess; each row is a different vertex.

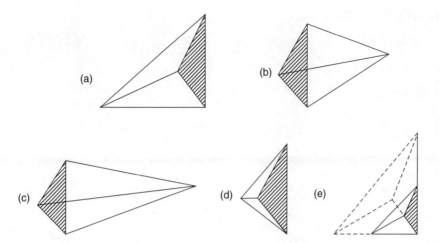

Figure 32.4 (a) Initial simplex, (b) reflection, (c) reflection and expansion, (d) contraction in one dimension and (e) multiple contractions to the highest point.

The subroutine `Ranking` finds the value of the function at, and the index of, the two lowest and the highest vertices. The subroutine `FindCenter` finds the center of the simplex, given by

$$\frac{1}{N+1}\sum_{i=1}^{N+1}\lambda_i.$$

The subroutine `Deform` reflects the lowest point in the opposite face of the simplex by an amount d. If $d = 1$ there is no reflection, if $d = 1/(N+1)$ then vertex goes to the center of the simplex and if $d = 0$ then the vertex goes into the opposite plane. Here we only use $d = -1$ for case (b), $d = 2$ for case (c) and $d = \frac{1}{2}$ for case (d).

```
Function maximize (FirstGuess As Object, N As Integer, _
  tol As Double, MaxIts As Integer)
ReDim outputarray (1 To N + 1, 1 To N + 1) As Double
ReDim y (1 To N + 1) As Double
ReDim p (1 To N + 1, 1 to N) As Double
ReDim ptemp (1 To N) As Double
Dim ranktest (1 To 3, 1 To 2) As Double

its = 0
rtol = 1

For i = 1 To N + 1
For j = 1 To N
p(i, j) = FirstGuess(i, j)
ptemp (j) = p(i, j)
Next j
y(i) = func(ptemp)
Next i

While (rtol > tol) And (its < MaxIts)

Call Ranking (y, N, ilo, ylo, inlo, ynlo, ihi, yhi)
```

SIMPLE OPTIMIZATION CODE (STOLEN FROM PRESS ET AL.)

```
Call Deform (y, p, N, ilo, ylo, -1)

ytest = ylo

If ytest >= yhi Then
    Call Deform(y, p, N, ilo, ylo, 2)
    ytest = ylo
Else
    If ytest <= ynlo Then
    ysave = y(ilo)
    Call Deform(y, p, N, ilo, ylo, 0.5)
    ytest = ylo
        If ytest <= ysave Then
            For i = 1 To N + 1
            If i <> ihi Then
                For j = 1 To N
                p(i, j) = 0.5 * (p(i, j) + p(ihi, j))
                ptemp(j) = p(i, j)
                Next j
            End If
            y(i) = func(ptemp)
            Next i
        End If
    End If
End If

    its = its + 1
    rtol = 2 * Abs (yhi - ylo) / (Abs(yhi) + Abs(ylo))
Wend

For i = 1 To N + 1
For j = 1 To N
outputarray(i, j) = p(i, j)
Next j
outputarray(i, N + 1) = y(i)
Next i

maximize = outputarray

End Function

Sub FindCenter(p, N As Integer, center)
For j = 1 To N
center(j) = 0
For i = 1 To N + 1
center(j) = center(j) + p(i, j) / (N + 1)
Next i
Next j
End Sub

Sub Ranking(y, N As Integer, ilo, ylo, inlo, ynlo, ihi, yhi)
ilo = 1
inlo = 1
ihi = 1
ylo = y(ilo)
ynlo = y(inlo)
yhi = y(ihi)
For i = 2 To N + 1
```

```
        If y(i) < ynlo Then
            If y(i) < ylo Then
                inlo = ilo
                ynlo = ylo
                ilo = i
                ylo = y(i)
            Else
                inlo = i
                ynlo = y(i)
            End If
        End If
        If y(i) > yhi Then
            ihi = i
            yhi = y(i)
        End If
    Next i
    End Sub

    Sub Deform(y, p, N) As Integer, ilo, ylo, d)
    ReDim ptest(1 To N) As Double
    ReDim center(1 To N + 1) As Double
    Call FindCenter(p, N, center)
    For j = 1 To N
        ptest(j) = d * p(ilo, j) + (1 - d) * (N + 1) / N * center(j) - p(ilo, j) / N)
    Next j
    ytest = func(ptest)
    If ytest > ylo Then
        y(ilo) = ytest
        ylo = ytest
    End If

    For j = 1 To N
    p(ilo, j) = ptest(j)
    Next j
    End Sub
```

32.10.2 Simulated Annealing

The main downside of the above maximization algorithm is that it will not necessarily find the global maximum. This is a problem with many optimization techniques. Currently, one of the most popular methods for overcoming this is **simulated annealing**. This gets its name from the form of crystallization that occurs as molten metals are cooled. In the optimization algorithm there is a degree of randomness in the search for a maximum, and as the number of iterations increases, so the randomness gets smaller and smaller. The anology is with the random motion of atoms in the liquid state of the metal; as the metal cools so the motion decreases. The hope is that the random motion will find the neighborhood of the global maximum, and as the 'temperature'/random motion decreases, the search will home in on the maximum.

One of the simplest simulated annealing techniques, related to the **Metropolis algorithm**, uses the addition or subtraction of a random number to the function value at each vertex, but is otherwise the same as the downhill simplex method. Suppose that we have a simplex and know the function values at each vertex. We now subtract a positive random number from the function value at each vertex. When we come to test a new replacement point we add a positive random number. In this way we still always accept a move to a better vertex, but occasionally we accept a move to a worse vertex. However, this worse vertex may actually be closer to the global maximum. How often we accept such a move depends on the size of the random variables. We must choose a distribution for the random variable and, importantly, its scale. As the number of iterations increases, so we decrease this scale so that it tends to zero as the number of iterations increases.

The scale of the random moves must decrease slowly enough for the simplex to have a good look around before converging on the global maximum.

32.11 OPTIMAL PORTFOLIOS FOR SPECULATORS

You are not restricted to looking for optimal portfolios just for static hedging and the elimination of model risk. You can also optimize within an arbitrage framework. For example, suppose you have a strong view on volatility that is not mimicked by the market: what's the best portfolio to hold to take advantage of this view? The techniques of optimization can be applied in this situation.

Since your view is different from the market's it is possible for you to believe that there is a model-dependent arbitrage opportunity. If that is the case, then when optimizing you must constrain your search range to finite quantities, otherwise your optimization algorithm will tell you to invest in infinite quantities to make an infinite profit.

32.12 SUMMARY

In this chapter I showed how to statically hedge a complex product with simpler contracts. We began with the classical static hedge, by setting up a portfolio having exactly the same theoretical value as our target contract at a specified set of asset/time points. The resulting hedge then reduces the risk from delta hedging. But it is not perfect; the hedge is still sensitive to the pricing model and its parameters.

Then I described the idea of static hedging for some nonlinear models. Because some of these models are less sensitive to parameter values (such as volatility) the resulting hedged contract is also insensitive. We also saw how to optimize this static hedge by maximizing the *marginal value* of the target contract, after allowing for the real cost of setting up the hedge.

FURTHER READING

- See Bowie & Carr (1994), Carr & Chou (1997), Carr, Ellis, & Gupta (1998) and Derman, Ergener & Kani (1997) for more details of static replicating portfolios.

- See Avellaneda & Parás (1996) for optimal static hedging as applied to uncertain volatility, and Oztukel (1996) for uncertain correlation.

- See Epstein & Wilmott (1997, 1998) for optimal static hedging in a fixed-income world.

- For a general description of optimization, with many references, algorithms and code, see the excellent Press *et al.* (1992).

- Avellaneda & Buff (1997) explain how to value large portfolios of barrier options in the uncertain volatility framework.

CHAPTER 33

the feedback effect of hedging in illiquid markets

In this Chapter...

* how delta hedging an option position or replication can influence and move the market in the underlying

33.1 INTRODUCTION

I have referred constantly to the 'underlying' asset. The implication of this is that the asset price leads and the option price follows; the option value is contingent upon the value and probabilistic properties of the asset, and not the other way around. In practice in many markets, however, the trade in the options on an asset can have a nominal value that exceeds the trade in the asset itself. So, can we still think of the asset price as leading the option?

In the traditional derivation of the Black–Scholes option pricing equation it is assumed that the replication trading strategy has no influence on the price of the underlying asset itself, that the asset price moves in a random way. Sometimes the justification for this is the action of **noise traders** or the random flow of information concerning the asset or the economy. Nevertheless, a significant number of trades are for hedging or replication purposes. And, crucially, these trades are for predictable amounts. What is their impact on the market?

Usually it is assumed that the effect of individual trading on the asset price is too small to be of any importance and is neglected when the strategy is derived. This seems justifiable if the market in question has many participants and a high degree of liquidity, which is usually true for modern financial markets. On the other hand, the portfolio insurance trading strategies are very often implemented on a large scale, and the liquidity of the financial markets is sometimes very limited. In the case of the October 1987 stock market crash some empirical studies and even the official report of the investigations carried out by the Brady commission suggest that portfolio insurance trading contributed to aggravate the effects of the crash.

In this chapter we are going to address the problem of the influence of these trading strategies on the price of the underlying asset and thus, in a feedback loop, onto themselves.

33.2 **THE TRADING STRATEGY FOR OPTION REPLICATION**

In theory, any simple option can be replicated by following the appropriate trading strategy. These trading strategies have enjoyed tremendous popularity among portfolio managers who use them to insure themselves against large movements in the share price. This strategy is called **portfolio insurance**. One of the most popular portfolio insurance strategies is the replication of a European put option. Any simple option having a value $V(S, t)$ can be *replicated* by holding

$$\Delta(S, t) = \frac{\partial V}{\partial S}(S, t) \tag{33.1}$$

shares at time t if the share price is S. (Contrast this with *hedging* an option for which we must hold this number short.) In particular, Black and Scholes found an explicit formula for the value $V(S, t)$ of the put option. As derived in Chapter 5, this formula is

$$V(S, t) = Ee^{-r(T-t)} N(-d_2) - S N(-d_1)$$

where

$$d_1 = \frac{\log\left(\dfrac{S}{E}\right) + \left(r + \dfrac{\sigma^2}{2}\right)(T - t)}{\sigma\sqrt{T - t}}$$

$$d_2 = d_1 - \sigma\sqrt{T - t}.$$

For the European put Equation (33.1) results in

$$\Delta(S, t) = N(d_1) - 1.$$

We can now think of $\Delta(S, t)$ as corresponding to a trading *strategy*. Put replication is one of the trading strategies, corresponding to one example of portfolio insurance, that we are going to analyze in detail. Figure 33.1 shows the delta of a European put at various times before expiry; the steeper the curve, the closer the option is to expiry.

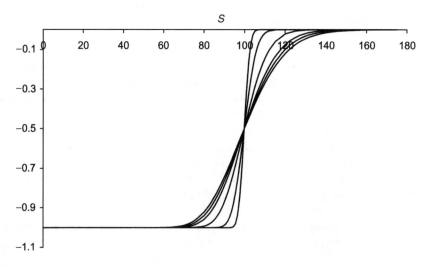

Figure 33.1 The delta for a European put at several times before expiry.

In the next section I propose a form for the reaction of the market to a trading strategy Δ and will derive the modification of the price process that takes account of the effect of trading.

33.3 THE EXCESS DEMAND FUNCTION

In order to quantify the effects of replication on the movement of the asset we need a model describing the relationship between supply of and demand for an asset and its price. Consider the difference between demand and supply in the market. This is the **excess demand** which we assume takes the form $\mathcal{X}(S, t, x)$ i.e. it is a function of price S, time t and, importantly, a random influence x. The random influence will ensure that our model does not stray too far from the classical lognormal, this will be a limiting case of our more complex model. We will model x as a random variable shortly. Its influence can be regarded as the effect of randomly arriving new information on the value of the underlying asset or the action of noise traders. Generally $(\partial/\partial S)\mathcal{X}$ is negative; rising price leads to falling demand.[1]

At any given time the equilibrium price $S_{Eq}(t)$ is the price for which demand is equal to supply, or excess demand is zero;

$$\mathcal{X}(S_{Eq}(t), t, x) = 0. \tag{33.2}$$

Typically, there is a strong tendency in any market to return to the equilibrium price if it has been disturbed. Surplus demand pushes the price up, towards the equilibrium price; excess supply makes the price drop, again towards the equilibrium price. The market equilibrium is stable. Note that this argument supposes that the excess demand function is negatively sloped; for a positive slope the equilibrium would still exist but it would be unstable. We will encounter locally positively sloped excess demand functions later on.

Disequilibrium is obviously possible but given the speed of the flow of information in these markets and the large number of professionals on the stock markets a full equilibrium in stocks and flows in modern financial markets is a good approximation. This does not mean that these markets are static: in our model both demand and supply can change in time because of the stochastic parameter x.

33.4 INCORPORATING THE TRADING STRATEGY

In this section we add to the original demand the extra demand resulting from the hedging of the put option. Not only will we have the random demand due to the noise traders but also a completely deterministic demand, due to the trading strategy Δ. With this additional demand of the form $\Delta(S, t)$ the equilibrium condition (33.2) becomes

$$\mathcal{X}(S, t, x) + \Delta(S, t) = 0 \tag{33.3}$$

(dropping the subscript on the equilibrium price) and the same must hold for the changes in $\mathcal{X}$ and Δ

$$d\mathcal{X} + d\Delta = 0.$$

---

[1] This is necessarily a simplification of the real story, but nevertheless a simplification often made in economic theory and one which holds the essence of truth.

We could consider arbitrary excess demand function $\mathcal{X}$, but for simplicity we assume from now on that

$$\mathcal{X}(S, t, x) = \frac{1}{\varepsilon}(x - S).$$

Here ε is a positive real number and x follows the stochastic process

$$dx = \mu_x \, dt + \sigma_x \, dX;$$

μ_x and σ_x can be functions of x and t. We can now think of x as being the 'intrinsic value' of the stock.

The parameter ε shows how strongly the excess demand function reacts to changes in the price. If the price changes by dS then the excess demand changes by $-dS/\varepsilon$. It gives an indication of the liquidity of the market. Liquid markets react very strongly to changes in the price, for such markets ε is small:

A *liquid market* is a market in which ε is *small*.
An *illiquid market* is a market in which ε is *large*.

As mentioned before most — but not all — financial markets are liquid markets.

If the parameter ε is assumed to be zero this reduces the price process S to the random walk x. We will not take this step but merely assume ε to be small. We will see later that in certain cases no matter how small ε is the individual effect on the excess demand cannot be neglected.

Now (33.3) becomes

$$(x - S) + \varepsilon \Delta(S, t) = 0. \tag{33.4}$$

In the undisturbed equilibrium, with appropriate choice of scalings, ε^{-1} is also equal to the **price elasticity of demand**.

Applying Itô's lemma to (33.4) we find that the stochastic process followed by S is

$$dS = \mu_S(S, t) \, dt + \sigma_S(S, t) \, dX \tag{33.5}$$

with μ_S and σ_S given by

$$\mu_S = \frac{\sigma_S}{\sigma_x} \left(\mu_x + \varepsilon \left(\frac{\partial \Delta}{\partial t} + \frac{1}{2} \sigma_S^2 \frac{\partial^2 \Delta}{\partial S^2} \right) \right)$$

and

$$\sigma_S = \frac{\sigma_x}{1 - \varepsilon \dfrac{\partial \Delta}{\partial S}}.$$

The details are left to the reader, the application of Itô's lemma is straightforward.

So far all analysis has been done for an arbitrary stochastic process for x. Usually the price process of share price is assumed to be lognormal, i.e.

$$dS = \mu S \, dt + \sigma S \, dX \tag{33.6}$$

with μ and σ constant. Our analysis of the model on the other hand yielded the modified price process (33.5). As the equilibrium price S is a known function of x we can choose

$$\mu_x = \mu S \quad \text{and} \quad \sigma_x = \sigma S$$

to achieve consistency between (33.5) and (33.6) in the sense that (33.5) reduces to (33.6) when $\varepsilon = 0$.

33.5 THE INFLUENCE OF REPLICATION

One of the most important portfolio insurance strategies Δ is put replication as explained in Section 33.2. Look again at Figure 33.1 to see the delta of the European put. As the expiry date of the option is approached Δ goes towards a step function with step from -1 to 0 at the exercise price.

Recalling Equation (33.4) the equilibrium condition can be written as

$$-x = -S + \varepsilon \Delta. \tag{33.7}$$

The effect of the trading strategy is a small (i.e. of order ε) perturbation added to the original demand function. Far from expiry the right-hand side of (33.7) is simply $-S$ since $\varepsilon \Delta$ is small. Close to expiry the Δ term becomes important and the shape of the demand curve alters dramatically, becoming as shown in Figure 33.2. Figure 33.2 shows the right-hand side of (33.7) for the put replication strategy; we are sufficiently close to expiry for the curve to no longer be monotonic.

As expiry approaches the sequence of events is as follows. Far from expiry there is hardly any deviation from the normal linear function. In an interval around the expiry price the demand

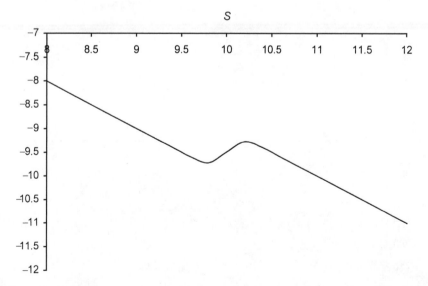

Figure 33.2 The sum of the original linear demand function and the extra demand due to put replication.

curve's slope becomes less and less negative until it is positively sloped. Thus there are times very close to expiry and close to the strike for which there is an unstable equilibrium.

Geometrically, Equation (33.7) can be interpreted as follows. For a given t, the market equilibria are defined as the points where the randomly moving horizontal line $-x$ intersects the function $-S + \varepsilon\Delta$. An example of this is shown in Figure 33.3. As long as $-S + \varepsilon\Delta$ is monotonically decreasing the equilibrium is unique for any value of x. But because Δ approaches a step function as expiry is approached, $-S + \varepsilon\Delta$ must become multi-valued at some point. From this time on the situation becomes more complicated.

Refer to Figure 33.3. In this figure the horizontal line denotes the left-hand side in the equilibrium condition (33.7) and the curve is the corresponding right-hand side for t close to expiry T. There are various possible situations depending on the value of x at that point in time:

1. x is sufficiently large or small as to be outside the critical region where there are several equilibria. The equilibrium asset price is unique, corresponding to the asset value where the horizontal line crosses the curve.

2. x is inside the critical region where there are three possible equilibria. This is the case shown in the figure. The middle equilibrium value for the asset is unstable, corresponding to a positively sloped demand function. The other two are stable.

3. The limiting case between cases 1 and 2. If the horizintal line just touches the curve, then we have two equilibria. One of these is stable, the other, corresponding to the asset value where the line just *touches* the curve, is stable only for movements in the random variable x of one sign. At this point either the equilibrium asset value will move continuously or will jump to the next equilibrium point: if a jump occurs there is a discontinuity in the price.

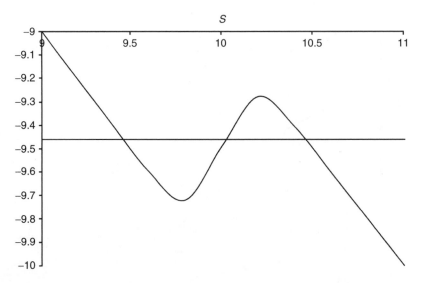

Figure 33.3 The intersection of the supply and demand curves. There are four cases depending on the relative position of the intersection and the maxima and minima.

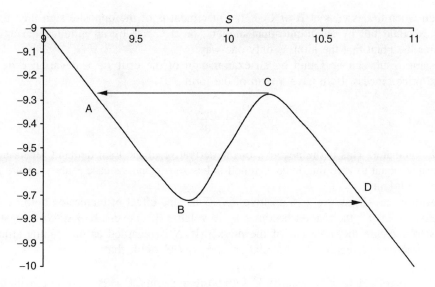

Figure 33.4 The positions of the four points A, B, C and D.

The third case is of particular interest. What happens to the equilibrium asset value is sketched in Figure 33.4. The points A, B, C and D are defined by

$$(A(t), -A(t) + \varepsilon\Delta(A(t), t)),$$

etc. as follows. $S = B(t)$ and $S = C(t)$ are the local extrema of the demand function, B is the local minimum and C is the local maximum. The definitions of these points are

$$-1 + \varepsilon\frac{\partial\Delta}{\partial S}(B, t) = 0 \tag{33.8}$$

$$-1 + \varepsilon\frac{\partial\Delta}{\partial S}(C, t) = 0 \tag{33.9}$$

and

$$\frac{\partial^2}{\partial S^2}\Delta(B, t) \geq 0 \geq \frac{\partial^2}{\partial S^2}\Delta(C, t). \tag{33.10}$$

$S = A(t)$ and $S = D(t)$ are defined as:

$$A \leq B \leq C \leq D \tag{33.11}$$

$$[-S + \varepsilon\Delta](A, t) = [-S + \varepsilon\Delta](C, t) \tag{33.12}$$

$$[-S + \varepsilon\Delta](D, t) = [-S + \varepsilon\Delta](B, t). \tag{33.13}$$

Thus there are four different points of interest: the two extrema *from* which the jumps come and the two points *to* which the price jumps.

The arrows in the figure mean that the asset can jump from one point of the curve to another, that is from B to D and from C to A. But that only happens if the increments of x are of the right sign when the asset is at B or C. If the increment is of the opposite sign then the asset price at B could fall by an infinitesimal amount, or at C rise by an infinitesimal amount. I emphasize the point that the jump is only one way.

The same result can be found by an examination of the drift μ_S and variance σ_S of the modified price process. Both have a term of the form

$$1 - \varepsilon \frac{\partial \Delta}{\partial S}$$

in the denominator. This is the negative of the derivative of the total demand function. When this becomes equal to zero the demand function has zero slope, as case 3 above. Here μ_S and σ_S approach infinity.

But even when $\partial \Delta / \partial S < \varepsilon^{-1}$, a positive $\partial \Delta / \partial S$ has the effect of increasing both σ_S and the absolute value of μ_S: the market becomes more volatile. If conversely $\partial \Delta / \partial S$ is negative, its effect is to decrease the volatility of the market. If Δ is regarded as the trading strategy to replicate a derivative security V, the relation $\Delta = \partial V / \partial S$ yields that:

- Replication of a derivative security V with positive gamma $\Gamma = \partial^2 V / \partial S^2$, i.e. with concave payoff profile, destabilizes the market of the underlying.

Long positions in put and call options have positive gamma.

33.6 **THE FORWARD EQUATION**

We can analyze the new stochastic process (33.5) by examining the probability density function $p(S, t)$ which gives the probability density of the share price being at S at time t subject to an initial distribution. Here S and t are the forward variables, usually written as S' and t'.

The probability density function satisfies the Kolmogorov forward equation

$$\frac{\partial p}{\partial t}(S, t) = \frac{1}{2} \frac{\partial^2}{\partial S^2}(p(S, t)\sigma_S^2(S, t)) - \frac{\partial}{\partial S}(p(S, t)\mu_S(S, t)).$$

We will use a delta function initial condition, meaning that we know exactly where the asset starts out.

33.6.1 The Boundaries

As pointed out in Section 33.5 there are two pairs of jump boundaries in the area close to expiry (see Figure 33.4). Jumps[2] can occur from point B to point D and from point C to point A.

The positions of these points change in time. A typical graph for a put-replicating trading strategy is shown in Figure 33.5. For any continuous time-dependent trading strategy all four boundary points have to arise at the same point. This point is characterized as a point of inflexion of the full demand function $x - S + \varepsilon \Delta$, which obviously satisfies the conditions (33.8)

---

[2] Determinstic jumps are not allowed in classical models of assets since they lead to arbitrage opportunities. We discuss this point later.

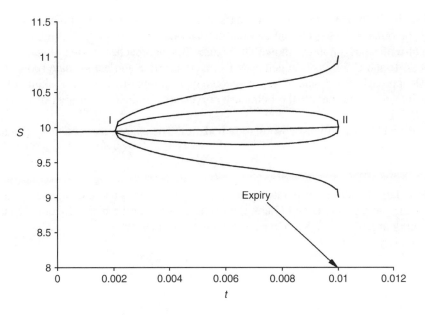

Figure 33.5 The time-dependent positions of the four points A, B, C and D for the put-replicating strategy.

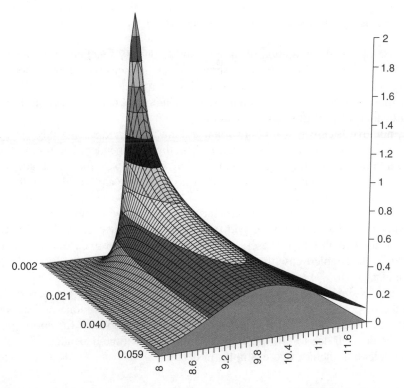

Figure 33.6 The probability density function for an asset with no feedback. The initial data is a delta function.

to (33.13). In Figure 33.5 this point is marked as I. Subsequently the points will constantly maintain the order $A < B < C < D$ and fan out as shown.

The put-replicating strategy shown in Figure 33.4 approaches a step function as expiry approaches. B and C disappear together at the exercise price and expiry date (point II). Points A and D approach $E - \varepsilon$ and $E + \varepsilon$ respectively resulting in a curve looking like a tulip lying on its side. Note that all the boundary curves are confined to an area of order $(\varepsilon \times \varepsilon^2)$ around (E, T).

Points B and C act locally as absorbing boundaries. The appropriate boundary conditions are therefore

$$p(B, t) = 0 = p(C, t).$$

The boundary conditions for points A and D are not so obvious because μ_S and σ_S have singularities as the jump points B and C are approached. They can be derived from a conservation of probability argument, see Schönbucher (1994) for further details.

The boundary conditions at $S = 0$ and $S \to \infty$ are

$$p(0, t) = 0 \quad \text{and} \quad \lim_{S \to \infty} p(S, t) = 0.$$

33.7 NUMERICAL RESULTS

In this section I present numerical results for the solution of the forward equation in three cases. The parameters as follows (unless otherwise stated) are: $E = 10$, $r = 0.1$, $\mu_x = 0.2$, $\sigma_x = 0.2$, and $\varepsilon = 1$.

The first example is the evolution of the probability density function in the absence of any feedback. Thus $\varepsilon = 0$. In Figure 33.6 is shown a three-dimensional plot of the probability density function against asset price and time. The starting condition is a delta function at $S = 10$. As time increases the curve flattens out as a lognormal density function. In Figure 33.7 is shown a contour map of this same function.

The first non-trivial case is a time-invariant trading strategy that is strong enough to give rise to the four new boundaries A, B, C and D. The time-independent addition to the demand, Δ, was taken to be that from a put-replicating trading strategy with time fixed at $t = T - 0.05$, shortly before expiry. The initial price is assumed to be known as $S = 11$, thus the probability density function contains a delta function at $S = 11$. This case helps to visualize the effects of the boundaries. This may perhaps model the effect of a large number of investors all delta hedging using the delta of an option with the same strike but each with a different expiry date such that the overall effect of these hedgers is to add a time-independent function to the demand. This is the simplest case to consider since the coefficients in the forward equation are time-independent.

The final case is simple put replication with only one investor moving the market. The trading strategy is the replication strategy for a European put option. This is one of the most popular trading strategies used by portfolio managers and of direct relevance for option pricing. The addition to the demand due to this one large investor is now time-dependent.

Let's take a look at the last two examples in detail.

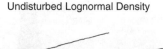

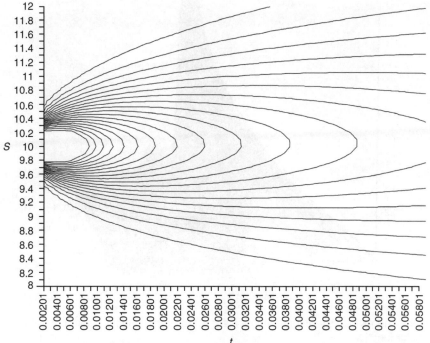

Figure 33.7   The contour plot of Figure 33.6.

33.7.1   Time-independent Trading Strategy

Time-invariant hedging schemes are mainly used to maintain the general performance of the portfolio without the need to satisfy any precise conditions at fixed points in time (such as the potential liabilities from writing an option). Here we will use the delta of a European put at time 0.05 before expiry.

Figure 33.8 shows the full development of the probability density function according to the feedback model as time proceeds. At $t = 0$ the probability density function is a delta function at $S = 11$, meaning that the price at $t = 0$ is known to be 11. Later the probability density function spreads out. Figure 33.9 shows the contour plot for the probability density function. The contour plot shows very clearly the unattainable region, the 'corridor', between asset values of approximately 9.6 and 10.3 which the asset price can never reach. This region is that between the two points previously labeled B and C. Since the replication strategy is time-independent this corridor does not change shape. Even though the starting value for the asset ($S = 11$) is above this region the asset can still reach values less than 9.6 by reaching the barrier C and jumping across to the point A. For more realistic values of ε this corridor is very narrow and, away from the corridor, is effectively only a small perturbation to the usual lognormal probability density function as shown in Figures 33.6 and 33.7.

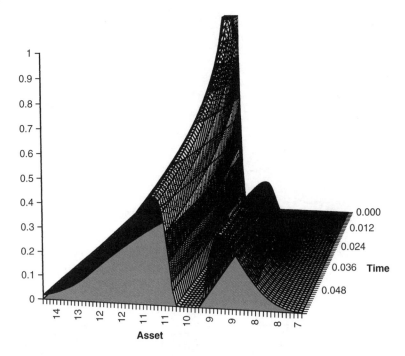

Figure 33.8 The probability density function for a time-independent trading strategy. The initial data is a delta function.

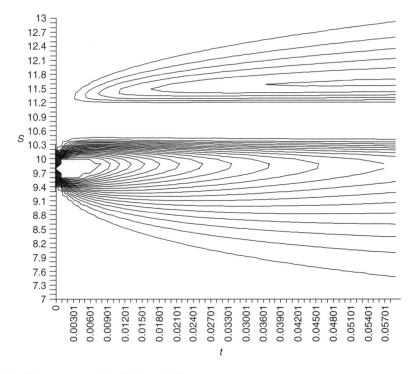

Figure 33.9 The contour plot of Figure 33.8.

33.7.2 Put Replication Trading

The more interesting case is the time-dependent put-replicating trading and the development in the area around the 'Tulip curve'.

The probability density function is shown in Figure 33.10, the corresponding contour plot is shown in Figure 33.11. Since this replication strategy is genuinely time-dependent the corridor that we saw in the above example is now the tulip shape shown in Figure 33.5. The unattainable barren region (the center of the 'Tulip curve') is most easily seen on the contour plot.

In this example, the asset price starts off at $S = 10$ and then evolves. The effect of the replication strategy is felt immediately but the tulip curve itself does not appear until about $t = 0.03$. At this time there appears the barren region around the exercise price (10) which the asset price avoids. This zone is shown most clearly in the contour plot of Figure 33.11. After expiry of the replicated option, $t = 0.04$, the barren zone disappears and all values of S are attainable.

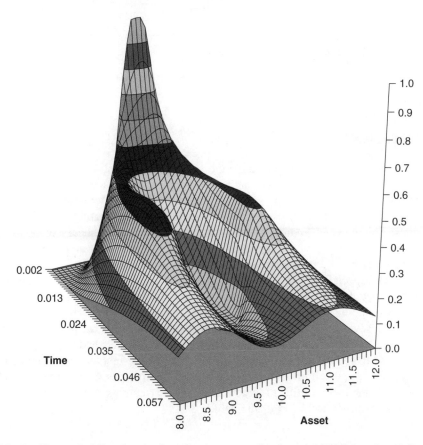

Figure 33.10 The probability density function for put replication; the initial data is a delta function.

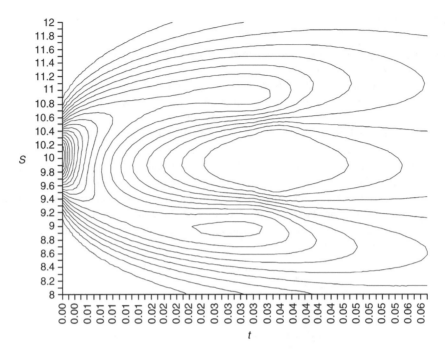

Figure 33.11 The contour plot of Figure 33.10.

33.8 SUMMARY

The influence of trading strategies has been the subject of much discussion but — apart from empirical studies — little theoretical research. In this chapter we have seen a way to formally incorporate trading strategies into the stochastic process followed by the underlying asset.

Many trading strategies that are used today are derived from replicating strategies for derivative securities. This class of trading strategies is also central to the theory of option pricing. We found that the securities whose replication destabilizes the market and their effects on the market can be described by their payoff profile and its gamma.

The effects are especially strong in markets with low liquidity and can even induce discontinuities in the price process. Such price discontinuities are not allowed in classical models of asset prices since they lead to arbitrage. The present model does allow such arbitrage and this can be justified in several ways. First, such effects as I describe do occur in practice and traders with a knowledge of the positions of other market players and their hedging requirements can take advantage of this knowledge. Second, this is just a simple first attempt at modeling feedback. It may be possible to remove certain arbitrage opportunities by incorporating elasticity in the response of the market price to large trades. The barren zone that we have found above may still be unstable, but no longer unattainable.

FURTHER READING

- See Schönbucher (1993), Schönbucher & Wilmott (1995a, b), Frey & Stremme (1995) and Sircar & Papanicolaou (1996) for further examples and analysis of feedback models.

- The book by O'Hara (1995) discusses many of the issues in market microstructure.
- Taleb (1997) discusses the reality of options markets, and feedback, market manipulation and discontinuous price paths are part of that reality. A particularly interesting account of the behavior of asset prices close to the barriers of knock-out options is given here.
- Porter & Smith (1994) describe some laboratory experiments in which simulated stock market trading led to interesting bubbles and crashes.
- See Korn (1997) for a very readable description of these and other models.

CHAPTER 34
utility theory

In this Chapter...

- utility definitions and axioms
- risk aversion
- certainty equivalent wealth
- maximization of expected utility
- ordinal and cardinal utility
- von Neumann–Morgenstern utility functions

34.1 INTRODUCTION

Most of derivatives theory is about hedging and the elimination of risk. For a variety of reasons, some of which we have discussed, perfect hedging is not always possible. In that case we must decide how to 'value' the unhedged residual risk. One way is to simply ignore it, and just concern oneself with real expectations of outcomes. Another is to examine both the average outcome and the standard deviation. We've seen both of these approaches in previous chapters.

Another approach is to consider the **utility** of each outcome; that is, we associate a number measuring the 'happiness' (in a loose sense) that each outcome gives us. Although $10 billion is twice as much as $5 billion, we wouldn't necessarily be twice as happy with the former; we'd be happier, probably, but since both numbers are so staggeringly large and unimaginable, the difference in our happiness would not be that great. We'll see how this fits into quantitative finance theory in later chapters. In this chapter we set the framework.

34.2 RANKING EVENTS

As a general rule people have (at least roughly) a ranking of preferences for various 'commodities'. An E-type Jaguar is preferable to any BMW, for example. Sometimes it is impossible to decide between two commodities, peanut butter cheesecake versus (my own) pumpkin pie. The commodities don't have to be physical quantities, they could be events, or outcomes.

Definition

Let's suppose we have commodities or events A, B, C, ... We write

$$A \succ B$$

if A is strictly preferred to B. Note that this is a personal choice.[1] If we can't decide between A and B then we write

$$A \sim B.$$

We can also write

$$A \succeq B$$

if A is weakly preferred to, or is at least as good as, B.

We must have some axioms, to ensure that the relationship between events makes sense. The usual axioms are as follows.

Axiom 1: Completeness

Given two events A and B, one of the following three must hold.

$$\text{Either} \quad A \sim B \quad \text{or} \quad A \succ B \quad \text{or} \quad B \succ A.$$

Axiom 2: Reflexivity

$$A \sim A.$$

Axiom 3: Transitivity

$$\text{If} \quad A \succeq B \quad \text{and} \quad B \succeq C \quad \text{then} \quad A \succeq C.$$

These axioms are sufficient to give us a ranking between various events. But they do not ensure the existence of a utility function. For this we need another definition and a final axiom.

Definition

Let A and B be two events and let $0 \leq \phi \leq 1$. By

$$\{\phi A + (1 - \phi)B\}$$

we mean the lottery having the two possible outcomes, A with probability ϕ and B with probability $1 - \phi$.

Axiom 4: Continuity

Let A, B and C be events such that

$$A \succ C \succ B$$

then there exists some $0 \leq \phi \leq 1$ such that

$$\{\phi A + (1 - \phi)B\} \sim C.$$

34.3 THE UTILITY FUNCTION

A **utility function** represents the 'worth' or 'happiness' associated with events or outcomes or, from now on, levels of wealth. The last is measured in units of currency, dollars, say. Although rarely used in practice they are often found in the financial and economic literature.

---

[1] Some people like BMWs.

I will denote the utility function by $U(W)$ where W is the wealth, measured at some time horizon. Generally speaking, you would expect utility functions to have the following properties. Here $'$ denotes differentiation with respect to W.

- The function $U(W)$ can vary among investors; each will have a different attitude to risk for example.

- $U'(W) \geq 0$: More is preferred to less. If it is a strict inequality then satiation is not possible, the investor will always like more than he has. This slope measures the marginal improvement in utility with changes in wealth.

- Usually $U''(W) < 0$: The utility function is strictly concave. Since this is the rate of change of the marginal 'happiness', it gets harder and harder to increase happiness as wealth increases. An investor with a concave utility function is said to be **risk averse**. This property is often referred to as the law of diminishing returns.

34.4 **RISK AVERSION**

The **absolute risk aversion function** is defined as

$$A(W) = -\frac{U''(W)}{U'(W)}.$$

The **relative risk aversion function** is defined as

$$R(W) = -\frac{W U''(W)}{U'(W)} = WA(W).$$

34.5 **SPECIAL UTILITY FUNCTIONS**

When it comes to choosing particular utility functions there are some popular choices.

Constant Absolute Risk Aversion (CARA)

The choice

$$U(W) = -\frac{1}{\eta}e^{-\eta W} \quad \text{with } \eta > 0$$

is a **Constant Absolute Risk Averse (CARA)** utility function. The absolute risk aversion function is just

$$A(W) = \eta, \quad \text{a constant.}$$

Constant Relative Risk Aversion (CRRA)

The choice

$$U(W) = \frac{W^\gamma - 1}{\gamma} \quad \text{with } \gamma < 1 \text{ and } \gamma \neq 0$$

is a **Constant Relative Risk Averse (CRRA)** utility function. The function $U(W) = \log W$ is also a member of this family; it's the limit of the above as $\gamma \to 0$.

The relative risk aversion function is just

$$R(W) = 1 - \gamma, \quad \text{a constant.}$$

Hyperbolic Absolute Risk Aversion (HARA)

The choice

$$U(W) = \frac{1 - \gamma}{\gamma} \left(\frac{\beta W}{1 - \gamma} + \eta \right)^{\gamma}$$

is a **Hyperbolic Absolute Risk Averse (HARA)** utility function. It is only valid for $\beta W / (1 - \gamma) + \eta > 0$ but is otherwise a very broad family.

34.6 CERTAINTY EQUIVALENT WEALTH

THE C.E.W. PUTS RISKY INVESTMENTS ON THE SAME FOOTING

When the end of period wealth is uncertain, and all outcomes can be assigned a probability, one can ask what amount of certain wealth has the same utility as the expected utility of the unknown outcomes. In other words, solve

$$U(W_c) = E[U(W)].$$

The quantity of wealth W_c that solves this equation is called the **certainty equivalent wealth**. One is therefore indifferent between the average of the utilities of the random outcomes and the guaranteed amount W_c.[2]

Investor 1 is more risk-averse than Investor 2 if for every portfolio

$$W_c|_1 < W_c|_2.$$

In words, the certainty equivalent of 1 is less than the certainty equivalent of 2 for every portfolio.

The concavity of the utility function ensures that

$$E[U(W)] < U(E[W]).$$

It follows that

$$U(W_c) < U(E[W])$$

and since $U'(W) > 0$

$$W_c < E[W].$$

i.e., if possible always accept the average outcome in place of a random one.

Example

Faced with winning or losing one dollar on the toss of a coin what should you, a risk-averse person, do?

---

[2] U(A bird in the hand) $= \sum_{i=1}^{2}$ Prob(Catching bird i) U(Bird $i \in$ bush).

Assuming that you start off without any money at all, the expected utility after the coin toss would be

$$\tfrac{1}{2}U(1) + \tfrac{1}{2}U(-1).$$

This is less than the utility of the expected wealth $U\left(\tfrac{1}{2} - \tfrac{1}{2}\right)$ which in this case also corresponds to not taking part in the contest in the first place. If you want to know how much the bet is (certainly) equivalent to, you must solve

$$U(W_c) = \tfrac{1}{2}U(1) + \tfrac{1}{2}U(-1).$$

The answer will be less than the average outcome of zero.

Let's do that calculation again but starting with a wealth of W and with a very small bet of ε (so that we can do a Taylor series approximation).

The expected utility after the coin toss is

$$\tfrac{1}{2}U(W + \varepsilon) + \tfrac{1}{2}U(W - \varepsilon) \sim U(W) + \tfrac{1}{2}\varepsilon^2 U''(W) + \cdots.$$

(The two first derivative terms cancel each other.) This is less than the utility of the average which is just $U(W)$.

The certainty equivalent is given by

$$U(W_c) = U(W) + \tfrac{1}{2}\varepsilon^2 U''(W) + \cdots.$$

It follows that[3]

$$W_c \sim W + \tfrac{1}{2}\varepsilon^2 \frac{U''(W)}{U'(W)} + \cdots$$

Of course, this analysis can easily be extended to arbitrary distribution of outcomes as long as the Taylor series is still valid...

The expected utility after a 'bet/investment' having a probability density function of $p(W)$ is

$$\int p(w)U(w)\,dw \approx \int p(w)\left(U(W) + (w - W)U'(W) + \tfrac{1}{2}(w - W)^2 U''(W) + \cdots\right)dw$$

$$= U(W) + \left(\int wp(w)\,dw - W\right)U'(W) + \tfrac{1}{2}U''(W)\int (w - W)^2 p(w)\,dw + \cdots. \quad (34.1)$$

The utility of the average is just

$$U\left(\int wp(w)\,dw\right) = U\left(\int (w - W)p(w)\,dw + W\right) \approx U(W) + \left(\int wp(w)\,dw - W\right)U'(W)$$

$$+ \tfrac{1}{2}U''(W)\left(\int wp(w)\,dw - W\right)^2 + \cdots. \quad (34.2)$$

(Under what conditions are these expansions valid?)

Comparing (34.1) and (34.2) the first two terms of each are the same, while the third term makes the former expression smaller. And the certainty equivalent?

The certainty equivalent is given by

$$U(W_c) = U(W) + \left(\int wp(w)\,dw - W\right)U'(W) + \tfrac{1}{2}U''(W)\int (w - W)^2 p(w)\,dw + \cdots.$$

---

[3] So that's where the absolute risk aversion function comes from.

It follows that

$$W_c \approx \int w\,p(w)\,dw + \frac{1}{2}\frac{U''(W)}{U'(W)}\int (w-W)^2\,p(w)\,dw + \cdots.$$

The first term in this is just the average future wealth and the second term is the correction due to the risk aversion of the investor.

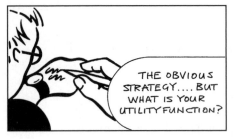

THE OBVIOUS STRATEGY.... BUT WHAT IS YOUR UTILITY FUNCTION?

34.7 **MAXIMIZATION OF EXPECTED UTILITY**

One of the main uses of utility theory is in choosing the optimal investment from a choice of investments. Typically these investments will be risky, having unknown outcomes at the end of the time horizon.

Example

If there are N assets, each having a random return R_i, then one way to optimize our portfolio is to choose the weights w_i of each asset such that the expected utility

$$E\left[U\left(W_0\sum_{i=1}^{N}w_iR_i\right)\right]$$

is maximized. Here W_0 is the wealth initially invested. This must be maximized subject to the budget constraint

$$\sum_{i=1}^{N}w_i = 1.$$

If we add an $N+1$th risk-free asset earning interest of r over the horizon then the optimization problem becomes

$$\max_{w_1,\ldots,w_n} E\left[U\left(W_0\left(r+\sum_{i=1}^{N}w_i(R_i-r)\right)\right)\right].$$

This incorporates the budget constraint.

34.7.1 Ordinal and Cardinal Utility

If we are only concerned with the ranking of events then we could equally use as our utility function $\Phi(U(W))$ where Φ is any strictly monotonically increasing function. Sometimes, the actual value of the utility function is important.

Two investors having different utility functions may rank events in the same order (this is **ordinal utility**) but when faced with uncertain outcomes/investments they may decide differently from each other (they have different **cardinal utility**).

We've just seen how to maximize expected utility to solve problems such as finding an optimal portfolio. If we are interested in finding this maximum then the new function

$$aU(W)+b \quad \text{with } a > 0$$

will do just as well since it has a maximum at the same value of W as the original $U(W)$. For an individual, the utility function is arbitrary up to a positive affine transformation (i.e. multiply by a positive constant and translate by a constant). For this reason specifying either the function $A(W)$ or $R(W)$ is sufficient information to rank an investor's choice of investments in terms of their cardinal utility.

Example

Let's see how the ordinal/cardinal distinction works in practice. An investor has the utility function $U(W) = W$. (We should have $U'' < 0$ but I want to keep things as simple as possible.) He has a choice between entering a lottery with outcomes of 0 or 9, with equal probabilities, or receiving 4. The expected utility of the first choice is

$$\tfrac{1}{2} \times 0 + \tfrac{1}{2} \times 9 = 4.5.$$

This beats the utility of the second choice which is 4; he takes the gamble.

Now consider a different utility function, which is in a sense just a montonic transformation of the first function, such as $U(W) = W^{1/2}$. The outcomes 0, 4, 9 are ranked in that order still, but what happens to the utility when considering expectations? The first choice has expected utility

$$\tfrac{1}{2} \times 0 + \tfrac{1}{2} \times 3 = 1.5.$$

This is now not as good as the second choice which has utility of 2. He takes the certain outcome.

The cardinal utility function or **von Neumann–Morgenstern utility function** attaches special meaning to the numerical value of the utility, and is more than just a mechanism for ranking certain outcomes.

We will be seeing more of expected utility maximization in later chapters.

34.8 **SUMMARY**

Utility theory is a useful framework for valuing risk. It's not popular outside of academic, and in particular, economic, circles but it is a subject than I am warming to. It gets a lot more complicated than what I've shown you here, but this will serve our purposes.

FURTHER READING

- For a more general discussion of utility theory see Ingersoll (1987) (from whom some of the above examples have been borrowed) and Owen (1995).

CHAPTER 35

more about American options and related matters

In this Chapter...

- why some people might exercise American options at non-optimal times
- the effect of non-optimal exercise on the profit of the option writer

35.1 INTRODUCTION

In 1998 Dr Hyungsok Ahn, then my postdoctoral researcher at Oxford University, now one of my employees, and I wrote a paper on the exercise of American options. The paper was never published in any learned journal.

In this chapter I reproduce the paper with just a few minor changes. However, to start with, here's what the nice *Derivatives Week* published by us on the subject of early exercise.

35.2　**WHAT *DERIVATIVES WEEK* PUBLISHED** © 1999 Institutional Investor, Inc., (article reproduced with kind permission)

8

DERIVATIVES WEEK

JANUARY 11, 1999
www.iinews.com

LEARNING CURVE® EXERCISING AMERICAN OPTIONS

INTRODUCTION

In this article we explain the ideas behind the valuation of options with early exercise features, so-called American options. We also aim to clarify some popular misconceptions about when an American option should be exercised.

HOW TO PRICE AMERICAN OPTIONS

If an option can only be exercised at expiry it is called European. If it can be exercised at any time prior to expiry, it is called American. Bermudan options have prespecified exercise dates which may be particular days or whole periods of time. Because they give the holder more rights, the American option is at least as valuable as an equivalent Bermudan option which in turn is at least as valuable as an equivalent European contract.

The idea behind valuing options with early exercise is to decide when the option should be exercised. Is there, in some sense, a best or optimal time for exercise? To correctly price American options we must place ourselves in the shoes of the option-writer. We must be clear about the principles behind his strategy. From the modeling point of view we assume that the writer of the option also is hedging his option position by trading in the underlying asset. The hedging strategy is dynamic and referred to as delta hedging. The position in the underlying asset is maintained delta neutral so as to be insensitive to movement of the asset. By maintaining such a hedge, the writer does not care about the direction in which the underlying moves, he eliminates all asset price risk. However, he does remain exposed to the exercise strategy of the option holder. If the writer makes an assumption about when the holder will exercise his option and this assumption turns out to be incorrect, this will have an impact on the writer's profit. Since the writer cannot possibly know what the holder's strategy will be, how can the writer reduce his exposure to this strategy?

The answer is simple. The writer assumes that the holder exercises at the worst possible time for the writer. He assumes that the option is exercised at the moment that gives the writer the least profit. This is often referred to as the optimal stopping time, although as far as the writer is concerned it is the last thing he wants to happen. So, out of all the possible exercise strategies we must find the one that gives the option the least value to the writer or the highest value to the holder. This sounds very complicated but anyone who has implemented the binomial method knows that it is just a matter of adding one line of code to the program. That line of code simply tests at each node in the tree whether the theoretical option value is greater than the payoff, if it is not then the payoff is used instead, and this corresponds to a time at which the option should be exercised.

To summarize this section, the assumptions are that the option writer is delta hedging and prices the option at the highest possible value over all exercise strategies.

WHEN SHOULD THE HOLDER EXERCISE?

The holder of the option rarely delta hedges. Perhaps he has bought the option as a static hedge for the rest of his portfolio, or perhaps as a speculative investment. Either way it is unlikely that he is insensitive to the direction of the underlying asset. The initial assumption concerning the writer of the option does not apply to the holder. Should the holder therefore act in the optimal way that follows from the two assumptions summarized at the end of the previous section?

Consider the simplest scenario. You buy a call because you believe the underlying asset is going to rise significantly. If you are correct you will make a substantial return. If there are no dividends on the underlying then it is 'theoretically' never optimal to exercise before expiry. We put the word theoretically between inverted commas because the 'theory' is only relevant to someone who is delta hedging. Now suppose that the stock does indeed rise, but the economic situation makes you believe that a sudden fall is imminent. What should you do? The obvious solution is to sell the option and lock in your profit. But this may not be possible, for example if the option is over-the-counter. The only way of locking in the profit may be to exercise the option early. The theory says don't exercise, but if the stock does fall then you lose the profit. At this stage it is important to remember that the theory is not relevant to you.

The writer and the holder of the option have different priorities, what is optimal to one is not necessarily optimal

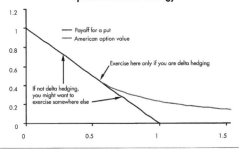

'Optional' Exercise Strategy

Payoff for a put
American option value

Exercise here only if you are delta hedging

If not delta hedging, you might want to exercise somewhere else

DERIVATIVES WEEK

JANUARY 11, 1999

www.iinews.com

9

to the other. The holder of the option simply may have a gut feeling about the stock and decide to exercise. That is perfectly valid. Or he may have a stop-loss strategy in place. He may even have a complex utility maximization strategy. Anyway, it is highly unlikely his exercise time will correspond to that calculated by the writer of the option.

The figure shows the payoff and theoretical value for an American put option some time before expiry. If you are delta hedging there is an optimal asset value at which to exercise.

HOW DOES THE WRITER FEEL ABOUT EXERCISE?

The writer has received a sum of money in exchange for the option. That sum of money was calculated assuming that the option holder exercises at a certain optimal time. This optimal exercise strategy gives the option its highest theoretical value. The writer receives this maximum amount even though the holder may exercise at any time.

It is clear that the writer can never lose. The worst that can happen to him is that the option is exercised at this theoretical optimal time. But this has already been priced into the premium he received. On the other hand, if the holder exercises at some other time he can only benefit. How does the writer feel about exercise? At worst he has no feelings, at best he is very happy!

CONCLUSION

Since there is a clever but complex theory behind the exercise of American options there has arisen some misunderstanding about the optimal time to exercise. It is commonly believed that there is only one optimal time. This is far from being true. Finally, we would like to make the observation that if you want to sell someone an option and early exercise doesn't add too much to the theoretical value then you should always make it American. This gives you the possibility of a surprise windfall profit if the holder exercises at an unexpected time.

*This week's Learning Curve was written by **Hyungsok Ahm**, a visiting academic at **Oxford University** and a Director of **Wonüng Financial Consultants** based in Korea, and **Paul Wilmott**, a Royal Society University Research Fellow at **Oxford University** and **Imperial College**, London.*

35.3 HOLD THESE THOUGHTS

I don't want you to get the wrong idea while reading this so I want you to keep the following in mind.

- The writer is in the Black–Scholes world, including no transaction costs
- The option holder can't delta hedge (why would he buy the option if he could?)
- The option cannot be sold, think of it as OTC

This means that the holder cannot close his position other than by exercising.

35.4 CHANGE OF NOTATION

Because this is the original paper I have kept the original notation. The main differences are

- For S_t and s read S

- For $v(t, s)$ read $V(S, t)$

- For v_t read $\dfrac{\partial V}{\partial t}$

- For v_s read $\dfrac{\partial V}{\partial S}$

- For v_{ss} read $\dfrac{\partial^2 V}{\partial S^2}$

- For K read E

35.5 AND FINALLY, THE PAPER...

I LIKE THIS A LOT

On Trading American Options

Hyungsok Ahn and Paul Wilmott
OCIAM, Oxford University

Abstract.   In this paper we consider the effect that early exercise has on the profit of the writer of an American option. The American option is correctly priced in the Black–Scholes framework assuming that the holder exercises at the worst possible time for the writer. This early exercise time is unique. But why should the holder exercise at this time? If the holder of the American option wants to both eliminate all risk by delta hedging and maximize his wealth then certainly he should exercise here. But if he is following this strategy why does he bother to buy the option? Complete markets make purchasing the option unnecessary. So let's assume that he is following some other strategy, as he is free to do. Now, clearly he would always be better off selling the option than exercising it, but what if the contract is OTC, and he can only close his position by exercise? If the option holder follows some strategy that results in early exercise at a time not given by the classical optimal free boundary then the writer makes more profit than might be expected. As examples, we assume that the holder exercises according to the maximization of his own utility function. We illustrate our results by applying them to several families of utility functions, namely the CARA, the HARA, and the expected return. While the option holder maximizes his utility, the issuer gains from the difference between the price maximizing exercise boundary and the exercise boundary of the option holder. We provide numerical results that describe the effect of the physical drift and the risk aversion on the issuer's expected profit.

I   INTRODUCTION

The American option is an option contract that allows the option holder to exercise before maturity if he is better off doing so. Because of the flexibility of choosing the exercise time, the price of the option is calculated as the value of the option in the worst case for the issuer among all feasible exercise strategies that the option holder may perform. Typically the price-maximizing exercise time, and hence the least favorable exercise time for the issuer, is described as an optimal stopping time, and the resulting pricing equation becomes a free-boundary partial differential equation (PDE). Although it may appear that it requires rather sophisticated mathematics to price an American option, the fundamental concept of the absence of arbitrage is still an integral part of determining the price. The issuer can construct a hedging portfolio involving the trading of the underlying assets in such a way that the value of the replicating portfolio (i.e., the upfront premium for the option plus the result of the trading) is not less than his liability even when his customer exercises the option at the least favorable time. Throughout this paper, we will assume that the American option is sold at this classical price and that the market is complete.

If he so desired, the option holder could delta hedge his position as well, constructing his portfolio exactly opposite to that of the issuer's and exercising his option at the price-maximizing exercise time. In this case, the balances of both the issuer and the option holder would be equal to zero. If this was what they intended, they could have chosen not to trade the option at the beginning and saved their effort in maintaining their hedge positions. In fact, option traders may choose not to construct a risk-free portfolio, especially when they use options as a means of investment instrument. This has nothing to do with market completeness: those who visit Las Vegas play slot machines even when they can sustain a risk-free wealth by not playing. Thus it is reasonable to assume that the option holder engages in some strategies. For example, he may adopt a stop-loss strategy: buy-and-hold the option until he decides to exercise it. Unlike the previous case, the option holder may gain or lose depending upon the behavior of the underlying asset price while the potential loss is not more than the premium he paid. The issuer gains unless the option holder exercises at the first time that the asset price reaches the price-maximizing exercise boundary. One of the questions we address in this paper is 'Should the option *holder* exercise at the price-maximizing exercise time?' Clearly, the option holder would always be better off selling the option than exercising it, but if the contract is OTC this possibility may not exist. We could, for example, assume that exercise is the only way available for him to close his position.

Trading an option is not a two-person zero-sum game because both the issuer and the holder can trade the underlying asset with other investors. For example, if the issuer maintains a risk-free portfolio by trading the underlying asset while the holder leaves the option unhedged, the issuer's balance will eventually become zero while the holder's depends upon where the underlying asset price lands at the exercise time. Hence, the worst case for the issuer is not necessarily the best case for the holder. It is not always true that the option holder is better off exercising the option at the price-maximizing exercise time. First, we consider the physical drift of the underlying asset. The price of an option depends upon the risk-neutral drift, not the physical drift of the underlying asset. The reason is that the presence of an option immediately allows one to construct locally riskless portfolios, and hence the risk-free rate is the only one that governs its price. As a result, the price-maximizing exercise boundary is also independent of the physical drift. Can we assert that the optimal exercise boundary for the option holder is not affected by the market direction? Almost certainly not. For example, it is well known that the price-maximizing exercise time for the American call is the maturity of the contract, provided that the underlying stock pays no dividend. If there is any evidence that the price of the underlying asset is expected to fall, however, a wise investor would exercise his call earlier before it expires worthless. Second, each investor has his own risk preference. Two different rational decision-makers may exercise differently, even when they agree on the probability distribution of outcomes. It is nonsense to argue that a single exercise strategy is the optimal strategy for every investor. As an aside, there is evidence (Overdahl and Martin (1994), for example) that a substantial proportion of all exercised American call options are exercised before their classical optimal stopping time. This is probably due to friction factors, errors or differences of opinion about parameter values. Our paper also applies in these situations.

This paper establishes the optimal exercise boundary provided that the option holder is a utility-maximizing investor. The optimal exercise boundary, or the utility-maximizing exercise boundary, depends upon the risk aversion and the physical drift. In Theorem 3.1 we confirm the following: (i) if the option holder is sufficiently risk-averse, early exercise is optimal even for a call; (ii) the optimal exercise time is a non-decreasing function of the physical drift, if the option is a call; (iii) if the option is a put, the optimal exercise time is a non-increasing function of the physical drift. We illustrate these results with several families of utility functions: the constant absolute risk averse (CARA), the hyperbolic absolute risk averse (HARA) and the linear utility (i.e. the expected return). Some of the highlights are:

(a) If the option holder's utility is of the CARA type, early exercise prevails for both call and put regardless of the absolute risk aversion parameter.

(b) Certain HARA utilities may yield two separate exercise boundaries.

(c) Upon the expected return criterion, a call option is exercised early only when the physical drift is surpassed by the risk-free rate.

Another result of this paper is the equation for the expected profit selling American options. As we stated earlier, the issuer gains from the difference between the price-maximizing exercise time and the exercise time chosen by his customer. The profit grows as the occupation time of the asset price in the region between the exercise boundaries of the price maximization and the utility maximization. The difference between the value of the option and the exercise value is the final piece of the profit. We provide numerical results on how the physical drift and the risk aversion affect the issuer's profit.

The paper is structured as follows. In the next section we review the classical results of pricing and hedging American options. In Section 3, we find the optimal exercise time for the utility maximizing investor. In Section 4, we analyze the effect of the option holder's optimal exercise strategy on the issuer's profit. Section 5 contains concluding remarks.

2 PRELIMINARY: PRICING AND HEDGING

The early exercise feature makes the valuation of the American option more intriguing than that of its European counterpart. The main concepts are the optimal stopping and the corresponding parabolic variational inequalities. Myneni (1992) surveyed the literature on the subject and summarized key results. Here we state the standing assumptions for the rest of the paper and review the variational inequalities.

The classical theory of option pricing is predicated on many assumptions for market completeness. We assume that the market is frictionless, that short-selling is allowed without restriction, that one can trade assets as frequently as one wishes, that all risk-free assets grow at the common rate r which is known *a priori*, and that there is a unique risk-neutral equivalent martingale measure. The last assumption becomes less abstract when we assume that the price of the underlying asset follows a geometric Brownian motion and that market participants are not capable of foreseeing the future. Thus, in what follows the price of the underlying asset evolves as:

$$dS_t = \sigma S_t \, dW_t + \mu S_t \, dt \tag{1}$$

where W is a standard Brownian motion. In addition the filtration is natural, meaning that the stream of information consists of the observations of the asset price only.

As shown in Harrison and Pliska (1981), the complete market assumption allows a trader to replicate the payoff of an arbitrary contingent claim by trading the underlying assets. We start by assuming that the issuer of the option maintains Δ shares of the underlying asset to hedge his position. In other words, the value of the issuer's portfolio is given by $\Delta S - v$ where v is the value of the option. This portfolio must grow at least at the risk-free rate r:

$$\Delta \, dS_t - dv \geq r(\Delta S_t - v) \, dt. \tag{2}$$

Because of the Markovian nature of the underlying asset price (1), the value of the option v at time t is a function of t and the asset price S_t. For the time being, we assume that v is continuously differentiable with respect to t and twice continuously differentiable with respect to s. Then we have

$$dv(t, S_t) = v_t(t, S_t)\, dt + v_s(t, S_t)\, dS_t + \tfrac{1}{2}\sigma^2 S_t^2 v_{ss}(t, S_t)\, dt. \tag{3}$$

which follows from Itô's formula. It is required for the issuer to pick $\Delta = v_s$ in order to fulfill (2) because the random growth dS_t is of the order $\sqrt{dt}$ and is much bigger than the dt terms. Rearranging (2) after replacing Δ by v_s yields:

$$\mathcal{L}v = v_t + \tfrac{1}{2}\sigma^2 s^2 v_{ss} + r(sv_s - v) \leq 0 \tag{4}$$

for each s. The value of the option will never fall below an immediate exercise value, otherwise the issuer loses. This yields the second condition for v:

$$v \geq \phi \tag{5}$$

where ϕ is the payoff of the option: $\phi(s) = \max(s - K, 0)$ for a call with strike K and $\phi(s) = \max(K - s, 0)$ for a put. At each time t, the option holder may or may not exercise his option. If $v > \phi$ at this moment, then exercising the option is not the least favorable outcome for the issuer because he can claim a non-zero profit $v - \phi$ instantly. In this case, $\mathcal{L}v = 0$ because the issuer has an arbitrage opportunity if $\mathcal{L}v$ were strictly less than zero. Therefore we obtain the third condition:

$$(\mathcal{L}v) \cdot (v - \phi) = 0. \tag{6}$$

The inequalities (4), (5), and (6) subject to $v(T, s) = \phi(s)$ form a parabolic obstacle problem. We refer to Friedman (1988) for the existence and the uniqueness of the solution to such problems. Jaillet, Lamberton, and Lapeyre (1990) showed that the solution of the parabolic variational inequalities (4), (5), and (6) has a continuous gradient at the free boundary (i.e. a smooth fit), and Van Moerbeke (1976) showed that the optimal stopping boundary is continuously differentiable. Thus Itô's formula (3) is valid at least in a weak sense: see San Martin and Protter (1993) for detail.

In the theory of optimal stopping, the space-time domain defined by $v > \phi$ is called the continuation region as the stopping is premature in this region and the graph of its boundary is called the optimal stopping boundary. In this paper we call this the price-maximizing exercise boundary, distinguishing it from the optimal stopping boundary from the utility maximization problem in the next section. We stress that the contents of this paper in no way conflict with the Black–Scholes–Merton theory of option pricing and hedging, that we are in a complete market and that there is thus only one fair value for an American option and this is given by the classical approach.

3 UTILITY MAXIMIZING EXERCISE TIME

We assume that the option holder possesses a utility function $U : \Re \to \Re$ that is strictly increasing and twice continuously differentiable. The investor, who purchases an American option at time 0, will select his exercise time by maximizing the expected utility of his discounted wealth. The class of feasible exercise times consists of stopping times that are less than or equal to T, the maturity of the option. This includes exercise times that are strategically selected based upon the price of the asset up to that date as well as pre-scheduled times (i.e., non-random). A feasible exercise time will be denoted by τ. If the option holder never exercises the option we set $\tau = T$. As before ϕ is designated for the payoff. Then, at time t, the option holder faces the following optimal stopping problem:

$$u(t, s) = \sup_{t \leq \tau \leq T} E^{t,s}[U(e^{-r\tau}\phi(S_\tau))] \qquad (7)$$

where $E^{t,s}$ is the conditional expectation given that $S_t = s$, τ is the option holder's exercise time, and ϕ is the payoff. The essential supremum is taken over all the feasible exercise times. Finally the expectation is governed by the physical measure, not the risk-neutral equivalent martingale measure. We consider only situations when (7) is well defined. A sufficient condition is that $U \circ \phi$ is bounded by a polynomial.

We could have defined u as the expected utility of $e^{-r\tau}\phi(S_\tau) - v(0, S_0)$, the discounted payoff minus the option price. In our definition, the option price is a part of the utility function U, since we treat the option price as a constant.

As in the case of the price maximization, the optimal stopping problem (7) is equivalent to a parabolic obstacle problem. Thus u satisfies a set of variational inequalities. We will describe the variational inequalities financially, omitting technical details. For notational convenience, we define $g(t, s) = U(e^{-rt}\phi(s))$. First we check that

$$u \geq g. \qquad (8)$$

This is because the maximum expected utility is not smaller than the utility of immediate exercise which is a special case of feasible stopping times. Next we will explain the following inequality for $t < T$:

$$u(t, s) \geq E[u(t + \delta, S_{t+\delta})] \qquad (9)$$

for each δ that makes $t + \delta$ a feasible exercise time. Note that the right side of (9) coincides with the expected utility when the option holder pursues the optimal stopping only after δ elapses. In other words, the option holder is dormant until time $t + \delta$ and he tries to find an optimal exercise time from then on. Thus the value of this expected utility cannot exceed the maximum expected utility which is on the left side of (9). The implication of (9) is the following inequality:

$$\mathcal{L}_\mu u = u_t + \mu s u_s + \tfrac{1}{2}\sigma^2 s^2 u_{ss} \leq 0 \qquad (10)$$

which is obtained by applying Itô's formula to u. If $\mathcal{L}_\mu u < 0$, then the maximum expected utility is expected to fall in an infinitesimal time, and hence the optimal strategy is to exercise the option immediately. That is, $u = g$. Therefore u must satisfy

$$(\mathcal{L}_\mu u) \cdot (u - g) = 0. \qquad (11)$$

The set of variational inequalities (8), (10), and (11) with terminal data $g(T, s)$ characterizes the maximum expected utility u. The optimal exercise time is the first time that the asset price S_t hits the free boundary of the inequalities.

Next we consider $h = e^{rt} U^{-1} \circ u$, the maximum expected certainty equivalence. U^{-1}, the inverse of U, is well defined as U is an increasing function of wealth. The merit of using this change of variable is that it facilitates us comparing the utility maximization to the price maximization. We confirm that h must satisfy the following variational inequalities:

$$h \geq \phi$$

$$\mathcal{D}h = h_t + \tfrac{1}{2}\sigma^2 s^2 \left(h_{ss} + \frac{U''}{U'}(e^{-rt}h)e^{-rt}(h_s)^2 \right) + \mu s h_s - rh \leq 0 \qquad (12)$$

$$(\mathcal{D}h) \cdot (h - \phi) = 0$$

subject to $h(T, s) = \phi(s)$. Therefore the utility-maximizing exercise boundary depends upon the physical drift and Pratt's measure of absolute risk aversion $-U''/U'$, and is different from the price-maximizing exercise boundary. The distortion in discount is caused by the nonlinearity of the utility function.

The utility-maximizing exercise time for an American option has the following properties:

(i)   If the absolute risk aversion is sufficiently large, then there is a positive probability of early exercise for both call and put.

(ii)   The exercise time is non-decreasing in μ, when the option is a call.

(iii)   The exercise time is non-increasing in μ, when the option is a put.

Proof.   Note that the exercise region coincides with the space-time domain of $\mathcal{D}h < 0$. If the absolute risk aversion $-U''/U'$ tends to infinity uniformly in its argument, then $\{(t, s) : \mathcal{D}h < 0, \; 0 \le t < T, \; s > 0\}$ is a set of a positive measures. Since the support of a non-degenerate geometric Brownian motion (i.e. $\sigma^2 > 0$) occupies the entire positive plane, the utility-maximizing exercise time can be less than the maturity with a positive probability. This proves (i). When the option is a call, h_s is positive. Thus $\mathcal{D}h$ becomes more negative when μ becomes smaller. If the option is a put, h_s is negative, and hence $\mathcal{D}h$ becomes more negative when μ becomes larger. Therefore we have (ii) and (iii). $\square$

Our next task is to locate the boundary when the time to maturity is arbitrarily close to zero. Note that the certainty equivalence h tends to ϕ as $t \to T$ and the utility-maximizing exercise boundary (as a function of time) is continuously differentiable. Thus when t is near T, the utility-maximizing exercise boundary is close to the boundary of $\mathcal{D}\phi < 0$. This is *the boundary at maturity*. If $\phi(s) = \max(s - K, 0)$ (i.e., a call option), then the boundary is above the strike K for each $t \in [0, T)$ and hence the boundary at maturity is

$$\partial \left[s > K : \tfrac{1}{2}\sigma^2 s^2 \frac{U''}{U'} \left(e^{-rT}(s - K) \right) e^{-rT} + (\mu - r)s + rK < 0 \right]. \tag{13}$$

Here, the symbol ∂ is used for indicating the boundary of a set. Similarly, if $\phi(s) = \max(K - s, 0)$ (i.e. a put option), the boundary at maturity is

$$\partial \left[s < K : \tfrac{1}{2}\sigma^2 s^2 \frac{U''}{U'} \left(e^{-rT}(K - s) \right) e^{-rT} - (\mu - r)s - rK < 0 \right]. \tag{14}$$

Sometimes (13) and (14) may contain more than one element. In such a case, we have more than one free boundary. In the remainder of this section, we provide an explicit expression for the boundary at maturity when the option holder's utility belongs to one of the following categories: the CARA, the HARA, and the expected return (i.e. the linear utility).

3.1   Constant Absolute Risk Aversion

This is the case when the absolute risk aversion is a constant regardless of the wealth of the investor. That is, $-U''/U' \equiv \lambda$ for a positive constant λ. Up to a constant, the utility is of the form $U(\omega) = -\alpha e^{-\lambda \omega}$ for a positive constant α.

First we consider a call option. We confirm that the boundary at maturity (13) reduces to

$$\max \left(K, \frac{1}{\lambda \sigma^2} \left(\mu - r + \sqrt{(\mu - r)^2 + 2\lambda \sigma^2 K r e^{-rT}} \right) e^{rT} \right). \tag{15}$$

Note that (15) tends to infinity as λ tends to zero. Hence as the risk aversion of the option holder vanishes, the utility-maximizing exercise time tends to the maturity which coincides with the price-maximizing exercise time. Next we consider a put option. The inequality in (14) is

$$-\tfrac{1}{2}\sigma^2\lambda e^{-rT}s^2 - (\mu - r)s - rK < 0. \tag{16}$$

If the physical drift is at least the risk-free rate ($\mu \geq r$), (16) is true for all positive s. Thus the boundary at maturity is K. Suppose that $\mu < r$. The quadratic inequality (16) is always satisfied if

$$d = (m - r)^2 - 2\lambda\sigma^2 K r e^{-rT} < 0.$$

In this case the boundary at maturity is also K. Now suppose that $d \geq 0$ as well as $\mu < r$. Solving the quadratic inequality (16), we obtain the boundary at maturity:

$$\min\left(K, \frac{1}{\lambda\sigma^2}\left(r - \mu + \sqrt{(r - \mu)^2 - 2\lambda\sigma^2 K r e^{-rT}}\right)e^{rT}\right).$$

3.2 Hyperbolic Absolute Risk Aversion

Merton (1990) provides a complete description of this family of utility functions. The hyperbolic absolute risk aversion means $-U''/U'(\omega) = \lambda/(\omega + \alpha)$ for a positive constant λ. This utility applies in the case when the wealth of the investor is bounded below $\omega + \alpha > 0$. Thus the richer the investor is, the less he is risk-averse. Up to a constant shift,

$$U(\omega) = \begin{cases} \dfrac{1}{\beta^\lambda} \dfrac{(\omega + \alpha)^{1-\lambda}}{1 - \lambda}, & \text{if } \lambda \neq 1 \\[2ex] \dfrac{1}{\beta} \log(\omega + \alpha), & \text{otherwise} \end{cases}$$

where $\beta > 0$. The parameter α is assumed positive as the option payoff could be zero.

Simple algebra reduces the inequalities in (13) and (14) to quadratic inequalities. For example, (13) is equivalent to

$$\partial[s > K : As^2 + Bs + C < 0] \tag{17}$$

where $A = \left(\mu - r - \tfrac{1}{2}\sigma^2\lambda\right)e^{-rT}$, $B = (\mu - r)(\alpha - e^{-rT}K) + re^{-rT}K$, and $C = rK(\alpha - e^{-rT}K)$. The continuation region and the exercise boundary depend upon the choice of parameters. An unusual case is when the parameters satisfy the following:

$$r + \tfrac{1}{2}\sigma^2\lambda < \mu < \tfrac{1}{2}\sigma^2\lambda\frac{e^{-rT}K}{\alpha}.$$

In this case, the continuation region near the maturity is separated by the exercise region:

$$[s > K : As^2 + Bs + C < 0] = \left[s : K < s < \frac{-B + \sqrt{B^2 - 4AC}}{2A}\right].$$

If the physical drift is sufficiently large, the option is very valuable to the holder when the option is very in-the-money. If not, the curvature reduces the holder's utility. Also note that there is no exercise boundary if

$$\mu > r + \tfrac{1}{2}\sigma^2\lambda \quad \text{and} \quad \alpha > e^{-rT}K.$$

This is the case when the physical drift is large while the risk aversion is not.

3.3 The Expected Return

This is a special case of $U(\omega) = \alpha\omega + \beta$ for a positive constant α. As U'' vanishes in this case, our analysis on the boundary at maturity becomes straightforward.

When the option is a call, the inequality in (13) becomes $(\mu - r)s + rK < 0$. This is never satisfied if $\mu \geq r$. Thus the utility-maximizing exercise time is the maturity when the physical drift is at least the risk-free rate. If $\mu < r$, on the other hand, the boundary at maturity is

$$\max\left(K, \frac{r}{r-\mu}K\right).$$

Next we consider a put option. If $\mu \geq r$, then the inequality in (14) is always satisfied. Thus the boundary at maturity is K in this case. If $\mu < r$, the boundary at maturity becomes

$$\min\left(K, \frac{r}{r-\mu}K\right).$$

4   PROFIT FROM SELLING AMERICAN OPTIONS

In the previous section, we observed that the option holder's exercise time could differ from the price-maximizing exercise time, when he optimizes his utility. When this happens, the issuer gains from the difference. In this section we examine the profit from selling American options to utility-maximizing investors.

The issuer charges $v(0, S_0)$ at time 0 as he sells an American option. He will hedge his short position as described in Section 2 until his customer exercises the option or the option expires. The discounted potential liability of the issuer is $e^{-r\tau}\phi(S_\tau)$ where τ is the actual time that his customer exercises. When the option holder never exercises, $\tau = T$ by convention. Thus the present value of the issuer's profit becomes:

$$P = v(0, S_0) + \int_0^\tau e^{-rt}\Delta(dS_t - rS_t\,dt) - e^{-r\tau}\phi(S_\tau) \tag{18}$$

The second term in the right side of (18) is the result of delta hedging with the cost of carry. First we add and subtract $e^{-r\tau}v(\tau, S_\tau)$ from the profit P. Applying Itô's formula to v yields:

$$v(0, S_0) + \int_0^\tau e^{-rt}\Delta(dS_t - rS_t\,dt) - e^{-r\tau}v(\tau, S_\tau) = -\int_0^\tau dt\,e^{-rt}\mathcal{L}v$$

where $\mathcal{L}$ is the Black–Scholes differential operator defined in (4). Thus we may rewrite the profit (18) as

$$P = -\int_0^\tau dt\,e^{-rt}\mathcal{L}v + e^{-r\tau}(v(\tau, S_\tau) - \phi(S_\tau)). \tag{19}$$

We define the expected profit at time t as

$$\psi(t, s) = E^{t,s}\left[-\int_t^\tau dt\,e^{-rt}\mathcal{L}v + e^{-r\tau}(v(\tau, S_\tau) - \phi(S_\tau))\right] \tag{20}$$

We will show that ψ satisfies a diffusion equation with a moving boundary which is known *a priori*. Recall that h is the maximum expected certainty equivalence of the option holder and its free boundary gives the optimal exercise time τ. Let $\mathcal{H}$ and $\mathcal{V}$ be the domains defined by $h > \phi$ and $v > \phi$, respectively. These are the regions of continuation for the utility maximization and the price maximization. We also define $\mathcal{G} = \mathcal{H} \backslash \mathcal{V}$; see Figure 1. Since $\mathcal{L}v$ vanishes on $\mathcal{V}$, the expected profit ψ satisfies

$$\psi_t + \mu s \psi_s + \tfrac{1}{2}\sigma^2 s^2 \psi_{ss} - e^{-rt}\mathcal{L}v\mathcal{I}_\mathcal{G} = 0 \tag{21}$$

subject to $\psi(T, s) = 0$ and $\psi = e^{-rt}(v - \phi)$ on $\partial\mathcal{H}$, the utility-maximizing exercise boundary. The indicator $\mathcal{I}_\mathcal{G}$ is one if (t, s) belongs to $\mathcal{G}$ and zero otherwise. If the option is a call, the left side of (21) vanishes because $\mathcal{G}$ is empty. If the option is a put, then $v = \phi$ on the complement of $\mathcal{V}$, and therefore

$$e^{-rt}\mathcal{L}v\mathcal{I}_\mathcal{G} = e^{-rt}\mathcal{L}\phi\mathcal{I}_\mathcal{G} = -re^{-rt}K\mathcal{I}_\mathcal{G}$$

where K is the strike price. Here we have used the fact that the price-maximizing exercise boundary is not above the strike when the option is a put.

Figure 2 shows the expected profit from selling an at-the-money American put to an investor who maximizes his expected return. In this case, the option holder's criterion in choosing the exercise time is free from risk aversion, and hence the outcome can be considered as the marginal effect of the physical drift to the issuer's expected gain. The initial asset price is 50, the asset volatility is 20% per annum, the maturity of the option is six months, and the risk-free rate is 8% per annum. When the physical drift coincides with the risk free rate, the exercise boundary that maximizes the expected return coincides with the price-maximizing exercise boundary, and hence there is no profit for the issuer. When the physical drift surpasses the risk-free rate, the holder's exercise boundary is inside the price-maximizing exercise boundary. In this case, $\mathcal{G}$ is empty and the only source of the issuer's profit is the difference between the value of the option and the exercise value (i.e., the value of ψ on the moving boundary $\partial\mathcal{H}$). If the physical drift is less than the risk-free rate, then the holder's exercise boundary is outside of the price-maximizing exercise boundary, and hence the issuer's profit grows with the

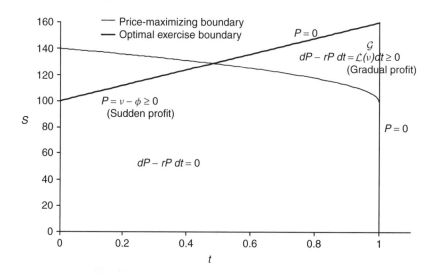

Figure 1   Overlapping exercise boundaries.

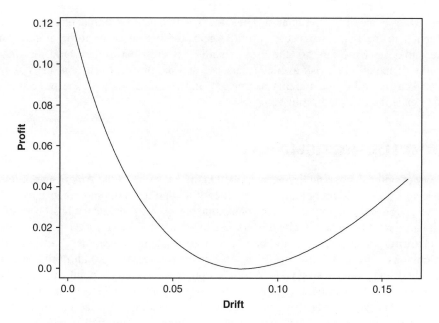

Figure 2 The effect of the physical drift.

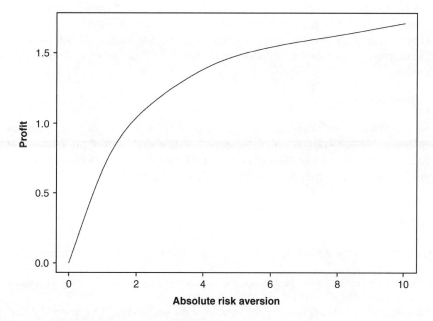

Figure 3 The effect of the absolute risk aversion.

occupation time of the asset price in between the two boundaries. This explains the asymmetry in the figure.

Figure 3 is the issuer's expected profit as a function of the absolute risk aversion. The option holder's exercise time maximizes the expected CARA utility, while the physical drift coincides with the risk-free rate of 8% per annum. Thus, the outcome is the marginal effect of the absolute risk aversion to the issuer's expected profit. Again, the option is an at-the-money American put, the initial asset price is 50, the asset volatility is 20% per annum, and the maturity is six months. If the absolute risk aversion vanishes and the physical drift and the risk-free rate coincide, then the utility-maximizing exercise boundary coincides with the price maximizing one, and hence the expected profit vanishes.

5 CONCLUDING REMARKS

The theory of optimal stopping has been applied to the valuation of the American option. People are prone to using the terminologies of the theory of optimal stopping when they talk about American options. For example, the price-maximizing exercise boundary has been referred to the optimal exercise boundary, while it is optimal to neither the issuer nor the option holder. This causes confusion to students, practitioners, and even academic researchers in the field. There are two obvious sources of confusion. First, financial software packages almost invariably value contracts, and find exercise strategies, from the option writer-hedger point of view. Rarely do they have anything to say about what is optimal for the contract holder. Some research papers even suggest that we price an American option by estimating the price-maximizing exercise boundary from the empirical data of exercise boundaries. In this paper, we have explained why these ideas are wrong. The holder of an American option should pursue his own profit maximization and choose the right exercise time for himself regardless of the price-maximizing exercise boundary, unless he is determined that the issuer should not gain. As the exercise times of market participants are affected by the market direction, their risk aversion factors, and their financial structures, an estimate of the price-maximizing exercise boundary from empirical data is also invalid.

Acknowledgements

We would like to thank Professor ____ _____, the editor of the _____ of _____, for persevering with us after reading our original rather incomprehensible first draft. Paul Wilmott would like to also thank the Royal Society for their support. Hyungsok Ahn would like to thank David Epstein for his service in improving the readability of the text.

REFERENCES

Friedman, A 1988 *Variational Principles and Free Boundary Problems*, Robert E. Kreiger Publishing, New York
Harrison, JM and Pliska, SR 1981 Martingales and stochastic integrals in the theory of continuous trading, *Stoch. Proc. and Their Appl.*, **11**, pp 215–260
Jaillet, P, Lamberton, D and Lapeyre, B 1990 Variational inequalities and the pricing of American options. *Acta Appl. Math.* **21** 263–289
Merton, RC 1990 *Continuous-Time Finance*. Blackwell, Oxford
Myneni, R 1992 The pricing of the American option. *Annals Appl. Probab.* **2**, (1) 1–23
Overdahl, JA & Martin, PG 1994 The exercise of equity options: theory and empirical tests. *Journal of Derivatives* Fall 38–51
San Martin, J & Protter, P 1993 General change of variables formulas for semimartingales in one and finite dimensions. *Probab. Theory Rel. Fields* **97** 363–381

Van Moerbeke, P 1976 On optimal stopping and free boundary problems, *Arch. Rational Mech. Anal.*, **60**, pp 101–148.

Wilmott, P., Dewynne, JN and Howison, SD 1993 *Option Pricing: Mathematical Models and Computation*, Oxford Financial Press, Oxford.

6 FAQ

IT'S IMPORTANT TO UNDERSTAND WHEN/WHY THE CONCEPT APPLIES

Q How can there be another value for American options in a Black–Scholes world?

A We are definitely not saying that there is another value. In a world where perfect delta hedging is possible and everyone agrees on the volatility of the underlying which follows a lognormal random walk there should be only one value for an American option. Concentrate, man.

Q Wouldn't the owner always be better off selling the option than exercising it?

A Yes, he would. But this is not always possible. Perhaps there just isn't the necessary liquidity in the market. Perhaps the contract is OTC and the original writer will not let the option holder off the hook without paying a stiff penalty.

Q Well, wouldn't the holder simply delta hedge until maturity?

A Again, yes, he would if he could. But if he can delta hedge why is he buying the option in the first place? Not everyone has access to the underlying asset sufficiently free of transaction costs to allow them to delta hedge.

Q Do people exercise options at 'non-optimal' times in practice?

A Yes, definitely with OTC contracts and any contracts in which the owner has to make complex decisions. Even vanilla options are sometimes exercised at odd times, and they can definitely be sold instead of exercised.

35.6 ANOTHER SITUATION WHERE THE SAME IDEA APPLIES: PASSPORT OPTIONS

Remember the passport option? This is a contract that pays off the positive part of a trader's trading account. If he has lost money at the expiry of the contract then he is not liable. If he has made money he keeps it. We saw in Chapter 20 how this turned into a stochastic control problem with the trader having to make decisions about how to invest. From the writer's point of view we have to assume that the trader invests in such a way as to give the option its highest value. In practice he will do whatever he thinks best to make the most money. This is very, very similar to the US option situation above. Before we go on with the analysis it might be best if you read through Chapter 20 again briefly to familiarize yourself with the basic problem and the mathematics involved.

35.6.1 Recap

To value the passport option we introduce a new variable π that is the value of the trading account. This quantity satisfies the following

$$d\pi = r(\pi - qS)\,dt + q\,dS,$$

where q is the amount of stock held at time t. I will restrict the size of the position in the stock by insisting that $|q| \le 1$.

The contract pays off

$$\max(\pi, 0)$$

at time T. This will be the final condition for our option value $V(S, \pi, t)$. Note that the option value is a function of three variables.

Now let us hedge this option:

$$\Pi = V - \Delta S.$$

...usual Itô stuff here...

Since $d\pi$ contains a dS term the correct hedge ratio is

$$\Delta = \frac{\partial V}{\partial S} + q\frac{\partial V}{\partial \pi}.$$

From the no-arbitrage principle follows the pricing equation

$$\frac{\partial V}{\partial t} + \tfrac{1}{2}\sigma^2 S^2 \frac{\partial^2 V}{\partial S^2} + q\sigma^2 S^2 \frac{\partial^2 V}{\partial S\,\partial \pi} + \tfrac{1}{2}q^2\sigma^2 S^2 \frac{\partial^2 V}{\partial \pi^2} + rS\frac{\partial V}{\partial S} + r\pi\frac{\partial V}{\partial \pi} - rV = 0.$$

If we are selling this contract then we should assume that the holder acts optimally, making the contract's value as high as possible. The highest value for the contract occurs when q is chosen to maximize the q terms in the above:

$$\max_{|q|\leq 1}\left(q\sigma^2 S^2 \frac{\partial^2 V}{\partial S\,\partial \pi} + \tfrac{1}{2}q^2\sigma^2 S^2 \frac{\partial^2 V}{\partial \pi^2}\right).$$

Call this strategy q^*.

The sentence in italics is crucial. In practice the option holder *will not* act in this fashion. Let's suppose that he acts to maximize his expected utility; this is a nice framework to analyze this problem, but not necessarily realistic.

35.6.2 Utility Maximization in the Passport Option

In this section we examine how the option holder trades the underlying and how much the issuer gains by selling the option. The investor who owns a passport option may construct his trading strategy to maximize his utility, trying to predict the market movement. When the physical trend of the market significantly differs from the risk-neutral drift, the option holder will benefit as long as he has a correct view on the market. At the same time, the issuer will gain from the difference between the price-maximizing trading strategy and the trading strategy performed by his customer.

First we assume that the option holder finds his strategy by solving the value of the maximum expected utility of the payoff:

$$u(S, \pi, t) = \max_{|q|\leq 1} E^{S,\pi,t}[e^{-r(T-t)}U(\max(\pi(T), 0))] \tag{35.1}$$

where E is the expectation under the real drift and U is the option holder's utility function which is increasing in its argument. u satisfies the following Bellman equation:

$$-\frac{\partial u}{\partial t} = r\pi\frac{\partial u}{\partial \pi} + \mu S\frac{\partial u}{\partial S} - ru + \max_{|q|\leq 1}\left(qS(\mu - r)\frac{\partial u}{\partial \pi} + \tfrac{1}{2}\sigma^2 S^2\left(q^2\frac{\partial^2 u}{\partial \pi^2} + 2q\frac{\partial^2 u}{\partial \pi \partial S} + \frac{\partial^2 u}{\partial S^2}\right)\right),$$

$$u(S, \pi, T) = U(\max(a, 0)) \tag{35.2}$$

where μ is the physical drift of the underlying asset.

If $U(x) = x$, $\partial^2 u/\partial \pi^2$ stays positive and hence the utility maximizing strategy has its value at either ± 1. The interpretation of the linear utility is that the option holder maximizes real expected return. A motivation for studying such utility is that the investor's portfolio is already insured by the passport option he owns and that it is tractable. In this case u has a similarity solution of the form $u(S, \pi, t) = Sh(\pi/S, t)$. Furthermore $h(y, t)$ satisfies the following equation:

$$-\frac{\partial h}{\partial t} = (\mu - r)\left(h - y\frac{\partial h}{\partial y}\right) + \max_{|q| \leq 1}\left(\tfrac{1}{2}\sigma^2(y - q)^2\frac{\partial^2 h}{\partial y^2} + q(\mu - r)\frac{\partial h}{\partial y}\right) \tag{35.3}$$

with the terminal data $\max(y, 0)$. From this, we obtain the option holder's trading strategy:

$$q = \text{sign}\left(\frac{\mu - r}{\sigma^2} \cdot \frac{\dfrac{\partial h}{\partial y}}{\dfrac{\partial^2 h}{\partial y^2}} - y\right). \tag{35.4}$$

When μ coincides with the risk-free rate r, (35.3) agrees with the price-maximizing value function for the option and q in (35.4) coincides with the price-maximizing strategy q^*. If μ differs from r, then the option holder's choice will be different from the price-maximizing strategy.

Next we discuss the issuer's hedging strategy, Δ. In Chapter 20, I explained that the hedging strategy must be in tune with the *actual* trading strategy performed by the option holder, and that it is given by

$$\Delta = \frac{\partial V}{\partial S} + q\frac{\partial V}{\partial \pi}.$$

Then the profit of the issuer becomes:

$$P = V(S_0, 0, 0) + \int_0^T e^{-rt}\Delta(dS - rS\,dt) - e^{-rT}V(S(T), \pi(T), T). \tag{35.5}$$

The first term is the price of the option he collects in cash, the second is the result of the delta hedging, and the third is the present value of the potential liability.

Applying Itô's formula to V yields:

$$P = -\int_0^T e^{-rt}\left(\frac{\partial V}{\partial t} + \tfrac{1}{2}\sigma^2 S^2\left(\frac{\partial^2 V}{\partial S^2} + 2q\frac{\partial^2 V}{\partial S\partial \pi} + q^2\frac{\partial^2 V}{\partial \pi^2}\right)\right)dt$$

$$= \tfrac{1}{2}\sigma^2 \int_0^T dt\, e^{-rt}S^2 \cdot \left((q^{*2} - q^2)\frac{\partial^2 V}{\partial \pi^2} + 2(q^* - q)\frac{\partial^2 V}{\partial S\partial \pi}\right) \tag{35.6}$$

where q^* is the price-maximizing strategy and q is the strategy performed by the option holder.

Recall that $V(S, \pi, t)$ has a similarity solution $SH(\pi/S, t)$. In particular, we have

$$\frac{\partial^2 V}{\partial S\partial \pi} = -\frac{\pi}{S^2}\frac{\partial^2 H}{\partial \xi^2} \quad \text{and} \quad \frac{\partial^2 V}{\partial \pi^2} = \frac{1}{S}\frac{\partial^2 H}{\partial \xi^2},$$

where $\xi = \pi/S$. Also recall that $q^* = -\text{sign}(\xi)$. Thus we have a further reduction in the integrand of (35.6):

$$S^2 \cdot \left((q^{*2} - q^2)\frac{\partial^2 V}{\partial \pi^2} + 2(q_\star - q)\frac{\partial^2 V}{\partial S\partial \pi}\right) = \frac{\partial^2 H}{\partial \xi^2} \cdot (2|\pi| + 2q\pi + (1 - q)^2 S) \tag{35.7}$$

Now suppose that the option holder finds his strategy by maximizing the expected return. Then, as we computed earlier in (35.4), the option holder's strategy q depends on π and S only through the ratio $\pi/S = \xi$ and has its value at either ± 1. Hence the last term in (35.7) drops out, and the profit of the issuer becomes:

$$P = \sigma^2 \int_0^T e^{-rt} S(|\xi| + q(\xi, t)\xi) \frac{\partial^2 H}{\partial \xi^2} \, dt. \tag{35.8}$$

To obtain the expected profit $E[P]$ of the issuer, we define

$$g(S, \pi, t) = \sigma^2 E^{S,\pi,t}\left[\int_t^T e^{-r\tau} S(|\xi| + q\xi) \frac{\partial^2 H}{\partial \xi^2} d\tau\right].$$

Again we observe that g has a similarity solution of the form $g(S, \pi, t) = S\psi(\pi/S, t)$ and that $\psi(\xi, t)$ satisfies the following:

$$-\frac{\partial \psi}{\partial t} = (\mu - r)(q - \xi)\frac{\partial \psi}{\partial \xi} + \mu\psi + \tfrac{1}{2}\sigma^2(q - \xi)^2 \frac{\partial^2 \psi}{\partial \xi^2} + \sigma^2 e^{-rt}(|\xi| + \xi q)\frac{\partial^2 H}{\partial \xi^2}$$

subject to $\psi(\xi, T) = 0$. To solve this equation, we need to obtain H from (20.3) and q from (35.3).

Figure 35.1 shows the expected gain by the issuer as a function of μ, the physical drift, that is $\psi(0, 0)$ against μ. The asset volatility is 20% per annum and the maturity of the option is six months. We calculate the profit $100 \cdot \psi$ at different values of the physical drift from zero to 16%. When the drift coincides with the risk-free rate of 8% per annum, the gain vanishes. As explained earlier, the issuer gains more as the gap between the drift and the risk-free rate become larger.

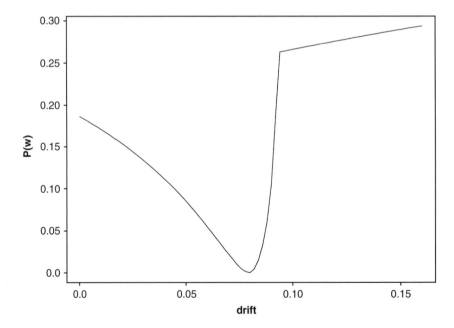

Figure 35.1 Issuer's expected gain versus the drift of the underlying asset.

35.13 **SUMMARY**

I'd like to hear what you have to think on this subject and I'll keep a poll of people's views. Email me on paul@wilmott.com. If you really hate this chapter please tear out the pages and send them to me.

FURTHER READING

- More details about the passport option are contained in Ahn, Penaud & Wilmott (1998).

CHAPTER 36
stochastic volatility and mean-variance analysis

In this Chapter...

- how to analyze risk when volatility is stochastic
- mean-variance analysis

36.1 INTRODUCTION

I tend not to like any model that requires the input of a market price of risk. The main reason is that this quantity is not directly observable. At best it can be deduced from the prices of derivatives, so called 'fitting'. But this is far from adequate, since the fitting will only work if those who set the prices of derivatives are using the same model and they are consistent in that the fitted market price of risk does not change when the model is refitted a few days later. To see what I mean, think back to Chapter 25. In that chapter I showed how to find the local deterministic volatility surface that is consistent with all quoted prices of vanilla options. If a few days later we find that this surface has changed, still being chosen to match market prices, then the model was wrong. Exactly the same problem occurs when we have stochastic volatility and we have fitted to find the market price of volatility risk.[1]

Whether we have a deterministic volatility surface or a stochastic volatility model with prescribed or fitted market price of risk, we will always be faced with how to interpret refitting. Was the market wrong before but is now right, or was the market correct initially and now there are arbitrage opportunities? We won't be faced with awkward questions like this if we don't expect our model, whatever it may be, to give unique and correct values. In this chapter we'll see how to estimate probabilities for prices being correct. We do this by only delta hedging and not dynamically vega hedging. Instead we look at means and variances for option values.

---

[1] I call this market-price-of-risk risk.

36.2 THE MODEL FOR THE ASSET AND ITS VOLATILITY

We are going to work with the very classical model

$$dS = \mu S\, dt + \sigma S\, dX_1$$

and

$$d\sigma = p(S, \sigma)\, dt + q(S, \sigma)\, dX_2$$

with a correlation of $\rho(S, \sigma)$. We'll only consider a non-dividend-paying asset, the modifications needed to allow for dividends are the usual.

We are going to examine the statistical properties of a portfolio that tries to replicate as closely as possible the original option position. We will not hedge the portfolio dynamically with other options so our portfolio will not be risk-free. Instead we will examine the mean and variance of the value of our portfolio as it varies through time.

With Π representing the value of the position $-\Delta$ in the asset,

$$\Pi = -\Delta S.$$

Thus

$$d\Pi = -\Delta\, dS,$$

and

$$d\Pi - r\Pi\, dt = -\Delta\, dS + r\,\Delta S\, dt = -(\mu - r)S\Delta\, dt - \sigma S\Delta\, dX_1.$$

36.3 ANALYSIS OF THE MEAN

Define the mean m of the portfolio value at any time by

$$m(S, \sigma, t) = E[\Pi(S, \sigma, t)|\text{state of the world at time } t].$$

Since

$$E[dm - rm\, dt] = E[d\Pi - r\Pi\, dt] = -(\mu - r)S\Delta\, dt,$$

we have

$$\frac{\partial m}{\partial t} + \tfrac{1}{2}\sigma^2 S^2 \frac{\partial^2 m}{\partial S^2} + \rho\sigma Sq \frac{\partial^2 m}{\partial S \partial \sigma} + \tfrac{1}{2}q^2 \frac{\partial^2 m}{\partial \sigma^2} + \mu S \frac{\partial m}{\partial S} + p \frac{\partial m}{\partial \sigma} - rm = -(\mu - r)S\Delta. \quad (36.1)$$

We still have to decide on Δ. We will choose it to minimize the variance locally, so we can't choose it until we've analyzed the variance in the next section. Note also that the final condition for (36.1) will be the payoff for our original option that we are trying to replicate.

This equation for m was easy to derive, the equation for the variance is a bit harder.

36.4 ANALYSIS OF THE VARIANCE

Define the variance $v(S, \sigma, t)$ by

$$v(S, \sigma, t) = E[(\Pi(S, \sigma, t) - m(S, \sigma, t))^2|\text{state of the world at time } t].$$

Thus

$$v(S + dS, \sigma + d\sigma, t + dt) = E[(\Pi(S + dS, \sigma + d\sigma, t + dt) - m(S + dS, \sigma + d\sigma, t + dt))^2|$$
$$\text{state of the world at time } t + dt]$$

$$= E[(\Pi(S, \sigma, t) + d\Pi - m(S, \sigma, t) - dm)^2|$$
$$\text{state of the world at time } t + dt].$$

Taking expectations over what happens from t to $t + dt$

$$E[v(S, t, t) + dv] = E[(\Pi(S, \sigma, t) - m(S, \sigma, t))^2] + E[(\Pi(S, \sigma, t) - m(S, \sigma, t))(d\Pi - dm)]$$
$$+ E[(d\Pi - dm)^2].$$

I've been a bit loose with my notation here, and I've also neglected any discounting. I'll put the latter back in a mo.

The middle term on the right-hand side of the above is zero from the definition of m. The third term becomes

$$E\left[\left(-\sigma S \Delta \, dX_1 - \frac{\partial m}{\partial \sigma} q \, dX_2 - \frac{\partial m}{\partial S} \sigma S \, dX_1\right)^2\right].$$

I'll leave this for you to simplify, but the end result is, for an arbitrary Δ,

$$\frac{\partial v}{\partial t} + \tfrac{1}{2}\sigma^2 S^2 \frac{\partial^2 v}{\partial S^2} + \rho\sigma Sq \frac{\partial^2 v}{\partial S \partial \sigma} + \tfrac{1}{2}q^2 \frac{\partial^2 v}{\partial \sigma^2} + \mu S \frac{\partial v}{\partial S} + p \frac{\partial v}{\partial \sigma} - 2rv$$
$$+ \sigma^2 S^2 \left(\frac{\partial m}{\partial S}\right)^2 + q^2 \left(\frac{\partial m}{\partial \sigma}\right)^2 + 2\rho\sigma Sq \frac{\partial m}{\partial S} \frac{\partial m}{\partial \sigma}$$
$$+ \sigma^2 S^2 \Delta^2 + 2\Delta \left(\sigma^2 S^2 \frac{\partial m}{\partial S} + \rho\sigma Sq \frac{\partial m}{\partial \sigma}\right) = 0. \tag{36.2}$$

Observe that I've put the discounting back in.

36.5 **CHOOSING Δ TO MINIMIZE THE VARIANCE**

Only the last two terms in (36.2) contain Δ. We therefore choose Δ to minimize this quantity, to ensure that the variance in our portfolio is as small as possible. This gives

$$\Delta = -\frac{\partial m}{\partial S} - \frac{\rho q}{\sigma S} \frac{\partial m}{\partial \sigma}. \tag{36.3}$$

36.6 **THE MEAN AND VARIANCE EQUATIONS**

Substituting (36.3) into (36.1) and (36.2) we get

$$\frac{\partial m}{\partial t} + \tfrac{1}{2}\sigma^2 S^2 \frac{\partial^2 m}{\partial S^2} + \rho\sigma Sq \frac{\partial^2 m}{\partial S \partial \sigma} + \tfrac{1}{2}q^2 \frac{\partial^2 m}{\partial \sigma^2} + rS \frac{\partial m}{\partial S} + \left(p - (\mu - r)\frac{\rho q}{\sigma}\right) \frac{\partial m}{\partial \sigma} - rm = 0 \tag{36.4}$$

and

$$\frac{\partial v}{\partial t} + \tfrac{1}{2}\sigma^2 S^2 \frac{\partial^2 v}{\partial S^2} + \rho\sigma Sq \frac{\partial^2 v}{\partial S \partial \sigma} + \tfrac{1}{2}q^2 \frac{\partial^2 v}{\partial \sigma^2} + \mu S \frac{\partial v}{\partial S} + p \frac{\partial v}{\partial \sigma} + q^2(1 - \rho^2) \left(\frac{\partial m}{\partial \sigma}\right)^2 - 2rv = 0. \tag{36.5}$$

The final conditions for these are obviously the payoff, for $m(S, \sigma, T)$, and zero for $v(S, \sigma, T)$. If the portfolio contains options with different maturities, the equations must satisfy the corresponding jump conditions as well.

Since the final condition for v is zero and the only 'forcing term' in (36.5) is $(\partial m/\partial \sigma)^2$, Equation (36.5) shows that the only way we can have a perfect hedge is for either q to be zero, i.e. deterministic volatility, or to have $\rho = \pm 1$. In the latter case the asset and volatility (changes) are perfectly correlated. The solution of (36.4) is then different from the Black–Scholes solution.

Equation (36.4) is very much like the pricing equation for stochastic volatility in a risk-neutral setting. It's rather like having a market price of volatility risk of $(\mu - r)\rho/\sigma$. But, of course, the reasoning and model are completely different in our case.

The system of equations is nonlinear (actually two linear equations, coupled by a nonlinear forcing term). We are going to exploit this fact shortly.

36.7 HOW TO INTERPRET AND USE THE MEAN AND VARIANCE

Take an option position in a world with stochastic volatility, and delta hedge as proposed above. Because we cannot eliminate all the risk we cannot be certain how accurate our hedging will be. Think of the final value of the portfolio together with accumulated hedging as being the 'outcome'. The distribution of the outcome will generally not be Normal. The shape will depend very much on the option position we are hedging. But we have calculated both the mean and the variance of the hedged portfolio. If we made the assumption that the distribution was not too far from Normal then the mean and the variance are sufficient to describe the probabilities of any outcome. If we wanted to be 95% certain that we would make money then we would have to sell the option for

$$m + 1.644853 v^{1/2}$$

or buy it for

$$m - 1.644853 v^{1/2}.$$

The 1.644853 comes from the position of the 95th percentile assuming a Normal distribution.

We'll use this idea below, but with a requirement that we are within one standard deviation of the mean, i.e. we make money 84% of the time.

36.8 STATIC HEDGING AND PORTFOLIO OPTIMIZATION

If we use as our option (portfolio) 'price' the following

$$\text{mean} - (\text{variance})^{1/2} = m - v^{1/2}$$

then we have a nonlinear model. Everything that was said in Chapter 32 about nonlinear pricing models applies here, in particular the possibility of optimal static hedging.

36.9 EXAMPLE: VALUING AND HEDGING AN UP-AND-OUT CALL

In this section, we consider the pricing and hedging of a short up-and-out call. Throughout this section, our choice of mean-variance combination is:

$$m - v^{1/2}. \tag{36.6}$$

First consider a single up-and-out call with barrier located at S_u. In this case, we solve Equations (36.4) and (36.5) subject to:

(a) $m(S_u, \sigma, t) = v(S_u, \sigma, t) = 0$ for each $(\sigma, t) \in (0, \infty) \times (0, T)$ where T is maturity;

(b) $m(S, \sigma, T) = -\max(S - E, 0)$ for each $(S, \sigma) \in (0, X) \times (0, \infty)$ where E is the strike;

(c) $v(S, \sigma, T) = 0$ for each (S, σ).

The discontinuity of the payoff at the knock-out barrier makes this position particularly difficult to hedge. In fact this can be easily seen from our equations. Figure 36.1 and Figure 36.2 are the pictures of calculated mean and variance respectively with the following specifications:

- Strike at 100, barrier at 110, and expiry 30 days;

- $p(\sigma) = 0.8(\sigma^{-1} - 0.2)$ and $q(\sigma) = 0.5$.

Near the barrier, $(\partial m/\partial \sigma)^2$ is huge (see Figure 36.1) and this feeds the variance, being the source term in (36.5). If the spot S is 100, and the current spot volatility σ is 20% per annum, the mean is -1.1101 and the variance is 0.3290. Thus if there is no other instrument available in the market, one would price this option at \$1.6836 to match with Equation (36.6).

36.9.1   Static Hedging

Suppose that there are six 30-day vanilla call options available in the market with the following specifications:

Option	1	2	3	4	5	6
Strike	96.62	100.00	104.17	108.70	112.36	116.96
Bid Price	4.6186	2.6774	1.1895	0.4302	0.1770	0.0557
Ask Price	4.6650	2.7043	1.2014	0.4345	0.1788	0.0562

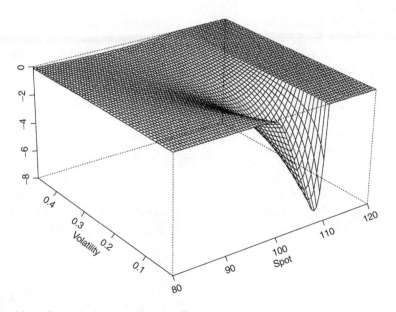

Figure 36.1   Mean for a single up-and-out call.

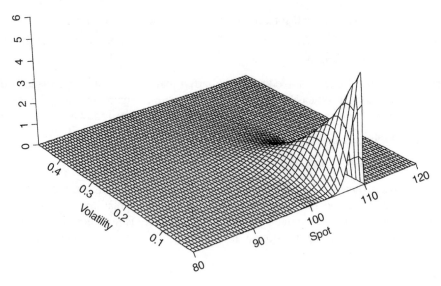

Figure 36.2 Variance for a single up-and-out call.

Aside

These hypothetical market prices were generated by computing the mean of each call option, with

$$d\sigma = \left(\frac{1}{\sigma} - 0.2\right) dt + 0.5 \, dX \tag{36.7}$$

where X is a Brownian motion with respect to the risk-neutral measure. Then 0.5% bid-ask spread was added.

Now we employ the optimal static vega hedge. Suppose we trade $(q_1, \ldots, q_6)$ of the above instruments and let E_i be the strikes among the payoffs. Furthermore, let $(m^{(0)}, v^{(0)})$ be the mean variance pair after knock-out and $(m^{(1)}, v^{(1)})$ be that before knock-out. Then $(m^{(i)}, v^{(i)})$, $i = 0, 1$, satisfies Equations (36.4) and (36.5) subject to:

(a) $m^{(1)}(110, \sigma, t) = m^{(0)}(110, \sigma, t)$ and $v^{(1)}(110, \sigma, t) = v^{(0)}(110, \sigma, t)$ for each (σ, t) in $(0, \infty) \times (0, T)$;

(b) $m^{(0)}(S, \sigma, t) = \displaystyle\sum_{i=1}^{6} q_i \max(S - E_i, 0)$ for each $(S, \sigma) \in (0, \infty) \times (0, \infty)$;

(c) $m^{(1)}(S, \sigma, T) = \displaystyle\sum_{i=1}^{6} q_i \, \max(S - E_i, 0) - \max(S - 100, 0)$ for each (S, σ) in $(0110) \times (0, \infty)$;

(d) $v^{(1)}(S, \sigma, T) = v^{(0)}(S, \sigma, T) = 0$ for each (S, σ) in $(0, \infty) \times (0, \infty)$.

Thus $m^{(1)}(S, \sigma, 0)$ stands for the mean of the cashflows excluding the upfront premium. We find a $(q_1, \ldots, q_6)$ that maximizes:

$$m^{(1)}(S, \sigma, 0) - \sum_{i=1}^{6} p(q_i) - \sqrt{v^{(1)}(S, \sigma, 0)}$$

where $p(q_i)$ is the market price of trading q_i shares of strike E_i. In the case of $S = 100$ and $\sigma = 0.2$, our optimal choice for vega hedge is given by:

Option	1	2	3	4	5	6
Strike	96.62	100.00	104.17	108.70	112.36	116.96
Quantity	0.0000	−1.1688	1.0207	3.1674	−3.6186	0.8035

The cost of this hedge position is \$1.1863. Figures 36.3 and 36.4 are the pictures of $m^{(1)}$ and $v^{(1)}$ after the optimal static vega hedge. After the optimal static vega hedge, the mean is 0.0398 and the variance is reduced to 0.0522. Thus the price for the up-and-out call that matches with our mean-variance combination (36.6) is \$1.3752 ($1.1863 - 0.0398 + \sqrt{0.0522}$). In the risk-neutral set-up (36.7), the price for this up-and-out call is \$1.1256. The difference mainly comes from the standard deviation term (variance$^{1/2}$) in (36.6) which is $\sqrt{0.0522} = 0.2286$.

36.10 SUMMARY

Constructing a risk-neutral model to fit the market prices of exchange traded options consistently over a reasonable time period is a difficult task. Putting aside the fundamental question of

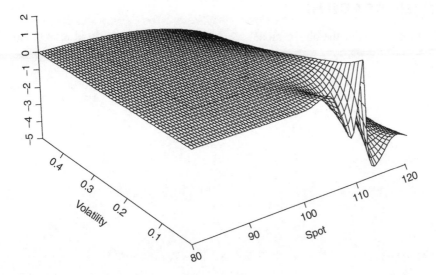

Figure 36.3   Mean of portfolio after optimal static vega hedging.

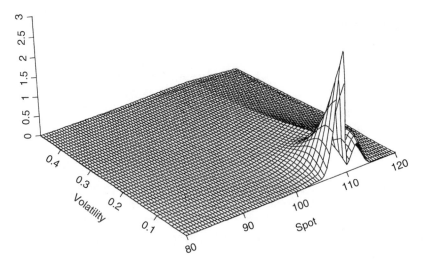

Figure 36.4 Variance of portfolio after optimal static vega hedging.

whether the axiomatic risk-neutral model for stochastic volatility is legitimate or not, we must face potential financial losses due to recalibration. In this chapter we take another approach. We first evaluate the mean and variance of the discounted future cashflow and then find market instruments that optimally reduce, the volatility risk.

I've set this problem up in a mean-variance framework but it could easily be extended to a more general utility theory approach.

FURTHER READING

- For all the details of the above model see Ahn, Arkell, Choe, Holstad & Wilmott (1999).

CHAPTER 37
advanced dividend modeling

In this Chapter...

- the importance of dividend modeling
- the stochastic dividend model
- uncertainty in dividend amount and timing

37.1 INTRODUCTION

We've already seen in Chapter 8 how to incorporate dividends into option models. But that supposes that we have a decent knowledge of the timing and the amount of the dividend. I relaxed the requirement for perfect knowledge of the amount of the dividend in Chapter 27 when we looked at best and worst cases. Here I present some more sophisticated models.

37.2 WHY DO WE NEED DIVIDEND MODELS?

The primary reason why the financial literature extensively analyzes dividends on equity is the perfect market hypothesis, i.e. that stocks are valued as their discounted dividend stream in perpetuity. To a large extent, equity analysts use this model and its variations to estimate a stock's fair value and forecast its future performance. Derivatives analysts, however, in order to value and hedge a portfolio of options on a stock or an index, normally require as standard parameters the current spot price, exercise price, risk-free rate, expiry time, volatility and, often, the stock's dividend yield over the life of the option. For many practitioners the dividend yield is considered to be relatively unimportant compared with the other parameters.

The amount of literature available on dividend models applied to derivatives is tiny. But if we compare the sensitivity ratios for different types of options, it becomes apparent that this might not describe the real world accurately. Compare, for example, option price sensitivity to the dividend yield with the sensitivity to the volatility, the vega. Figure 37.1 shows the ratio of these two sensitivities against volatility and dividend yield. The option is a European call with one year to expiry, struck at 98, the risk-free rate is 5%. The underlying asset is 100. We can see

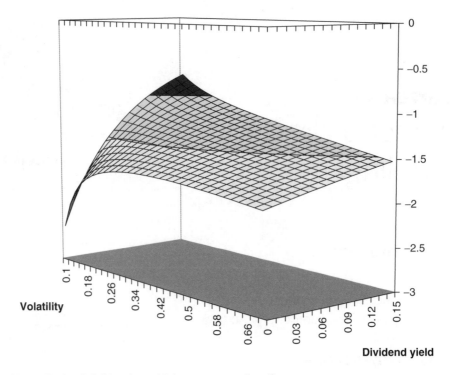

Figure 37.1   Ratio of dividend sensitivity to vega, call option.

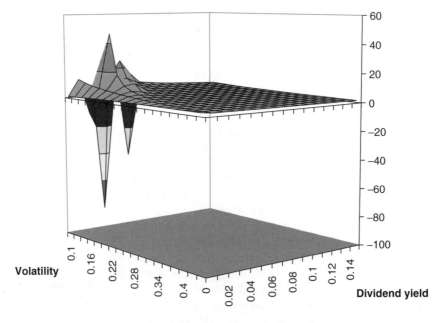

Figure 37.2   Ratio of dividend sensitivity to vega, binary option.

that for most parameter values this ratio is greater than one. In other words, the sensitivity to the dividend is more important than the sensitivity to the volatility. This is even more pronounced for a binary call option; see Figure 37.2.

In this chapter I attempt to give a concise summary of the relation between dividends on equity and the value of derivatives. You'll see various approaches to modeling the dividend process of a security or an index, elaboration of existing models and some new ideas. Choices normally have to be made as to whether to use deterministic dividends or stochastic dividends and discrete dates or a continuous flow. We'll see all of these here.

37.3 EFFECTS OF DIVIDENDS ON ASSET PRICES

In most financial theory the standard assumption is that the spot price S is lognormally distributed, satisfying a stochastic differential equation. The paths of the random walk are continuous almost everywhere, but smooth almost nowhere. However, such a model is only valid for non-dividend paying stocks. If the underlying stock is paying discrete dividends there are two points in time where the stock and/or the underlying asset might jump discontinuously. The two dates to examine closely are the date that the dividend is announced and the date when the asset goes ex-dividend.

Figure 37.3 shows the paths of a stock and option price, with dividend announcement date and ex-dividend date marked.

Consider first the ex-dividend date t_i. If markets are efficient and frictionless, the stock price drops discontinuously by exactly the, possibly time- and spot-dependent, dividend D;

$$S(t_i^+) = S(t_i^-) - D.$$

However any option written on the underlying stock stays unchanged in its price due to no-arbitrage arguments, thus

$$V(S, t_i^-) = V(S - D, t_i^+)$$

We saw this in Chapter 8.

The other key date t_a, which itself may be random, is the time when the company announces the dividend size and ex-dividend date on its stock. Firstly, the value of the derivative may

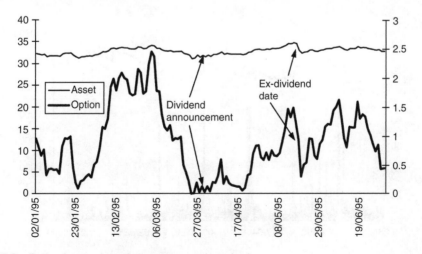

Figure 37.3 Path of a stock and an option on that stock.

jump discontinuously because of a discrepancy between implied and actual dividend and/or the exact ex-dividend date. The implied values, in practice, are inferred from prices of traded futures. We'll come back to implied dividends shortly. Across the dividend announcement date the price of the derivative can change. The path of the derivative will stay continuous only if the size of the dividend and its ex-dividend date are as expected by the market.

On the other hand, in practice, the stock price itself may change in light of this new information, but it is not always clear in which direction. A growing dividend may signal good earnings prospects and a dividend decline, financial distress; but not necessarily. However, on a more theoretical level, the famous theorem by Modigliani & Miller states that stock returns should not change with the dividend policy, hence there should be no discontinuous jump in S across t_a.

37.3.1   Market Frictions

In reality the jump in stock price across the ex-dividend date is not exactly the amount of the dividend due to frictions in the market. The error term is mostly attributed to different taxation of dividends and capital gains.

Some empirical tests conclude that due to short-term trading the tax effects are removed. That means, because of low transaction costs, market makers who unlike other investors have nearly equal tax rates, buy stocks with high dividend yield. Indeed, the trading volume in stocks around t_i is exceptionally high; see Figure 37.4.

A second market friction is introduced if the dividend payout date is not the same as the ex-dividend date. Typically, German companies pay their dividend on the ex-dividend date, UK companies five weeks after the ex-dividend date and other European companies up to six months after the ex-dividend date. Hence the stock price should jump by an amount that discounts the dividend accordingly.

We've seen the theory behind deterministic dividends in Chapter 5. We'll take a quick look at the term structure of dividends and then concentrate on random or uncertain dividend models.

37.3.2   Term Structure of Dividends

For individual stocks and short-dated index instruments practitioners rely on fundamental research when estimating the dividend correction to an option value. That means they try and

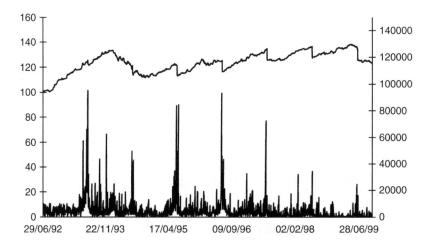

Figure 37.4   Price and volume.

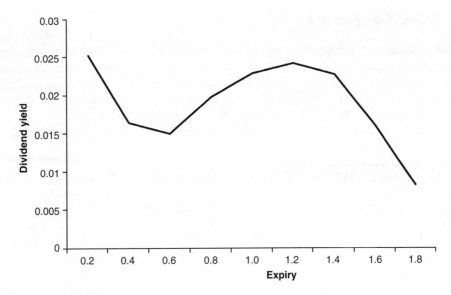

Figure 37.5 Term structure of dividend yield.

estimate every single dividend amount, timing and the anticipated percentage drop. However for long-dated derivatives or for large indices it may be more efficient to rely on implied dividend yields. The term-structure of dividends can be inferred from traded futures and the interest rate yield curve. The relationship between the forward price F and the spot price S is

$$F = Se^{(r-D(t;T))(T-t)}.$$

Here r is the interest rate, which I've assumed to be constant for all time, and $D(t;T)$ the dividend yield over the period t to T.

Typically, there are only a limited number of liquid futures and therefore the implied dividend term structure will consist of only a discrete number of actual points. If the dividend yield of a non-traded interim expiry is required it can be interpolated; see Figure 37.5.

37.4 **STOCHASTIC DIVIDENDS**

Let's now look at a more traditional stochastic model, this time for dividend yield.

$$dD = p(S, D, t)\, dt + q(S, D, t)\, dX_2$$

with a correlation of ρ.

If we want to obtain the value of a derivative on the dividend-yielding stock we will have to deal with an additional source of randomness that has to be hedged away by another instrument. To derive an equation we again set up a hedged portfolio, hedging one option with the underlying and another option.

Naturally, we end up with

$$\frac{\partial V}{\partial t} + \tfrac{1}{2}\sigma^2 S^2 \frac{\partial^2 V}{\partial S^2} + \sigma q S \frac{\partial^2 V}{\partial S \partial D} + \tfrac{1}{2}q^2 \frac{\partial^2 V}{\partial D^2} + (r - D)S \frac{\partial V}{\partial S} + (p - \lambda q)\frac{\partial V}{\partial D} - rV = 0$$

and the territory is familiar.

37.5 **POISSON JUMPS**

One phenomenon that sometimes happens during the life of an option is the announcement of special, i.e. non-recurring, dividends. These tend to be paid out when a firm is restructuring its financing or after mergers. These dividends lead to jumps in the option prices since they are not usually anticipated. These events can be incorporated as a Poisson jump process when evaluating the price of the option. This is exactly the same as modeling a stock market crash, which could be done in the Merton, Chapter 29, or worst-case, Chapter 30, sense.

37.6 **UNCERTAINTY IN DIVIDEND AMOUNT AND TIMING**

Instead of approximating the averages of a set of parameters or making assumptions about the respective stochastic processes that they follow, it may be more robust to come up with estimates of intervals that cannot be breached, then the option is valued in a worst-case scenario. We've seen this idea in Chapters 27 and 30 and will see it applied to interest rates later on. Indeed, in Chapter 27 we even saw how to apply the idea to uncertain dividend amounts or yields. Applied to dividends the simplest model is to assume that the dividend or the dividend yield lies within a range. This can be applied to a continuously- or discretely-paid dividend. See Figure 37.6 for a short time series of Deutsche Bank dividends.

Figure 37.7 shows the ex-dividend months for components of the Eurostoxx50, 1996–1998. We can see from this that companies are not necessarily consistent in the timing of their dividends, never mind the amounts.

A more realistic scenario is to assume that both dividends and ex-dividend dates are uncertain, both lying in ranges. Let's write the range for dividends as

$$D^- \leq D \leq D^+$$

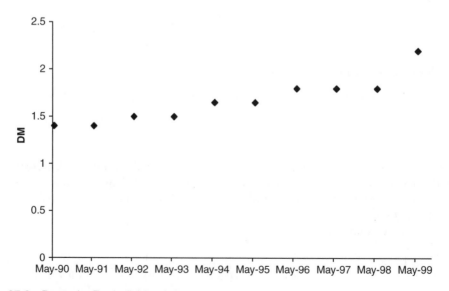

Figure 37.6 Deutsche Bank dividends.

1996

Jan	Feb	Mar	Apr	May	Jun	Jul	Aug	Sep	Oct	Nov	Dec
					elf						
			schneider	veba	telitalia						
			portel	unicredit	stgobain						
			nokia	rdutch	lvmh						
		vivendi	lufthansa	repsol	loreal		unilever				
		telefonic	endesa	petrofina	fortis		repsol				
		rhone	elsevier	electrabe	eni		ing	rdutch			
	siemens	philips	carrefour	deubank	endesa	generali	elsevier	kpn			
telefonic	paribas	ing	Bayer	ahold	alcatel	fiat	allianz	ahold	kpn	unilever	rwe
mannes	allirish	airliquid	akzonobel	aegon	abnamro	akzonobel	abnamro	aegon	BBV	lvmh	allirish

1997

Jan	Feb	Mar	Apr	May	Jun	Jul	Aug	Sep	Oct	Nov	Dec
				veba	elf						
				unicredit	telitalia						
				telefonic	stgobain	repsol					
				rdutch	lvmh	kpn	vivendi				
				petrofina	lufthansa	ing	nokia				
telefonic	siemens			paribas	francetel	generali	metro				
lvmh	schneider		portel	electrabe	eni	fiat	elsevier				
loreal	mannes	rhone	elsevier	deubank	deutel	endesa	ahold	unilever		unilever	
kpn	Bayer	philips	carrefour	airliquid	allirish	BBV	aegon	rdutch	BBV	fortis	
ing	allirish	endesa	akzonobel	aegon	alcatel	ahold	abnamro	abnamro	akzonobel	allianz	rwe

1998

Jan	Feb	Mar	Apr	May	Jun	Jul	Aug	Sep	Oct	Nov	Dec
				veba							
				unicredito							
				telefonic	elf						
				portel	telitalia						
				petrofina	stgobain						
				paribas	mannes	unilever					
				electrabe	lvmh	repsol				unilever	
		rhone		deutel	lufthansa	metro	ing			rwe	
	siemens	philips	kpn	deubank	francetel	generali	elsevier	rdutch		rdutch	
vivendi	endesa	nokia	carrefour	airliquid	eni	fiat	allirish	repsol	BBV	ing	
telefonic	endesa	kpn	Bayer	ahold	endesa	ahold	aegon	allianz	akzonobel	fortis	
lvmh	allirish	elsevier	akzonobel	aegon	alcatel	abnamro	abnamro			BBV	loreal

Figure 37.7 Ex-dividend months for components of the Eurostoxx50, 1996–1998.

and the time range

$$t_i^- \le t_i \le t_i^+.$$

We are now very much in the discretely-paid dividend world.[1]

Assuming that ranges for one ex-dividend date do not overlap ranges for another, we must introduce the functions $V_0(S, t)$ and $V_1(S, t)$ representing option values with dividends having been paid and not yet paid, respectively. These two functions exist during the time range containing the ex-dividend date. We also

UNCERTAINTY
IN TIMING, NEAT

---

[1] These definitions for $t_i^{+/-}$ are different from our usual definitions, encountered when there are jump conditions.

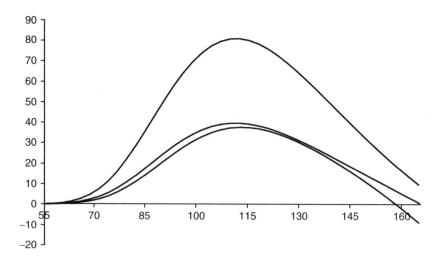

Figure 37.8 Best and worst cases when amount and timing of dividend are uncertain.

have the function $V(S, t)$ as the option value outside of these ranges. These functions all satisfy the basic Black–Scholes equation. They also satisfy the conditions

$$V_1(S, t) \leq \min_{D^- \leq D \leq D^+} (V_0(S - D, t)).$$

We must then have

$$V(S, t_i^-) = \min(V_1(S, t_i^-), \min_{D^- \leq D \leq D^+} (V_0(S - D, t_i^-)))$$

and

$$V_0(S, t_i^+) = V_1(S, t_i^+) = V(S, t_i^+).$$

In Figure 37.8 are shown the best and worst cases for a portfolio of long two calls struck at 80 and short three struck at 110. There are six months to expiry. This is the same example as in Chapter 27. There we looked at a single dividend paid just before expiry lying in the range zero to five. Here we increase the uncertainty by making the dividend date lie between 0.49 and 0.51 years.

Since the expiry of the portfolio is in 0.5 years, the ex-dividend date could be before or after expiry. The best-worst range can be so big that static hedging is necessary to reduce the spread.

37.7 **SUMMARY**

Dividends can have a major impact on the price of derivatives. Knowledge about the size and timing of dividends is often more important than having accurate volatility knowledge. Here we've seen some of the latest approaches to the subject of modeling dividends. Much of this is quite familiar by now, such as stochastic dividends. However, the uncertain timing of dividends is a new concept to us.

FURTHER READING

- There is a good description of dividends in Ingersoll (1987).

- For more on the term structure of dividends see Fabozzi & Kipnis (1984).

- The idea of stochastic dividends is due to Geske (1978). See also Hilliard & Reis (1998).

- Barone-Adesi & Whaley (1986) fail to reject the hypothesis that there is a tax-factor for dividends on US stocks. However, Kaplanis (1986) finds significant differences for implied dividends in option prices on UK stocks.

- For more details of the models described above see Bakstein & Wilmott (1999).